Radiopharmaceuticals for PET Imaging—Issue A

Radiopharmaceuticals for PET Imaging — Issue A

Editors

Anne Roivainen
Xiang-Guo Li

MDPI • Basel • Beijing • Wuhan • Barcelona • Belgrade • Manchester • Tokyo • Cluj • Tianjin

Editors
Anne Roivainen
Turku PET Centre,
University of Turku and
Turku University Hospital
Finland

Xiang-Guo Li
Turku PET Centre,
University of Turku
Finland

Editorial Office
MDPI
St. Alban-Anlage 66
4052 Basel, Switzerland

This is a reprint of articles from the Special Issue published online in the open access journal *Molecules* (ISSN 1420-3049) (available at: https://www.mdpi.com/journal/molecules/special_issues/radiopharmaceuticals_PET_imaging).

For citation purposes, cite each article independently as indicated on the article page online and as indicated below:

LastName, A.A.; LastName, B.B.; LastName, C.C. Article Title. *Journal Name* **Year**, *Article Number*, Page Range.

ISBN 978-3-03943-242-4 (Hbk)
ISBN 978-3-03943-243-1 (PDF)

© 2020 by the authors. Articles in this book are Open Access and distributed under the Creative Commons Attribution (CC BY) license, which allows users to download, copy and build upon published articles, as long as the author and publisher are properly credited, which ensures maximum dissemination and a wider impact of our publications.

The book as a whole is distributed by MDPI under the terms and conditions of the Creative Commons license CC BY-NC-ND.

Contents

About the Editors vii

Preface to "Radiopharmaceuticals for PET Imaging—Issue A" ix

Masoud Sadeghzadeh, Barbara Wenzel, Daniel Gündel, Winnie Deuther-Conrad, Magali Toussaint, Rareş-Petru Moldovan, Steffen Fischer, Friedrich-Alexander Ludwig, Rodrigo Teodoro, Shirisha Jonnalagadda, Sravan K. Jonnalagadda, Gerrit Schüürmann, Venkatram R. Mereddy, Lester R. Drewes and Peter Brust
Development of Novel Analogs of the Monocarboxylate Transporter Ligand FACH and Biological Validation of One Potential Radiotracer for Positron Emission Tomography (PET) Imaging
Reprinted from: *Molecules* **2020**, *25*, 2309, doi:10.3390/molecules25102309 1

Magali Toussaint, Winnie Deuther-Conrad, Mathias Kranz, Steffen Fischer, Friedrich-Alexander Ludwig, Tareq A. Juratli, Marianne Patt, Bernhard Wünsch, Gabriele Schackert, Osama Sabri and Peter Brust
Sigma-1 Receptor Positron Emission Tomography: A New Molecular Imaging Approach Using (S)-(−)-[^{18}F]Fluspidine in Glioblastoma
Reprinted from: *Molecules* **2020**, *25*, 2170, doi:10.3390/molecules25092170 23

Bernhard Sattler, Mathias Kranz, Barbara Wenzel, Nalin T. Jain, Rareş-Petru Moldovan, Magali Toussaint, Winnie Deuther-Conrad, Friedrich-Alexander Ludwig, Rodrigo Teodoro, Tatjana Sattler, Masoud Sadeghzadeh, Osama Sabri and Peter Brust
Preclinical Incorporation Dosimetry of [^{18}F]FACH—A Novel ^{18}F-Labeled MCT1/MCT4 Lactate Transporter Inhibitor for Imaging Cancer Metabolism with PET
Reprinted from: *Molecules* **2020**, *25*, 2024, doi:10.3390/molecules25092024 41

Jessica Bridoux, Sara Neyt, Pieterjan Debie, Benedicte Descamps, Nick Devoogdt, Frederik Cleeren, Guy Bormans, Alexis Broisat, Vicky Caveliers, Catarina Xavier, Christian Vanhove and Sophie Hernot
Improved Detection of Molecular Markers of Atherosclerotic Plaques Using Sub-Millimeter PET Imaging
Reprinted from: *Molecules* **2020**, *25*, 1838, doi:10.3390/molecules25081838 53

Chiara Da Pieve, Ata Makarem, Stephen Turnock, Justyna Maczynska, Graham Smith and Gabriela Kramer-Marek
Thiol-Reactive PODS-Bearing Bifunctional Chelators for the Development of EGFR-Targeting [^{18}F]AlF-Affibody Conjugates
Reprinted from: *Molecules* **2020**, *25*, 1562, doi:10.3390/molecules25071562 63

Peter J. H. Scott, Robert A. Koeppe, Xia Shao, Melissa E. Rodnick, Alexandra R. Sowa, Bradford D. Henderson, Jenelle Stauff, Phillip S. Sherman, Janna Arteaga, Dennis J. Carlo and Ronald B. Moss
The Effects of Intramuscular Naloxone Dose on Mu Receptor Displacement of Carfentanil in Rhesus Monkeys
Reprinted from: *Molecules* **2020**, *25*, 1360, doi:10.3390/molecules25061360 75

Sofia Otaru, Surachet Imlimthan, Mirkka Sarparanta, Kerttuli Helariutta, Kristiina Wähälä and Anu J. Airaksinen
Evaluation of Organo [^{18}F]Fluorosilicon Tetrazine as a Prosthetic Group for the Synthesis of PET Radiotracers
Reprinted from: *Molecules* 2020, 25, 1208, doi:10.3390/molecules25051208 85

Patricia E. Edem, Jesper T. Jørgensen, Kamilla Nørregaard, Rafaella Rossin, Abdolreza Yazdani, John F. Valliant, Marc Robillard, Matthias M. Herth and Andreas Kjaer
Evaluation of a ^{68}Ga-Labeled DOTA-Tetrazine as a PET Alternative to ^{111}In-SPECT Pretargeted Imaging
Reprinted from: *Molecules* 2020, 25, 463, doi:10.3390/molecules25030463 103

Lars Jødal, Anne Roivainen, Vesa Oikonen, Sirpa Jalkanen, Søren B. Hansen, Pia Afzelius, Aage K. O. Alstrup, Ole L. Nielsen and Svend B. Jensen
Kinetic Modelling of [^{68}Ga]Ga-DOTA-Siglec-9 in Porcine Osteomyelitis and Soft Tissue Infections
Reprinted from: *Molecules* 2019, 24, 4094, doi:10.3390/molecules24224094 117

Brendan J. Evans, Andrew T. King, Andrew Katsifis, Lidia Matesic and Joanne F. Jamie
Methods to Enhance the Metabolic Stability of Peptide-Based PET Radiopharmaceuticals
Reprinted from: *Molecules* 2020, 25, 2314, doi:10.3390/molecules25102314 139

Guillaume Becker, Sylvestre Dammicco, Mohamed Ali Bahri and Eric Salmon
The Rise of Synaptic Density PET Imaging
Reprinted from: *Molecules* 2020, 25, 2303, doi:10.3390/molecules25102303 163

Christos Sachpekidis, Hartmut Goldschmidt and Antonia Dimitrakopoulou-Strauss
Positron Emission Tomography (PET) Radiopharmaceuticals in Multiple Myeloma
Reprinted from: *Molecules* 2020, 25, 134, doi:10.3390/molecules25010134 183

About the Editors

Anne Roivainen (Ph.D.) is Professor of Preclinical Imaging and Drug Research at the Turku PET Centre, University of Turku, Finland. She is medical biochemist with a background in molecular biology and immunology research of rheumatic diseases and, thereafter' 20 years' experience in molecular imaging, especially positron emission tomography and its modeling, tracer development, and preclinical and clinical applications.

Xiang-Guo Li (Ph.D.) is Adjunct Professor at the Turku PET Centre, University of Turku, Finland. He is a radiochemist and his research interests are in new radiopharmaceutical development, radiochemistry, and radiolabeling techniques in the field of positron emission tomography.

Preface to "Radiopharmaceuticals for PET Imaging—Issue A"

Positron emission tomography (PET) has become a clinical routine in public healthcare. To perform diagnosis with PET, suitable radiopharmaceuticals are needed. A variety of drug targets are present in different diseases. Even in the same disease (e.g., breast cancer), the profile of drug targets may vary greatly among patients. Therefore, new radiopharmaceuticals are needed as ever before to read out the target profile in different ways. This Special Issue "Radiopharmaceuticals for PET Imaging" is dedicated to presenting the fresh research results on radiopharmaceuticals. We are pleased to see that many research labs have been able to disseminate their results, and 12 papers are included in this edition, Issue A; Issue B on the same topic is currently under editing. We hope this issue will contribute to promoting scientific communication among its readers. We hope you enjoy reading.

Anne Roivainen, Xiang-Guo Li
Editors

Article

Development of Novel Analogs of the Monocarboxylate Transporter Ligand FACH and Biological Validation of One Potential Radiotracer for Positron Emission Tomography (PET) Imaging

Masoud Sadeghzadeh [1,*], Barbara Wenzel [1], Daniel Gündel [1], Winnie Deuther-Conrad [1], Magali Toussaint [1], Rareş-Petru Moldovan [1], Steffen Fischer [1], Friedrich-Alexander Ludwig [1], Rodrigo Teodoro [1], Shirisha Jonnalagadda [2], Sravan K. Jonnalagadda [2], Gerrit Schüürmann [3,4], Venkatram R. Mereddy [2], Lester R. Drewes [5] and Peter Brust [1]

[1] Department of Neuroradiopharmaceuticals, Institute of Radiopharmaceutical Cancer Research, Helmholtz-Zentrum Dresden-Rossendorf, Permoserstraße 15, 04318 Leipzig, Germany; b.wenzel@hzdr.de (B.W.); d.guendel@hzdr.de (D.G.); w.deuther-conrad@hzdr.de (W.D.-C.); m.toussaint@hzdr.de (M.T.); r.moldovan@hzdr.de (R.-P.M.); s.fischer@hzdr.de (S.F.); f.ludwig@hzdr.de (F.-A.L.); r.teodoro@hzdr.de (R.T.); p.brust@hzdr.de (P.B.)
[2] Department of Chemistry and Biochemistry, Department of Pharmacy Practice & Pharmaceutical Sciences, University of Minnesota, Duluth, MN 55812, USA; sgurrapu@d.umn.edu (S.J.); skjonnal@d.umn.edu (S.K.J.); vmereddy@d.umn.edu (V.R.M.)
[3] UFZ Department of Ecological Chemistry, Helmholtz Centre for Environmental Research, Permoserstraße 15, 04318 Leipzig, Germany; gerrit.schuurmann@ufz.de
[4] Institute of Organic Chemistry, Technical University Bergakademie Freiberg, Leipziger Straße 29, 09599 Freiberg, Germany
[5] Department of Biomedical Sciences, University of Minnesota Medical School Duluth, 251 SMed, 1035 University Drive, Duluth, MN 55812, USA; ldrewes@d.umn.edu
* Correspondence: m.sadeghzadeh@hzdr.de; Tel.: +49-341-2341794630; Fax: +49-341-2341794699

Academic Editors: Anne Roivainen and Xiang-Guo Li
Received: 26 March 2020; Accepted: 11 May 2020; Published: 14 May 2020

Abstract: Monocarboxylate transporters 1-4 (MCT1-4) are involved in several metabolism-related diseases, especially cancer, providing the chance to be considered as relevant targets for diagnosis and therapy. [^{18}F]FACH was recently developed and showed very promising preclinical results as a potential positron emission tomography (PET) radiotracer for imaging of MCTs. Given that [^{18}F]FACH did not show high blood-brain barrier permeability, the current work is aimed to investigate whether more lipophilic analogs of FACH could improve brain uptake for imaging of gliomas, while retaining binding to MCTs. The 2-fluoropyridinyl-substituted analogs **1** and **2** were synthesized and their MCT1 inhibition was estimated by [^{14}C]lactate uptake assay on rat brain endothelial-4 (RBE4) cells. While compounds **1** and **2** showed lower MCT1 inhibitory potencies than FACH (IC$_{50}$ = 11 nM) by factors of 11 and 25, respectively, **1** (IC$_{50}$ = 118 nM) could still be a suitable PET candidate. Therefore, **1** was selected for radiosynthesis of [^{18}F]**1** and subsequent biological evaluation for imaging of the MCT expression in mouse brain. Regarding lipophilicity, the experimental log D$_{7.4}$ result for [^{18}F]**1** agrees pretty well with its predicted value. In vivo and in vitro studies revealed high uptake of the new radiotracer in kidney and other peripheral MCT-expressing organs together with significant reduction by using specific MCT1 inhibitor α-cyano-4-hydroxycinnamic acid. Despite a higher lipophilicity of [^{18}F]**1** compared to [^{18}F]FACH, the in vivo brain uptake of [^{18}F]**1** was in a similar range, which is reflected by calculated BBB permeabilities as well through similar transport rates by MCTs on RBE4 cells. Further investigation is needed to clarify the MCT-mediated transport mechanism of these radiotracers in brain.

Keywords: monocarboxylate transporters (MCTs); FACH; ^{18}F-labeled analog of FACH; α-CCA; blood-brain barrier (BBB); positron emission tomography (PET) imaging

1. Introduction

Monocarboxylate transporters (MCTs), comprising 14 isoforms, are dedicated to the solute carrier 16 (*SLC16*) gene family [1,2]. Of all the MCT isoforms, MCT1-4 are well characterized and known as membrane-bound carriers that bidirectionally transport short-chain monocarboxylic acids, most notably L-lactate, pyruvate, and ketone bodies along with protons across the plasma membrane of mammalian cells [2]. The tissue distribution of the MCT isoforms is quite variable. Although MCT1 is ubiquitously distributed in the muscles, it is additionally expressed along with MCT4 in the brain and other peripheral organs like small intestine, liver, heart, kidney, and blood cells [1,3]. Aberrant expression such as, upregulation of MCT1 and MCT4 has been reported in a large number of tumors (e.g., neuroblastomas, high-grade gliomas, carcinomas of renal cells, breast epithelium, colorectal and squamous tissues, and cervical and lung cancers) where expression is correlated to poor outcomes. In these tissues, the MCTs serve to facilitate the shuttling of lactate between cells with different metabolic requirements [4–6]. Due to the metabolic reprogramming, considered as a hallmark of cancer, tumor cells indeed switch from glucose to lactate as a crucial energy supply, hence, their metabolism heavily relies on glycolysis and consequently the lactate efflux through MCT1 and MCT4 in order to prevent their own acidosis and to regenerate NAD$^+$ [7]. Accordingly, both transporters are attractive therapeutic and even diagnostic targets for the treatment and detection of human cancers [4,7,8].

Positron emission tomography (PET) is known as a powerful tool for non-invasive molecular detection of early metabolic changes in cancer progression [9]. [^{18}F]Fluorodeoxyglucose ([^{18}F]FDG), a radiolabeled glucose analog, is well known as a standard PET tracer used for diagnosis, staging and treatment monitoring in clinical oncology [10,11]. Considering the lack of specificity and sensitivity of [^{18}F]FDG for several types of tumors [10], there is still an unmet clinical need for cancer detection and therapy. Thus, the complementary concept based on components of the aerobic glycolysis and metabolism by malignant cells is more intriguing and potentially rewarding [2,12–14]. In this regard, [^{18}F]DASA-23, recently developed as a potent radiotracer for imaging tumor glycolysis by targeting pyruvate kinase M2, is currently in phase I clinical trials (ClinicalTrials.gov, NCT03539731) [15]. Because the metabolic reprogramming in cancer cells may also result in the overexpression of MCT1/MCT4 in many cancers [4,7], MCT-targeting PET studies provide an opportunity to achieve more accurate and useful understanding of certain aspects of the tumor-specific metabolism [8,16].

During the last decade, only a few ^{11}C- or ^{18}F-labeled substrates of MCTs such as [^{11}C]lactate, [^{11}C]pyruvate as well as their ^{18}F-labeled analogs were investigated for imaging of MCTs by PET [17–20]. To the best of our knowledge, only limited examples of the MCTs inhibitors were investigated as PET radiotracers for in vivo applications [21]. Although one of the best characterized inhibitors, α-cyano-4-hydroxycinnamic acid (α-CCA), possesses a 10-fold selectivity for MCT1 compared to other subtypes (Figure 1) [22], it also shows significant inhibitory potency towards the mitochondrial pyruvate carrier in isolated mitochondria [23]. Accordingly, very potent and more specific MCT1/MCT4 inhibitors have been developed based on a comprehensive structure-activity relationship study on a series of α-CCA derivatives [24,25]. Based on this approach, we recently developed and evaluated [^{18}F]FACH as the first ^{18}F-labeled inhibitor of MCTs (Figure 1) [26]. Along with a high inhibitory potency towards MCT1 (11.0 nM) and MCT4 (6.5 nM) [26], [^{18}F]FACH showed very promising pharmacokinetics in healthy mice, in particular in the kidneys, as organ with a high physiological expression of MCT1 [1,3,27].

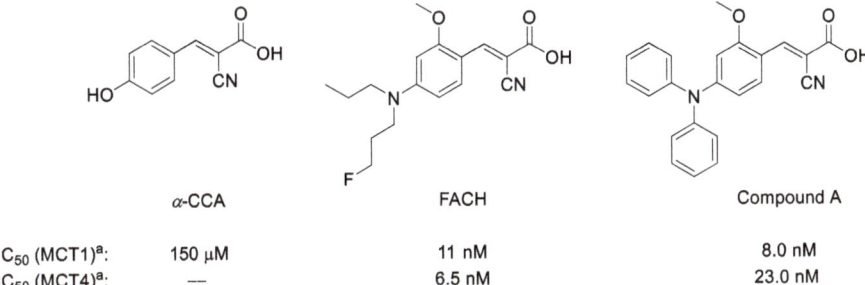

	α-CCA	FACH	Compound A
IC$_{50}$ (MCT1)[a]:	150 µM	11 nM	8.0 nM
IC$_{50}$ (MCT4)[a]:	---	6.5 nM	23.0 nM

Figure 1. Chemical structures of α-CCA and its novel derivatives as potent monocarboxylate transporters 1/4 (MCT1/MCT4) inhibitors; [a] Inhibition of [^{14}C]lactate uptake was determined in RBE4 cells (for MCT1) and in MDA-MB-231 cells (for MCT4) by measuring intracellular radioactivity after 60 min incubation without and with the respective inhibitor at 37 °C [25].

Nonetheless, [^{18}F]FACH showed only moderate brain uptake, which might be related to the rather hydrophilic features (log $D_{7.4}$ = 0.42) [28], assuming that [^{18}F]FACH could only passively enter the brain. This assumption would be in accordance with the fact that for the structural analog α-CCA, the site of its MCT inhibition has been demonstrated to be the extracellular surface [29,30]. Moreover, it is well established that MCT1 is also the prominent monocarboxylic acid transporter in the cerebral microvascular endothelium, facilitating the bidirectional transport of lactate through brain endothelial cells and the blood-brain barrier (BBB) [31]. Many brain tumors, such as gliomas and neuroblastomas, produce high amounts of lactic acid and consequently up-regulate MCT1, thus, inducing acidosis in the tumor microenvironment [5]. MCT1 is therefore proposed as a most likely therapeutic target for neuroblastomas and gliomas, and α-CCA has been able to suppress tumor growth via inhibition of MCT1 [5,32,33]. Accordingly, the development of MCT1-targeting radiotracers possessing sufficiently high brain permeability would be an important step forward toward brain imaging.

Although α-CCA as well as its new analog FACH (Figure 1) contain a Michael acceptor unit, their highly predominating carboxylate form under physiological conditions (ACD-calculated pK_a < 1 vs. pH 7.4; see Table 1 below) masks a respective electrophilic reactivity. This implies that these compounds are most likely not active as protein-attacking electrophiles. Moreover and as mentioned above, α-CCA has been reported to remain extracellular, inhibiting MCT from the outside in a competitive manner and without adverse effects under therapeutic concentrations [29,30].

In this context, it is interesting that for all respective cinnamic acid derivatives, the electron-withdrawing α-CN substituent is required for their MCT inhibition potency, which holds also for FACH [26]. As a possible explanation, we hypothesize that their non-covalent interaction with the MCT protein at the cellular surface may include an electrostatic Arg-carboxylate binding motif as primary anchor. This may facilitate a further complex stabilization through approaching a Cys-thiol by the Michael acceptor β-carbon that is activated further through the α-CN substitution (Figure 2, left). To balance the carboxylate anionic charge, a separate extracellular proton, required for the MCT action as respective symporter, could be attached temporarily at a His-nitrogen (not shown in Figure 2). In case of a successful carboxylate transport such as for the lactate efflux, a respective (additional) His-proton could be liberated to the interstitial compartment.

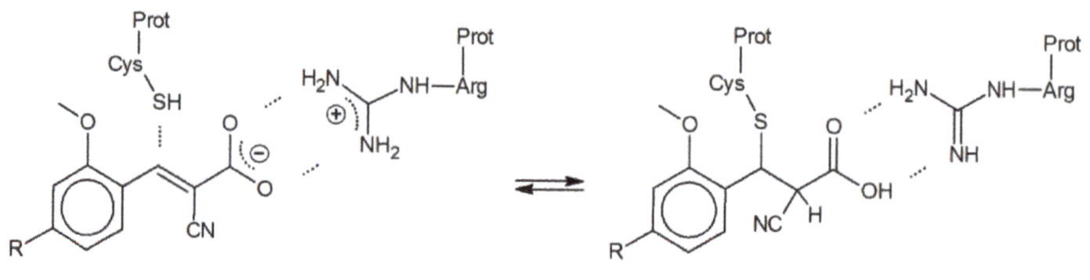

Figure 2. Hypothetical MCT-inhibitor binding at the extracellular surface (left) and at an interior protein site with no aqueous solvation (right). Michael addition may become active for the neutral carboxylic acid form (right), but would be reversible due to the α-CN substitution that enhances the retro-Michael reaction significantly (see, e.g., [34,35]).

In case the Arg-carboxylate interaction would result in a charge compensation sufficient to unmask the Michael acceptor reactivity, the Cys-thiol might add to the α,β-unsaturated unit, possibly following a proton transfer from Arg to the carboxylate (right part of Figure 2). In this case, however, the α-CN substituent enhances both the Michael and the retro-Michael reactivity, making this covalent reaction reversible through stabilization of the carbanion intermediate [34,35]. In conclusion, we hypothesize that despite the Michael acceptor unit common to all cinnamic acid derivatives, their mode of action is probably non-covalent or under water-poor/free conditions at least only temporarily covalent, with a correspondingly negligible risk to form permanent covalent bonds to nucleophilic protein sites. Results from a respective toxicity study will be reported in due course.

With the goal to improve the brain uptake by passive diffusion, we designed new analogs of FACH by replacing the less lipophilic propyl groups with more lipophilic aryl and heteroaryl moieties (Figure 3). Notably, the structurally modified analogs yet need to retain an acceptable inhibitory potency towards MCT1. On the basis of compound A, which was reported to exhibit high MCT1 inhibition (IC_{50} = 8.0 nM, Figure 1) [25], two fluorinated analogs were developed by introducing 2-fluoropyridinyl and phenyl groups (compounds **1** and **2**, Figure 3). Herein we describe the organic synthesis of the new compounds and their inhibitory potency for MCT1-mediated lactate transport. Furthermore, radiofluorination of **1** was performed and the resulting new radiotracer [^{18}F]**1** was investigated in mice to assess the impact of higher lipophilicity on the in vivo features compared to [^{18}F]FACH for imaging of MCT1 in mouse brain.

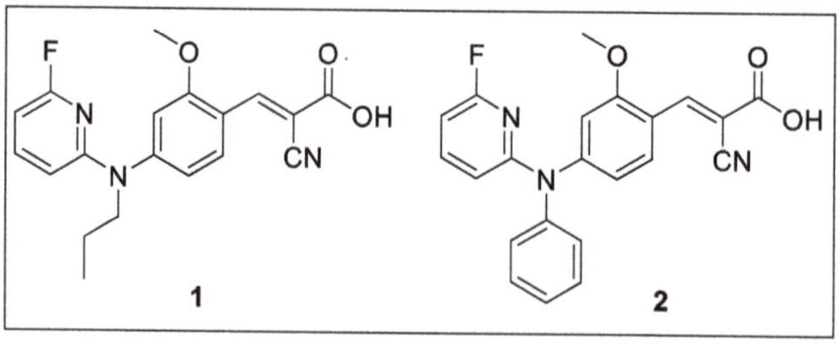

Figure 3. New analogs of FACH investigated in the current study.

2. Results and Discussion

2.1. Organic Chemistry and Monocarboxylate Transporter Inhibition

For developing the compounds **1** and **2**, the di-arylamine intermediate **5** was synthesized via the Buchwald-Hartwig aryl amination according to the previously reported procedures (Scheme 1) [36,37].

Alkylation of 6-fluoro-*N*-(3-methoxyphenyl)pyridin-2-amine **5** using 1-iodopropane and sodium hydride afforded **6** in 95% yield [38]. Compound **7** was obtained via a second Buchwald–Hartwig amination of **5** with phenyl bromide in negligible yield (< 10%). However, a stepwise addition of the palladium (Pd) catalyst and the phosphine ligand together with a longer reaction time led to the formation of **7** in moderate yield (46%). This might be related to the decreased electron density of the nitrogen atom due to the 2-fluoropyridinyl substituent and/or the steric hindrance effect. Both **6** and **7** were afterwards subjected to Vilsmeier–Haack formylation [39] to afford **8** and **9** with yields of 57% and 68%, respectively. Finally, Knoevenagel condensation of aldehydes **8** and **9** with cyanoacetic acid generated **1** and **2** in nearly quantitative yields (Scheme 1) [26].

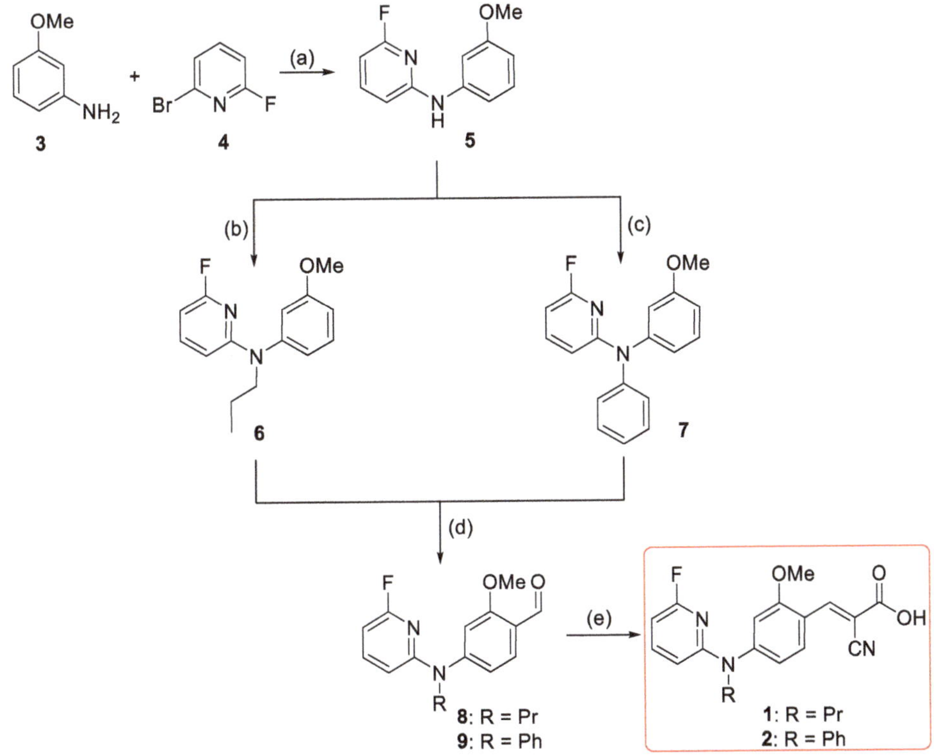

Scheme 1. Synthesis of **1** and **2**; reagents and reaction conditions: (a) Pd(OAc)$_2$ (5 mol %), Xantphos (5 mol %), Cs$_2$CO$_3$, 1,4-dioxane, Ar, 105 °C, 50 min, 96%; (b) 1-iodopropane, NaH (60% oil dispersion), DMF, Ar, r.t., 1.5 h, 95%; (c) PhBr, Pd(OAc)$_2$ (15 mol %), Xantphos (15 mol %), Cs$_2$CO$_3$, 1,4-dioxane, Ar, 105 °C, 24 h, 46%; (d) POCl$_3$, DMF, Ar, 80 °C, 2–4 h, 57% (for **8**) and 68% (for **9**); (e) i. cyanoacetic acid, piperidine, ACN, reflux; ii. HCl (6 M), r.t. 30 min, 95% (for **1**) and 98% (for **2**).

Inhibition of MCT1-mediated lactate transport of **1** and **2** was investigated by [^{14}C]lactate uptake assays using immortalized rat brain endothelial-4 cells (RBE4) [40] which express mainly MCT1 [24,25]. Both compounds dose-dependently inhibited the lactate uptake, with IC$_{50}$ values of 118 nM (**1**) and 274 nM (**2**). Accordingly, replacing the 1-fluoropropyl group of FACH by a 2-fluoropyridinyl group in **1** resulted in a 10-fold decrease of the inhibitory potency. When comparing **2** with compound A, the substitution of the phenyl ring by a 2-fluoropyridinyl ring in **2** (Figure 1) caused an even stronger reduction of the inhibitory potency [25]. We therefore decided to proceed with **1** for radiofluorination and biological evaluation.

In order to develop the new MCT1-targeting radiotracer [^{18}F]1, a precursor including a suitable leaving group was required for the nucleophilic aromatic substitution (S$_N$Ar) with [^{18}F]fluoride. A nitro precursor (15) with an unprotected carboxylic acid function (Scheme 2) was synthesized considering the good results obtained for the aliphatic nucleophilic substitution with unprotected precursor in the one-step radiosynthesis of [^{18}F]FACH [28]. Initial Buchwald–Hartwig aryl amination between 2-amino-6-nitropyridine and 3-bromoanisole provided the N-substituted anisidine 12 in 80% yield (Scheme 2) [36]. Alkylation of 12 under basic condition provided 13 with a yield of 93% [38]. Vilsmeier–Haack formylation of 13 gave aldehyde 14 in 92% yield [39]. It was followed by Knoevenagel condensation with cyanoacetic acid to provide 15 in 67% overall yield [26]. The chemical purity of the precursor 15 was > 98%, according to NMR and HPLC analyses.

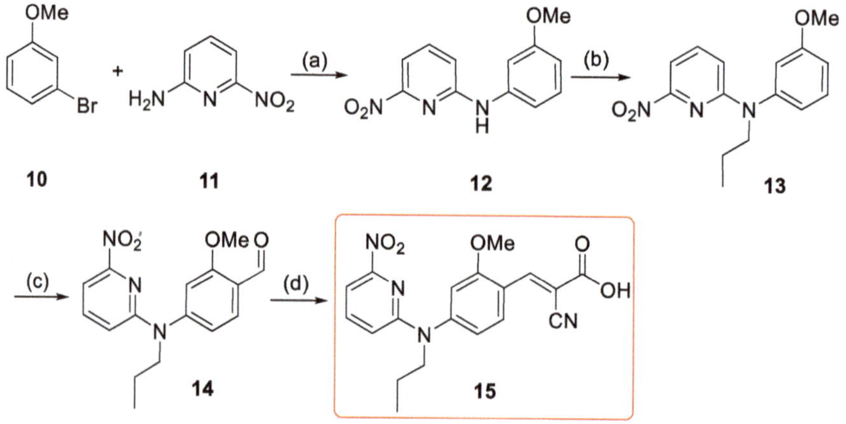

Scheme 2. Synthesis of the nitro precursor (15): Reagents and reaction conditions: (a) Pd(OAc)$_2$ (5 mol %), Xantphos (5 mol %), Cs$_2$CO$_3$, 1,4-dioxane, Ar, 105 °C, 2 h, 80%; (b) 1-iodopropane, NaH (60% oil dispersion), DMF, Ar, r.t., 1.5 h, 93%; (c) POCl$_3$, DMF, Ar, 80 °C, 1.5 h, 92%; (d) i. cyanoacetic acid, piperidine, ACN, reflux, 5 h; ii. HCl (6 M), r.t. 30 min, above 98%.

2.2. Radiosynthesis, Stability, and Determination of log D$_{7.4}$

As shown in Scheme 3, [^{18}F]1 was synthesized on the basis of an S$_N$Ar reaction via substitution of the NO$_2$ leaving group of 15 by [^{18}F]fluoride in the presence of Kryptofix® (K$_{2.2.2}$) and K$_2$CO$_3$. In dimethylsulfoxide (DMSO), the S$_N$Ar reaction proceeded smoothly and resulted in high radiochemical yields of 73 ± 12% (n = 4, non-isolated, radio-HPLC) for [^{18}F]1 after 15 min conventional heating at 130 °C. Besides unreacted [^{18}F]F$^-$, radioactive by-products accounted for less than 5%. [^{18}F]1 was isolated by semi-preparative HPLC (Supplementary Data, Figure S1A), trapped on a pre-conditioned Sep-Pak C18 light cartridge, eluted with ethanol, and formulated in isotonic saline containing 10% of EtOH (v/v) for better solubility. Analytical radio- and UV-HPLC analyses of the final product co-eluted with the reference 1, confirmed the identity of the radiotracer (Supplementary Data, Figure S1B). Finally, [^{18}F]1 was obtained with radiochemical yields of 51 ± 11% (n = 3, decay-corrected to the end of the bombardment) in a total radiosynthesis time of about 90 min, at a radiochemical purity of ≥ 98% and with molar activities in the range of 180–200 GBq/µmol (n = 3, end of synthesis) using starting activities of 2–3 GBq.

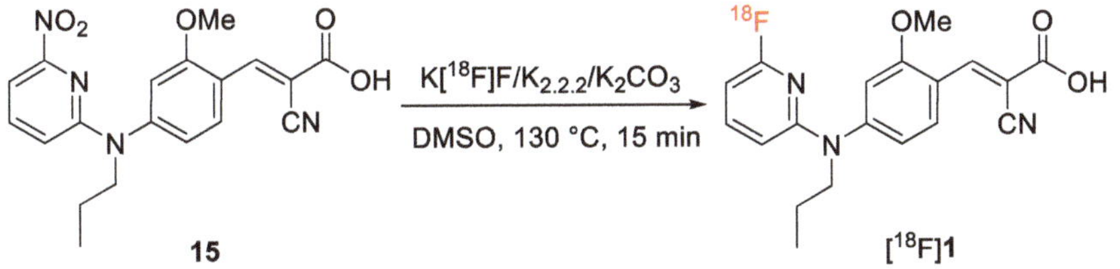

Scheme 3. Radiosynthesis of [^{18}F]1.

The stability of the radiotracer was investigated by incubation of [^{18}F]1 in *n*-octanol, saline, phosphate-buffered saline (PBS) and ethanol. Samples were analyzed by radio-thin-layer chromatography (TLC) and radio-HPLC and no degradation or defluorination was observed in any of the solvents after 60 min incubation at 40 °C.

A variety of physicochemical parameters affects the brain permeability of different brain-targeting radiotracers [41]. Lipophilicity, often, but not necessarily, correlates with the ability to cross the BBB, and is considered as an important physicochemical property. In Table 1, calculated bioavailability-related parameters are listed for α-CCA, FACH and selected structural analogs. Accordingly, the new derivatives **1** and **2** show the desired higher hydrophobicity (log K_{ow} = logarithmic *n*-octanol/water partition coefficient [42], log $D_{7.4}$ = log K_{ow} corrected for ionization at pH 7.4 [43,44]) as compared to FACH. Nevertheless, Table 1 also shows that the predicted brain-blood partition coefficients (log K_{BB} [43]) of **1** (−0.49) and **2** (−1.05) are below the one of FACH (−0.10).

Table 1. Calculated physicochemical parameters of our drug candidates and structurally related compounds.[1]

Compound	log K_{ow}	log $D_{7.4}$	f_u	pK_a	log K_{BB}
FACH	4.43	0.69	4.5×10^{-8}	0.05	−0.10
Compound A	5.41	1.66	2.5×10^{-7}	0.80	−0.57
1	4.83	1.08	7.4×10^{-8}	0.27	−0.49
2	5.07	1.32	2.1×10^{-7}	0.73	−1.05
Cinnamic acid	2.07	−0.91	8.7×10^{-4}	4.34	−0.14
α-CN cinnamic acid	2.27	−1.48	1.6×10^{-7}	0.60	−0.54
α-CCA	1.79	−1.96	2.8×10^{-7}	0.85	−0.45

[1] The decadic logarithms of the *n*-octanol/water partition coefficient (log K_{ow}) have been calculated with EPI Suite [42], and these values have been employed for predicting the pK_a-pH-corrected *n*-octanol/water distribution coefficients, log $D_{7.4}$. For the latter, the ACD approach log K_{ow} (ionized) = log K_{ow} (unionized) − 3.75 (applicable for the relevant range of log K_{ow} data) [43,44] with $D_{7.4} = f_u \times K_{ow} + (1 - f_u) \times K_{ow}$ (ionized) has been employed (see also [44]). Note that the ACD-calculated log K_{ow} data are lower by 0.5–1.5 log units except for a slightly larger value for cinnamic acid, resulting in correspondingly lower calculated log $D_{7.4}$ data. Moreover, f_u denotes the compound fraction unionized at pH 7.4 according to the Henderson-Hasselbalch relationship, and pK_a as well as the brain-blood partition coefficient (K_{BB}) have been calculated with the ACD software.

Note further that FACH is both more lipophilic and more BBB-permeable than α-CCA. Regarding the Michael-acceptor unit mentioned above, comparison of cinnamic acid and its α-CN derivative shows that the α-CN substitution decreases the pK_a value by 3.7 units, most likely because of its combined inductive and mesomeric electron-withdrawing effect. Accordingly, all α-CCA derivatives are significantly acidic with pK_a values below 1, indicating for all of them that the dissociated carboxylate form is prevalent under physiological conditions. Experimental investigation of the lipophilicity of [^{18}F]1 through employing the shake-flask method using *n*-octanol and PBS (pH 7.4) resulted in a log $D_{7.4}$ value of 0.820 ± 0.003 (n = 4). This value agrees pretty well with its calculated counterpart of 1.08 (Table 1), which holds correspondingly for the FACH log $D_{7.4}$ value (0.42 experimental [28] vs. 0.69 calculated) as well as for the respective difference in log $D_{7.4}$ values.

2.3. In Vitro and In Vivo Biological Validation of [^{18}F]1

It is well demonstrated that several members of the MCT family are highly expressed in mammalian kidney, where over 95% of the lactate reabsorption takes place [1,3,45,46]. MCT1 mRNA and protein have clearly been detected on both the human kidney derived cell line HK-2 and human kidney cortex. In HK-2 cells it was found exclusively on the basal membrane [45]. Therefore, the specific binding of [^{18}F]1 to MCTs was initially proven by in vitro autoradiography using cryosections of the mouse kidney. As reflected by the autoradiographic images presented in Figure 4, co-incubation of ~1 nM [^{18}F]1 with 10 µM α-CCA-Na resulted in significantly lower binding of the radiotracer. Therefore, the binding of [^{18}F]1 in mouse kidney in vitro is highly specific.

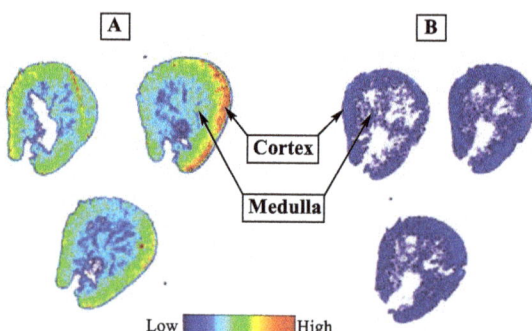

Figure 4. In vitro autoradiography of [^{18}F]1 in transversal cryosections of the mouse kidney. Total (**A**) and non-specific (**B**) binding of ~1 nM [^{18}F]1 obtained without and with co-incubation with 10 µM α-CCA-Na.

To investigate the stability of [^{18}F]1 in vivo, samples of plasma and brain homogenates obtained from CD-1 mice at 30 min after intravenous injection of the radiotracer were analyzed for radiometabolites by using reversed-phase and micellar (MLC) radio-HPLC. MLC allows a direct injection of the samples into the HPLC system without the elimination of the tissue matrix as already described [47,48]. In general, the results obtained with both methods are comparable and the analyses revealed solely intact radiotracer and no detectable radiometabolites in plasma (Figure 5A–B) and brain (Supplementary Data, Figure S2) samples. Notably, in both samples two peaks a/b were detected by analytical HPLC (Figure 5/Figure S2) which are supposed to represent the neutral and deprotonated form of the radiotracer ([^{18}F]1a/b). This finding suggests that the analytical HPLC conditions do not reflect the physiological milieu at which the neutral compound fraction would be negligible according to the Henderson–Hasselbalch equation. Research into the speciation of **1** under analytical-chemical conditions is subject to a future investigation.

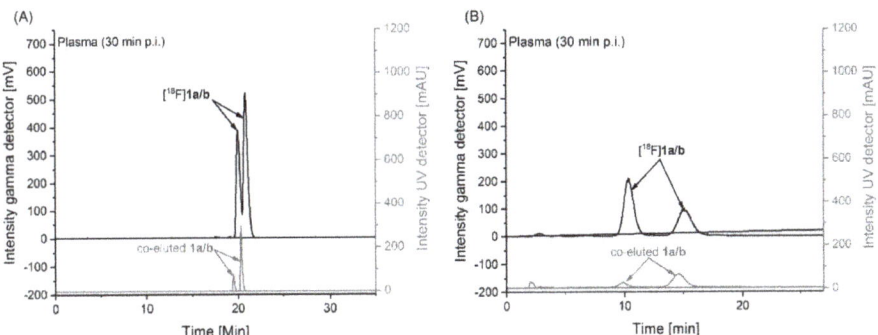

Figure 5. Analytical UV- and radio-HPLC chromatograms representing two peaks a/b which are supposed to reflect the neutral and deprotonated form of the radiotracer ([^{18}F]**1a/b**) in mouse plasma at 30 min p.i. measured under: (**A**) reversed phase (Reprosil-Pur C18-AQ, 250 × 4.6 mm, gradient with an eluent mixture of ACN/20 mM NH$_4$OAc (aq.), 370 nm, 1.0 mL/min), and (**B**) micellar conditions (Reprosil-Pur C18-AQ, 250 × 4.6 mm, isocratic mode with water containing 50 mM sodium dodecyl sulfate/10 mM NaHPO$_4$, 1.0 mL/min).

Pharmacokinetic studies of [^{18}F]**1** were performed by dynamic PET imaging in mouse using a dedicated small animal PET/MR camera. The target-specificity of [^{18}F]**1** was investigated by pre-administration of the blocking compound α-CCA-Na. Maximum intensity projections of PET studies from a representative control and α-CCA-Na treated animals and time-activity curves (TACs) from tissues of interest are presented in Figure 6. [^{18}F]**1** cleared rapidly from the blood with an initial TAC peak standardized uptake value (SUV) of 7.3 and a SUV of 1.5 after 10 min followed by a slow blood clearance to a SUV of 0.9 after 60 min in the control group (Figure 6B). Pre-administration of the MCT inhibitor α-CCA-Na resulted in an initial TAC peak SUV of 6.9 which was comparable to the control group, whereas a higher SUV of 3.5 after 10 min and a SUV of 1.5 was reached after 60 min p.i., reflecting higher availability of the radiotracer in the blood (Figure 6B). This is expected to be caused by blocking the uptake of [^{18}F]**1** in peripheral organs in vivo. In comparison to the control conditions, the pre-administration of α-CCA-Na significantly reduced the activity accumulation in the MCT1-expressing renal cortex [46] throughout the whole imaging period, which is shown by the SUV ratio (SUVR) of kidney cortex-to-blood (Figure 6D). Furthermore, the displacement study revealed 39.2% drop of the SUV, 20 min after i.v. injection of α-CCA-Na (Figure 6E), which implicates a reversible tissue uptake of [^{18}F]**1** in the kidney cortex. Nevertheless, further studies are needed to clarify the exact mechanism of the radiotracer uptake. Regarding liver, where the highly expressed MCT1 transports L-lactate into the parenchymal cells for gluconeogenesis [1], a constantly increasing accumulation of activity can be observed under both control and blocking conditions, although at lower values under pre-administration of α-CCA-Na (Figure 6C).

Taking into consideration the high activity concentrations persistently accumulated in the kidney and liver, the blocking effect of α-CCA-Na in both tissues will result in a strong increase in the fraction of available tracer in blood as reflected by the higher SUV in blood observed in the blocking experiments (Area Under the Curve (AUC)$_{0-60\ min}$ = 140 SUV × minutes) compared to the control experiments (AUC$_{0-60\ min}$ = 75 SUV × minutes).

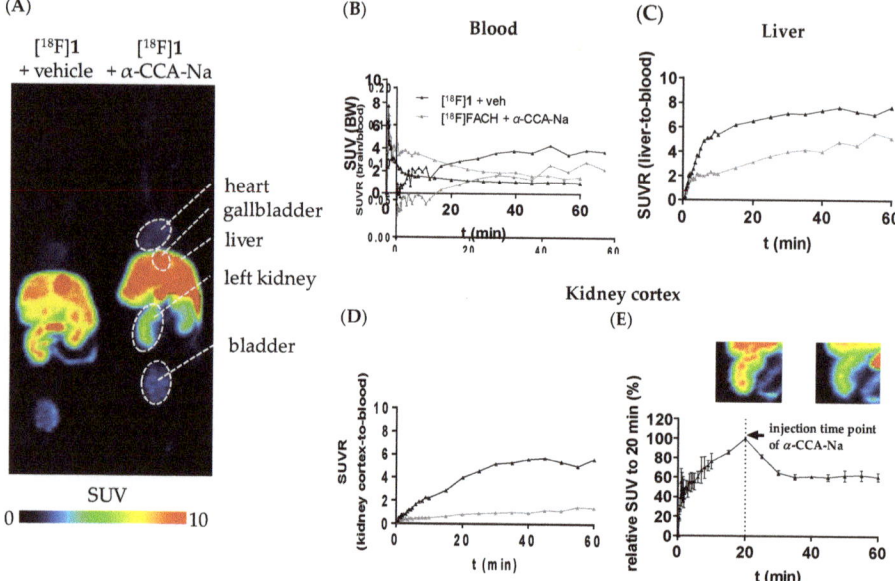

Figure 6. (**A**) Maximum intensity projections (MIPs) of small animal positron emission tomography (PET) images (45–60 min p.i., n = 1) of female CD-1 mice depicting the differential distribution of [^{18}F]1 in peripheral organs without and with pre-administration of α-CCA-Na; (**B–D**) Time-activity curves representing the SUV of blood, as well as the standardized uptake value (SUV) ratio (SUVR) from liver-to-blood and kidney cortex-to-blood; (**E**) displacement study: i.v. application of α-CCA-Na 20 min after tracer injection (n = 2), MIPs of the left kidney (left: 15–20 min p.i. and right: 45–60 min p.i.).

According to Figure 7A, [^{18}F]1 shows a similar brain uptake as [^{18}F]FACH despite a somewhat higher lipophilicity. Note, however, that this is in accordance with our ACD predictions regarding both log $D_{7.4}$ and log K_{BB} values (Table 1). The slightly higher lipophilicity (log K_{ow} and log $D_{7.4}$) of **1** is accompanied by a slightly lower BBB penetration (log K_{BB}), demonstrating further that passive brain uptake is not governed alone by log $D_{7.4}$. Moreover, a selective uptake into a particular brain region could not be verified for both radiotracers (Figure 7, C and D). The SUVR (brain-to-blood) of [^{18}F]**1** was reduced by α-CCA-Na by only 25% compared to the control animal (Figure 7B), which is much less than the reduction of kidney uptake and indicates that the uptake of [^{18}F]**1** into the brain is partially mediated by MCT1 expressed at the endothelial cells of the BBB [49,50], and partly by non-specific diffusion mediated by lipophilicity.

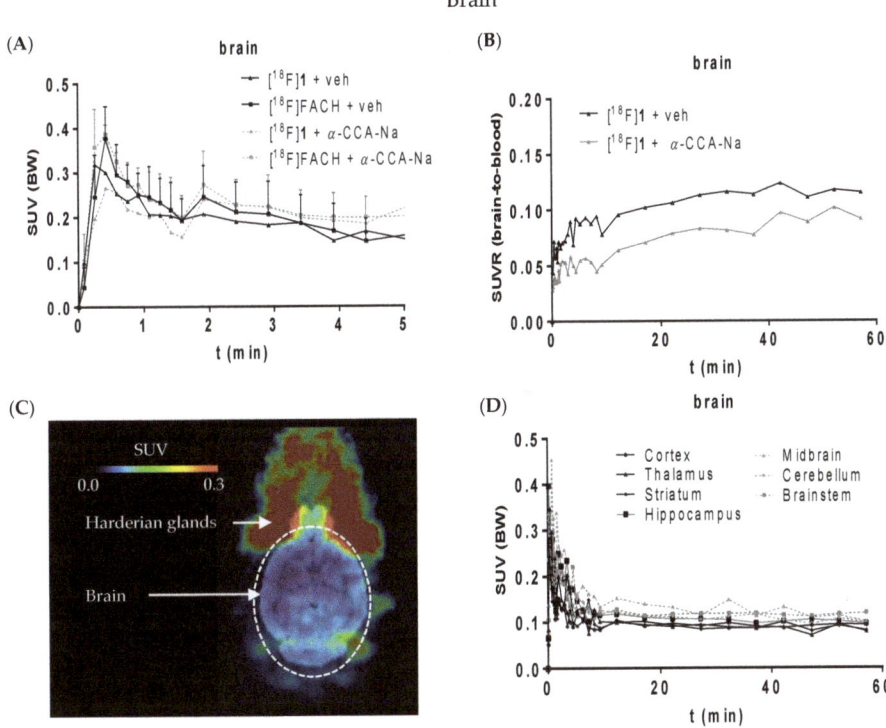

Figure 7. (**A**) Time-activity curve of [^{18}F]1 (n = 1) and [^{18}F]FACH (n = 3) representing the SUVs without and with pre-administration of α-CCA-Na; (**B**) The SUVR of [^{18}F]1 showing the brain to blood uptake ratio (n = 1); (**C**) horizontal section of representative PET image (averaged time frames from 0–60 min p.i.) of mouse head; (**D**) SUVs of different brain regions over time after injection of [^{18}F]1.

Brain capillary endothelial cells are tightly bound to each other and constitute the permeability barrier of the BBB/Neurovascular Unit (NVU), which serves to restrict the transport of compounds into the brain. The permeation of compounds across the BBB/NVU is determined not only by lipophilicity, ionic feature and molecular size, but also by various transporters expressed on the endothelial cell membrane [51]. For example, the radiotracer [^{11}C]choline has initially been demonstrated a comparably low brain accumulation with 0.08% of injected dose/g brain (%ID/g) 10 min after injection in mice [52] and 0.15% ID/g at 15 min after injection in rats [53]. This low uptake of [^{11}C]choline from blood is mediated by a choline-specific transport system of brain endothelial cell membranes [54]. Despite the low BBB/NVU permeability of [^{11}C]choline, it is a clinically relevant radiopharmaceutical for brain tumor imaging with high and specific accumulation in proliferating cancer cells [55,56]. Notably, in our control experiment using the MCT radioligand [^{18}F]1, a moderate brain accumulation of 0.4% ID/g at 10 min after injection was observed. It has been shown that MCT1 is expressed along with MCT4 in the cerebral microvascular endothelium cells suggesting a key role in transporting endogenous monocarboxylates into and out of the brain [50,57,58]. Together with the important role of MCT1 and MCT4 for glioma metabolism [5,32,33,59], it provides evidence that [^{18}F]FACH and/or [^{18}F]1 might be suitable for brain tumor imaging.

3. Materials and Methods

3.1. Organic Synthesis

The syntheses of **1**, **2** and **15** (nitro precursor) were implemented by slight modifications of the previously reported procedures [26,36–39]. All final compounds described here meet the purity requirements determined by HPLC, NMR, and HR-MS analyses.

3.1.1. General Information

Unless otherwise noted, moisture-sensitive reactions were conducted under dry nitrogen or argon. Pd(OAc)$_2$, Xantphos (4,5-bis(diphenylphosphino)-9,9-dimethylxanthene), and Cs$_2$CO$_3$ were purchased from Sigma-Aldrich (SIGMA-ALDRICH Chemie GmbH, Schnelldorf, Germany). Other chemicals and reagents were purchased from commercial sources and were used without further purification. For thin-layer chromatography (TLC), Silica gel 60 F254 plates (Merck KGaA, Darmstadt, Germany) were used. Flash chromatography (fc) was performed using Silica gel 60, 40–64 µm (Merck). Room temperature was 21 °C. ^{1}H, ^{13}C, and ^{19}F spectra were recorded on Varian "MERCURY plus 300/400" (Varian, Palo Alto, CA, USA) and Bruker Advance DRX-400 (Bruker, Billerica, MA, USA); δ in ppm related to tetramethylsilane; coupling constants (J) are given with 0.1 Hz resolution. Multiplicities of NMR signals are indicated as follows: s (singlet), d (doublet), t (triplet), q (quartet), m (multiplet), dd (doublet of doublets), ddd (doublet of doublet of doublets), td (triplet of doublets), dq (doublet of quartets) and h (hexet (sextet)). High-resolution mass spectra (HR-MS) were recorded on an FT-ICR APEX II spectrometer (Bruker Daltonics; Bruker Corporation, Billerica, MA, USA) using electrospray ionization (ESI) in positive ion mode.

3.1.2. General Procedures

General Procedure A: The Buchwald-Hartwig Aryl Amination Reaction.

Substituted halide (2.5 mmol, 1.0 eq.) was dissolved in dry dioxane (10 mL) in a dry Schlenk tube, under an argon atmosphere. Pd(OAc)$_2$ (28 mg, 0.125 mmol, 0.05 eq.), Xantphos (72 mg, 0.125 mmol, 0.05 eq.), and Cs$_2$CO$_3$ (2.04 g, 6.25 mmol, 2.5 eq.) were added afterwards and the mixture was stirred at 75 °C. After 20 min, amine (3.0 mmol, 1.2 eq.) was added, and the reaction mixture was conducted at 105 °C under an argon atmosphere. The reaction monitored by TLC and it was completed in 30 min. After cooling to room temperature, the reaction mixture was diluted with diethyl ether (Et$_2$O), the solids were filtered and washed by Et$_2$O. The solvents were evaporated under *vacuum* and the oily residue was then purified by column chromatography.

General Procedure B

To a solution of substituted amine (2.0 mmol, 1.0 eq.) in N,N-dimethylformamide (DMF, 10 mL), NaH (26.0 mmol of a 60% dispersion in mineral oil, 13.0 eq.) was added in small portions under argon. 1-Iodopropane (0.488 mL, 5.0 mmol, 2.5 eq.) was thereafter added to the mixture. After stirring for 1 h at room temperature, the mixture was slowly poured on an ice-water mixture and stirred for 5 min. The mixture was extracted with ethyl acetate (EtOAc, 2 × 25 mL), the extracts were combined, dried with anhydrous MgSO$_4$, and concentrated by evaporation of the solvents under vacuum. The residue was then purified by column chromatography.

General Procedure C

POCl$_3$ (1.03 mL, 11.0 mmol, 1.1 eq.) was added dropwise to DMF (4.6 mL, 60.0 mmol, 6.0 eq.) at 0 °C and the mixture was stirred 30 min at room temperature. To this solution, N,N-disubstituted aniline (10.0 mmol, 1.0 eq.) was added and the reaction mixture was heated up to 80 °C. After 2–4 h, the reaction was quenched by addition of a mixture of ice water, stirred for additional 5 min. The pH was thereafter adjusted to 6–7 by using aqueous 1 M NaOH. The residue was extracted with chloroform

(CHCl$_3$, 3 × 15 mL), and the organic phases were combined, washed with H$_2$O and brine. The extracts were dried with anhydrous MgSO$_4$ and the corresponding N,N-disubstituted benzaldehyde was purified by column chromatography after evaporation of the solvents.

To a solution of the substituted benzaldehyde (5.0 mmol, 1.0 eq.) in 20 mL acetonitrile (ACN), cyanoacetic acid (0.991 mL, 15.0 mmol, 3.0 eq.) and piperidine (0.494 mL, 5.0 mmol, 1.0 eq.) were added and refluxed overnight at 85 °C. Upon the completion of the reaction, as judged by TLC, the above solution was poured into a mixture of 3M HCl (10 mL) on ice. The solution was stirred for 30 min and the solids were filtered using a Büchner funnel and washed with ice-cold ACN. The solid was afterwards poured into adequate amount of n-hexane and stirred for 15 min to remove the remaining aldehyde. The final compounds were obtained in pure form after filtration of the solids and consequent washing with n-hexane.

6-Fluoro-N-(3-methoxyphenyl)pyridin-2-amine (5). The reaction was carried out according to the general procedure A. Column chromatography: silica, EtOAc/n-hexane, 1:2; Yellow oil: 96% yield; TLC: (silica gel, EtOAc/n-hexane, 2:1), R$_f$ = 0.8. ^{1}H NMR (300 MHz, CDCl$_3$) δ 7.56 (q, J = 8.0 Hz, 1H), 7.25 (m, 1H), 6.97 (t, J = 2.2 Hz, 1H), 6.90 (ddd, J = 8.0, 2.1, 0.8 Hz, 1H), 6.73 (dd, J = 8.1, 2.3 Hz, 1H), 6.66 (ddd, J = 8.3, 2.4, 0.8 Hz, 1H), 6.33 (dd, J = 7.8, 2.2 Hz, 1H), 3.83 (s, 3H); ^{13}C NMR (75 MHz, CDCl$_3$) δ 162.87 (d, J = 238.2 Hz), 160.49, 154.92 (d, J = 16.1 Hz), 142.12 (d, J = 8.4 Hz), 140.82, 130.03, 113.09, 108.87, 106.67, 104.30 (d, J = 4.2 Hz), 98.29 (d, J = 36.1 Hz), 55.27; ^{19}F NMR (282 MHz, CDCl$_3$) δ −69.08 (d, J = 8.1 Hz).

6-Fluoro-N-(3-methoxyphenyl)-N-propylpyridin-2-amine (6). The reaction was carried out according to the general procedure B. Column chromatography: silica, EtOAc/n-hexane, 1:3; Yellow oil: 95% yield; TLC: (silica gel, EtOAc/n-hexane, 1:3), R$_f$ = 0.75. ^{1}H NMR (400 MHz, CDCl$_3$) δ 7.35–7.27 (m, 2H), 6.95–6.65 (m, 3H), 6.12 (td, J = 7.6, 2.5 Hz, 2H), 3.85 (d, J = 7.6 Hz, 2H), 3.81 (s, 3H), 1.66 (m, 2H), 0.92 (t, J = 7.4 Hz, 3H); ^{13}C NMR (101 MHz, CDCl$_3$) δ 162.85 (d, J = 234.7 Hz), 160.87, 157.87 (d, J = 16.1 Hz), 145.82, 140.72 (d, J = 8.3 Hz), 130.50, 120.09, 113.64, 111.95, 104.78 (d, J = 4.1 Hz), 95.17 (d, J = 37.4 Hz), 55.30, 51.76, 21.05, 11.26; ^{19}F NMR (377 MHz, CDCl$_3$) δ −69.25 (d, J = 8.3 Hz).

6-Fluoro-N-(3-methoxyphenyl)-N-phenylpyridin-2-amine (7). The reaction was carried out according to the general procedure A. For this compound, higher amounts of Pd(OAc)$_2$ (84 mg, 0.375 mmol, 0.15 eq.) and Xantphos (217 mg, 0.375 mmol, 0.15 eq.) were added stepwise (3 × 0.125 mmol) to the reaction mixture over 24 h. Column chromatography: silica, EtOAc/n-hexane, 1:3; Milky oil: 46% yield; TLC: (silica gel, EtOAc/n-hexane, 1:3), R$_f$ = 0.85. ^{1}H NMR (400 MHz, CDCl$_3$) δ 7.48 (dt, J = 8.4, 7.9 Hz, 1H), 7.38–7.29 (m, 2H), 7.28–7.12 (m, 4H), 6.79 (ddd, J = 7.9, 2.0, 0.9 Hz, 1H), 6.77–6.69 (m, 2H), 6.50 (ddd, J = 8.1, 2.2, 0.5 Hz, 1H), 6.34 (ddd, J = 7.8, 2.9, 0.5 Hz, 1H), 3.75 (s, 3H); ^{13}C NMR (101 MHz, CDCl$_3$) δ 162.43 (d, J = 237.8 Hz), 160.49, 157.50 (d, J = 15.1 Hz), 146.32, 145.06, 141.55 (d, J = 8.1 Hz), 130.02, 129.39, 126.62, 125.23, 119.07, 112.63, 110.74, 109.10 (d, J = 4.5 Hz), 99.41 (d, J = 37.0 Hz), 55.29; ^{19}F NMR (377 MHz, CDCl$_3$) δ −67.46 (d, J = 7.3 Hz).

4-((6-Fluoropyridin-2-yl)(propyl)amino)-2-methoxybenzaldehyde (8). The reaction was carried out according to the general procedure C. Column chromatography: silica, EtOAc/petroleum ether (PE), 1:2; Yellow oil: 57% yield; TLC: (silica gel, EtOAc/PE, 1:2), R$_f$ = 0.55. ^{1}H NMR (300 MHz, CDCl$_3$) δ 10.35 (s, 1H), 7.82 (d, J = 8.4 Hz, 1H), 7.46 (dt, J = 8.5, 7.9 Hz, 1H), 6.88 (ddd, J = 8.4, 2.0, 0.7 Hz, 1H), 6.82 (d, J = 1.9 Hz, 1H), 6.55 (ddd, J = 8.1, 2.4, 0.5 Hz, 1H), 6.30 (ddd, J = 7.8, 2.9, 0.5 Hz, 1H), 3.98–3.89 (m, 2H), 3.88 (s, 3H), 1.69 (dq, J = 14.8, 7.4 Hz, 2H), 0.94 (t, J = 7.4 Hz, 3H); ^{13}C NMR (75 MHz, CDCl$_3$) δ 189.19, 163.05 (d, J = 244.3 Hz), 161.72, 159.68 (d, J = 15.1 Hz), 149.95, 140.12 (d, J = 8.1 Hz), 130.33, 118.77, 111.37 (d, J = 4.5 Hz), 106.62, 100.47 (d, J = 37.0 Hz), 101.76, 55.99, 49.89, 20.95, 11.38; ^{19}F NMR (282 MHz, CDCl$_3$) δ −68.09 (d, J = 8.6 Hz).

4-((6-Fluoropyridin-2-yl)(phenyl)amino)-2-methoxybenzaldehyde (9). The reaction was carried out according to the general procedure C. Column chromatography: silica, EtOAc/n-hexane, 1:3; Yellow oil: 68% yield; TLC: (silica gel, EtOAc/n-hexane, 1:3), R$_f$ = 0.60. ^{1}H NMR (300 MHz, CDCl$_3$) δ 10.32 (s, 1H), 7.73

(d, *J* = 8.5 Hz, 1H), 7.59 (q, *J* = 8.0 Hz, 1H), 7.46–7.37 (m, 2H), 7.33–7.25 (m, 1H), 7.24–7.16 (m, 2H), 6.78 (d, *J* = 2.0 Hz, 1H), 6.69 (ddd, *J* = 8.5, 2.0, 0.8 Hz, 1H), 6.60 (dd, *J* = 8.0, 1.7 Hz, 1H), 6.54–6.46 (m, 1H), 3.77 (s, 3H); ^{13}C NMR (75 MHz, CDCl$_3$) δ 188.37, 162.91 (d, *J* = 235.1 Hz), 162.63,156.61 (d, *J* = 14.9 Hz), 152.14, 144.37, 142.13 (d, *J* = 8.1 Hz), 130.06, 129.50, 127.66, 126.70, 120.65, 116.19, 111.59 (d, *J* = 4.6 Hz), 106.62, 101.86 (d, *J* = 36.6 Hz), 55.60; ^{19}F NMR (282 MHz, CDCl$_3$) δ −66.92 (d, *J* = 8.5 Hz).

(E)-2-Cyano-3-(4-((6-fluoropyridin-2-yl)(propyl)amino)-2-methoxyphenyl)acrylic Acid (1). The reaction was carried out according to the general procedure C. Yellow solid: 95% yield; TLC: (silica gel, CHCl$_3$/MeOH/acetic acid (AcOH), 95:5:0.1), R$_f$ = 0.60. ^{1}H NMR (400 MHz, CDCl$_3$/CD$_3$OD) δ 8.63 (s, 1H), 8.30 (d, *J* = 8.7 Hz, 1H), 7.46 (q, *J* = 8.2 Hz, 1H), 6.86 (dd, *J* = 8.7, 2.1 Hz, 1H), 6.74 (d, *J* = 2.1 Hz, 1H), 6.59 (dd, *J* = 8.1, 2.2 Hz, 1H), 6.31 (dd, *J* = 7.8, 2.8 Hz, 1H), 3.93–3.86 (m, 2H), 3.80 (s, 3H), 1.65 (h, *J* = 7.4 Hz, 2H), 0.90 (t, *J* = 7.4 Hz, 3H); ^{13}C NMR (101 MHz, CDCl$_3$/CD$_3$OD) δ 164.86, 162.72 (d, *J* = 237.5 Hz), 160.52, 156.27 (d, *J* = 15.2 Hz), 151.55, 148.32, 141.46 (d, *J* = 8.3 Hz), 133.13, 130.43, 116.63, 116.39, 108.28 (d, *J* = 4.3 Hz), 106.26, 100.01, 99.02 (d, *J* = 36.9 Hz), 55.72, 51.83, 21.14, 11.14; ^{19}F NMR (376 MHz, CDCl$_3$/CD$_3$OD) δ -68.23 (dt, *J* = 8.3, 2.2 Hz). HR-MS (ESI) *m/z*:calcd. for [C$_{18}$H$_{17}$FN$_3$O]$^+$ = 310.1350; found = 310.1332 [M–CO$_2$H]$^+$.

(E)-2-Cyano-3-(4-((6-fluoropyridin-2-yl)(phenyl)amino)-2-methoxyphenyl)acrylic Acid (2). The reaction was carried out according to the general procedure C. Orange solid: 98% yield; TLC: (silica gel, CHCl$_3$/MeOH/AcOH, 95:5:0.1), R$_f$ = 0.75. ^{1}H NMR (400 MHz, CDCl$_3$) δ 8.75 (s, 1H), 8.37 (d, *J* = 8.8 Hz, 1H), 7.65 (q, *J* = 8.0 Hz, 1H), 7.47 (t, *J* = 7.7 Hz, 2H), 7.41–7.17 (m, 3H), 6.80 (d, *J* = 1.9 Hz, 1H), 6.73 (dd, *J* = 8.8, 1.9 Hz, 1H), 6.69–6.61 (m, 1H), 6.58 (dd, *J* = 7.9, 2.7 Hz, 1H), 3.79 (s, 3H); ^{13}C NMR (101 MHz, CDCl$_3$) δ 167.91, 162.31 (d, *J* = 240.5 Hz), 160.76, 156.25 (d, *J* = 14.6 Hz), 152.17, 149.60, 143.99, 142.34 (d, *J* = 8.2 Hz), 130.35, 130.24, 127.87, 127.16, 116.26, 115.94, 115.58, 112.29 (d, *J* = 4.5 Hz), 105.47, 102.54 (d, *J* = 36.5 Hz), 97.31, 55.82; ^{19}F NMR (377 MHz, CDCl$_3$) δ -66.88 (d, *J* = 8.2 Hz). HR-MS (ESI) m/z: calcd. for [C$_{21}$H$_{15}$FN$_3$O]$^{+.}$ = 344.1194; found = 344.1180 [M–CO$_2$H]$^+$.

N-(3-Methoxyphenyl)-6-nitropyridin-2-amine (12). The reaction was carried out according to the general procedure A. Column chromatography: silica, EtOAc/*n*-hexane, 1:1; Yellow solid: 80% yield; TLC: (silica gel, EtOAc/*n*-hexane, 1:1), R$_f$ = 0.65. ^{1}H NMR (400 MHz, CDCl$_3$) δ 7.78–7.70 (m, 1H), 7.63 (dd, *J* = 7.6, 0.6 Hz, 1H), 7.31–7.24 (m, 1H), 7.23–7.19 (m, 1H), 7.14 (dd, *J* = 8.2, 0.6 Hz, 1H), 6.88 (ddd, *J* = 7.9, 2.0, 0.8 Hz, 1H), 6.71 (ddd, *J* = 8.3, 2.5, 0.8 Hz, 1H), 3.86 (s, 3H); ^{13}C NMR (101 MHz, CDCl$_3$) δ 160.29, 155.31, 155.06, 140.46, 139.94, 129.79, 114.77, 112.21, 109.24, 107.35, 105.65, 55.16.

N-(3-Methoxyphenyl)-6-nitro-N-propylpyridin-2-amine (13). The reaction was carried out according to the general procedure B. Column chromatography: silica, EtOAc/*n*-hexane, 1:3; Yellow oil: 93% yield; TLC: (silica gel, EtOAc/*n*-hexane, 1:3), R$_f$ = 0.70. ^{1}H NMR (400 MHz, CDCl$_3$) δ 7.53–7.29 (m, 3H), 6.88 (dd, *J* = 8.3, 2.3 Hz, 1H), 6.82 (d, *J* = 7.9 Hz, 1H), 6.79–6.68 (m, 1H), 6.59 (d, *J* = 8.2 Hz, 1H), 3.95 (t, *J* = 7.4 Hz, 2H), 3.82 (s, 3H), 1.69 (h, *J* = 7.4 Hz, 2H), 0.95 (t, *J* = 7.4 Hz, 3H); ^{13}C NMR (101 MHz, CDCl$_3$) δ 161.44, 157.92, 156.25, 145.36, 139.22, 131.23, 120.24, 114.28, 113.93, 112.86, 105.78, 55.72, 52.38, 21.21, 11.68.

2-Methoxy-4-((6-nitropyridin-2-yl)(propyl)amino)benzaldehyde (14). The reaction was carried out according to the general procedure C. Column chromatography: silica, EtOAc/PE, 1:2; Yellow oil: 92% yield; TLC: (silica gel, EtOAc/PE, 1:2), R$_f$ = 0.60. ^{1}H NMR (400 MHz, CDCl$_3$) δ 10.40 (s, 1H), 7.89 (d, *J* = 8.3 Hz, 1H), 7.63–7.53 (m, 2H), 6.94–6.89 (m, 2H), 6.88 (d, *J* = 1.9 Hz, 1H), 4.03 (t, *J* = 7.4 Hz, 2H), 3.91 (s, 3H), 1.74 (h, *J* = 7.4 Hz, 2H), 0.97 (t, *J* = 7.4 Hz, 3H); ^{13}C NMR (101 MHz, CDCl$_3$) δ 188.37, 163.12, 156.66, 155.98, 151.33, 139.56, 130.32, 122.69, 118.05, 115.38, 108.86, 107.32, 55.84, 52.30, 21.08, 11.38.

(E)-2-Cyano-3-(2-methoxy-4-((6-nitropyridin-2-yl)(propyl)amino)phenyl)acrylic Acid (15). The reaction was carried out according to the general procedure C. Yellow solid: >98% yield; TLC: (silica gel, CHCl$_3$/MeOH/AcOH, 95:5:0.1), R$_f$ = 0.70. ^{1}H NMR (400 MHz, CDCl$_3$/CD$_3$OD) δ 8.62 (s, 1H), 8.30 (d, *J* = 8.5 Hz, 1H), 7.58 (t, *J* = 7.9 Hz, 1H), 7.51 (d, *J* = 7.6 Hz, 1H), 6.92 (d, *J* = 8.3 Hz, 1H), 6.89 (dd, *J* = 8.6, 1.7 Hz, 1H), 6.80 (d, *J* = 1.5 Hz, 1H), 3.98 (t, *J* = 7.4 Hz, 2H), 3.82 (s, 3H), 1.68 (h, *J* = 7.4 Hz, 2H), 0.91

(t, J = 7.3 Hz, 3H); ^{13}C NMR (101 MHz, CDCl$_3$/CD$_3$OD) δ 164.39, 160.51, 156.43, 155.70, 150.30, 148.24, 139.62, 130.62, 118.14, 117.66, 116.35, 115.78, 107.90, 107.54, 101.63, 55.82, 52.13, 20.95, 11.16. HR-MS (ESI) *m/z* calcd. for [C$_{18}$H$_{17}$N$_4$O$_3$]$^{+\cdot}$= 337.1295; found = 337.1283 [M–CO$_2$H]$^{+\cdot}$.

3.2. Assessment of the MCT1 Inhibition by [^{14}C]lactate Uptake Assay

Inhibition of the MCT1-mediated lactate transport was determined using the rat brain endothelial cell line (RBE4, a gift from F. Roux research group [40]) as previously described [26]. PCR analysis and Western Blot demonstrated that only the MCT1 isoform is expressed by RBE4 cells.

3.3. Radiochemistry

3.3.1. General Methods

No-carrier-added [^{18}F]fluoride was produced via the [^{18}O(p,n)^{18}F] nuclear reaction by irradiation of an [^{18}O]H$_2$O target (Hyox 18 enriched water, Rotem Industries Ltd., Arava, Israel) on a Cyclone 18/9 (IBA RadioPharma Solutions, Lourain-La-Neuve, Belgium) with fixed energy proton beam using Nirta [^{18}F]fluoride XL target. Radio thin layer chromatography (radio-TLC) was performed on silica gel (ALUGRAM SIL G/UV254, Machery-Nagel, Düren, Germany) pre-coated plates with a mixture of dichloromethane (DCM)/MeOH (4:1) as eluent. The plates were exposed to storage phosphor screens (BAS IP MS 2025 E, GE Healthcare Europe GmbH, Freiburg, Germany) and recorded using the Amersham Typhoon RGB Biomolecular Imager (GE Healthcare Life Sciences). Images were quantified with the ImageQuant TL8.1 software (GE Healthcare Life Sciences).

Analytical chromatographic separations were performed on a JASCO LC-2000 system, incorporating a PU-2080Plus pump, AS-2055Plus auto injector (100 μL sample loop), and a UV-2070Plus detector coupled with a gamma radioactivity HPLC flow detector (Gabi Star, raytest Isotopenmessgeräte GmbH, Straubenhardt, Germany). Data analysis was performed with the Galaxy chromatography software (version 1.10.0.5590, Agilent Technologies Deutschland GmbH, Waldbronn, Germany) using the chromatograms obtained at 254 and 370 nm. Reprosil-Pur C18-AQ column (250 × 4.6 mm; 5 μm; Dr. Maisch HPLC GmbH; Ammerbuch-Entingen, Germany) with an eluent mixture of ACN/20 mM NH$_4$OAc (aq., pH 6.8) and a flow of 1.0 mL/min were used (gradient: eluent A 10% ACN/20 mM NH$_4$OAc (aq.); eluent B 90% ACN/20 mM NH$_4$OAc (aq.); 0–5 min 100% A, 5–20 min up to 62% B, 20–21 min up to 100% B, 21–26 min 100% B, 26–27 min up to 100% A, 27–35 min 100% A; isocratic: 30% MeOH/20 mM NH$_4$OAc (aq.), flow: 0.75 mL/min, UV detection at 370 nm).

Semi-preparative HPLC separation was performed by using the HPLC system implemented in the TRACERlab FX2 N synthesizer (GE Healthcare). A Reprosil-Pur 120 CN column (250 × 20 mm, 10 μm, Dr. Maisch HPLC GmbH, Germany) was used with 45% MeOH/20 mM NH$_4$OAc (aq.) as eluent mixture and a flow rate of 7.0 mL/min. The ammonium acetate concentration stated as aq. 20 mM NH$_4$OAc, corresponds to the concentration in the aqueous component of an eluent mixture.

The molar activities were determined based on a calibration curve carried out under isocratic HPLC conditions (Reprosil-Pur C18-AQ, 250 × 4.6 mm, 30% MeOH/20 mM NH$_4$OAc (aq.), flow: 0.75 mL/min) using chromatograms obtained at 370 nm as an appropriate maximum of UV absorbance.

3.3.2. Radiosynthesis of [^{18}F]1

No carrier added [^{18}F]fluoride in 1.5 mL water was trapped on a Sep-Pak Accell Plus QMA Carbonate Plus light cartridge (Waters GmbH, Eschborn, Germany). The activity was eluted with 400 μL of an aqueous solution of potassium carbonate (K$_2$CO$_3$, 1.8 mg, 13 μmol) into a 4 mL V-shape vial prefilled with Kryptofix® 2.2.2 (K$_{2.2.2}$, 11 mg, 29 μmol) in 1 mL ACN. The aqueous [^{18}F]fluoride was azeotropically dried under vacuum and nitrogen flow within 7–10 min using a single mode microwave (75 W, at 50–60 °C, power cycling mode; Discover PETWave from CEM GmbH, Kamp-Lintfort, Germany) [60]. Two aliquots of ACN (2 × 1.0 mL) were added during the drying procedure and the final complex was obtained in a dried form. A solution of 1.0 mg of the nitro precursor **15** in 750 μL

DMSO was added, and the ^{18}F-labeling was performed at 130 °C. To determine the radiochemical yields of the labeling process, samples were taken at different time points (5, 10, 15, and 20 min) and analyzed by radio-HPLC and radio-TLC.

After cooling to < 30 °C, the reaction mixture was diluted with 2.0 mL aqueous NH_4HCO_2 (adjusted to pH 4 with formic acid) and 2.0 mL MeOH/water (1:/1) and directly applied to an isocratic semi-preparative RP-HPLC for isolation of [^{18}F]1. The collected radiotracer fraction was diluted with 40 mL water to perform final purification by sorption on a Sep-Pak® C18 light cartridge (Waters, GmbH, Eschborn, Germany) and successive elution with 1.3 mL of ethanol. The ethanolic solution was concentrated under a gentle argon stream at 70 °C to a final volume of 10–50 µL. Afterwards, the radiotracer was diluted in isotonic saline to obtain a final product containing 10% of EtOH (v/v).

3.3.3. Determination of In Vitro Stability and log $D_{7.4}$ of [^{18}F]1

In vitro stability was investigated by incubation of about 5 MBq of [^{18}F]1 in phosphate-buffered saline (PBS, pH 7.4), n-octanol, saline and ethanol (500 µL) at 40 °C up to 60 min. Samples were taken at 60 min and the radio-TLC and radio-HPLC analyses were performed.

The lipophilicity of [^{18}F]1 was experimentally determined by partitioning a small amount of the radiotracer between n-octanol and PBS (pH 7.4) at room temperature using the conventional shake-flask method. The radiotracer (15 µL, ~800 kBq) was added to the tubes containing a mixture of n-octanol and PBS (3.0 mL, n = 4). After shaking for 20 min using the mechanical shaker (HS250 basic, IKA Labortechnik GmbH & Co. KG, Staufen, Germany), the samples were centrifuged for 5 min at 5000 rpm, followed by separation of the phases. Aliquots were taken from the organic and the aqueous phases (1 mL of each) and activity was measured in a γ-counter (PerkinElmer Wallac Wizard 1480 Gamma Counter, manufactured by WALLAC, Turku, Finland). For the second extraction, another 1 mL aliquot of the organic layer was mixed with 2.0 mL n-octanol and 3.0 mL of PBS and was subjected to the same procedure. The distribution coefficient (D) was calculated as [activity (cpm/mL) in n-octanol]/[activity (cpm/mL) in PBS, pH 7.4] stated as the decade logarithm (log $D_{7.4}$).

3.4. Animal Studies

3.4.1. General Information

All experimental work including animals has been conducted in accordance with the national legislation on the use of animals for research (Tierschutzgesetz (TierSchG), Tierschutz-Versuchstierverordnung (TierSchVersV)) and has been approved by the responsible research ethics committee (TVV 18/18, DD24.1-5131/446/19, Landesdirektion Sachsen, 20 June, 2018). Female CD-1 mice, 10–12 weeks, were obtained from the Medizinisch-ExperimentellesZentrum at Universität Leipzig. For the time of the experiments, the animals were kept in a dedicated climatic chamber with free access to water and food under a 12:12 h dark:light cycle at a constant temperature of 24 °C.

3.4.2. In Vitro Autoradiographic Analysis of Binding Sites of [^{18}F]1 in Mouse Kidney

The kidneys was isolated after cervical dislocation from one CD-1 mouse, frozen rapidly in isopentane at −25 °C for 5 min, and stored at −25 °C until the sectioning. Cryosections (transversal, 10 µm) were obtained using a microtome (MICROM HM560, Thermo Scientific Microm, Fisher Scientific GmbH, Schwerte, Germany), mounted on microscopy slides (SuperFrost, Thermo Scientific Menzel, Fisher Scientific GmbH, Schwerte, Germany), dried for ~2 h at room temperature, and stored at −25 °C until the autoradiography study. For the experiment, the slides were taken out from the freezer, the cryosections dried under a stream of cold air, and pre-incubated with PBS for 15 min at room temperature. The incubation solution was decanted, the slices dried again under a stream of cold air, and covered afterwards with the incubation solution ([^{18}F]1, 198 kBq/mL PBS or 1.19 nM at the time of incubation, without (total binding) or with co-incubation with 10^{-5}, 10^{-7}, and 10^{-9} M α-CHC-Na). Incubation at room temperature was terminated after 60 min, the slides were washed twice in 50 mM

TRIS-HCl, pH 7.4 at 4 °C, on ice for two minutes each followed by dipping in ice-cold demineralized water for 5 s and rapid drying under a stream of cold air. Afterwards, the slides were exposed to a phosphor imager plate (BAS-IP TR 2025, FujiFilm Corporation, Tokyo, Japan) along with standards obtained by pipetting and drying 1 µL of each concentration of a serial dilution of the radiotracer solution on to a microscopic slide.

3.4.3. Analysis of Radiometabolites of [^{18}F]1 in Mouse

The radiotracer [^{18}F]1 (32.84 MBq in 200 µL isotonic saline) was injected in an awake female CD-1 mouse (12 weeks old, 31 g) via the tail vein. After 30 min, the animal was slightly anesthetized with isoflurane, and blood was collected from retro-orbital bleeding followed by cervical dislocation. Blood plasma was separated by centrifugation of the blood sample at 10,000 rpm for 2 min. Brain was also isolated, cleaned by pouring PBS and homogenized in demineralized water (~2 mL/g tissue) in a borosilicate glass with a PTFE plunger by 10 strokes at 1000 rpm (POTTER S, Homogenizer, B. Braun Biotech, Sartorius AG, Göttingen, Germany). Blood plasma and brain homogenate were provided for preparation of samples for radio-HPLC analyses as described in the following.

RP-HPLC: Protein precipitation was performed by addition of an ice-cold ACN/H$_2$O mixture (9:1) to the plasma and brain samples in a ratio of 4:1 (*v/v*: solvent/tissue sample, n = 2). The samples were vortexed for 2 min, incubated on ice for 3 min, and the suspensions were centrifuged at 10,000 rpm at 4 °C for 5 min. For the second extraction, the precipitates were re-dissolved in 100 µL of the solvent mixture, vortexed for 3 min, incubated on ice for 5 min and subjected to the same centrifuging procedure. The combined supernatants (total volume between 1–2 mL) were concentrated at 70 °C under argon flow to a final volume of approximately 100 µL and were analyzed by analytical radio-HPLC. A Reprosil-Pur C18-AQ column (250 mm × 4.6 mm; 5 µm) was used as stationary phase and elution was done using the gradient mode as described in Section 3.3.1. To determine the percentage of radioactivity in the supernatants compared to total activity, aliquots of each step as well as the precipitates were quantified by gamma counting.

MLC: Preparation of the plasma and brain samples was performed as already described [47,48]. A Reprosil-Pur C18-AQ column (250 × 4.6 mm, particle size: 10 µm) coupled with a pre-column of 10 mm length was used. Separations were performed by using an isocratic mode with an aqueous eluent containing 50 mM sodium dodecyl sulfate and 10 mM Na$_2$HPO$_4$ at a flow rate of 1.0 mL/min.

3.4.4. Small Animal Micro-PET/MR Studies

The dynamic biodistribution of the radiotracers was assessed by small animal PET (nanoscan, Mediso, Hungary) over 60 min PET with a subsequently T1 weighted MR. Anaesthetized (2% isofluran, carrier gas mixture of 40% air and 60% O$_2$) female CD-1 mice (bodyweight = 30.3 ± 1.1 g) were kept during imaging on a heated animal bed to sustain body temperature and were pretreated with vehicle (0.9% saline) and α-CHC-Na (25 mg/kg bodyweight) prior to tracer application ([^{18}F]1: 5.8 ± 0.2 MBq, 1.1 fmol/g bodyweight and [^{18}F]FACH: 5.9 ± 0.5 MBq, 3.7 ± 1.8 fmol/g), whereby all injections were administered intravenously. The acquisitions were performed in normal mode and a coincidence Mode 1–5. For subsequent dynamic reconstructions, list mode data were sorted into sinograms (12 × 10 s, 6 × 30 s, 5 × 60 s, and 10 × 300 s). The frames were reconstructed by Ordered Subset Expectation Maximization applied to 3D sinograms (OSEM3D) with an attenuation correction with 4 iterations, 6 subsets and a voxel size of 0.4 mm^3 (Nucline v2.01, Mediso, Hungary). Analyses of reconstructed studies were performed with PMOD software (v4.005, PMOD Technologies LLC, Zurich, Switzerland) and results are expressed in Standardized Uptake Value (SUV).

4. Conclusions

In summary, two new analogs of FACH, **1** and **2**, were synthesized and the former, with moderate MCT1 inhibition, was regarded to be a good PET candidate and therefore chosen for labeling with fluorine-18. Although the partition coefficient log $D_{7.4}$ of [^{18}F]1 was 2-fold higher than the one of [^{18}F]FACH, the brain accumulation of both radiotracers was in a similar moderate range. This demonstrates that log $D_{7.4}$ alone does not govern passive diffusion into the brain, which is also reflected by comparing log $D_{7.4}$ with calculated log K_{BB} (brain-blood partition coefficient). Nevertheless, the high uptake of [^{18}F]1 in kidney and other peripheral MCT-expressing organs together with the strong inhibition by specific drugs provide evidence that this radiotracer is suitable for future investigation of MCT imaging with PET. Moreover, these results suggest that further structural modifications towards improving MCT-mediated transport might result in higher brain uptake in vivo.

5. Patents

The nitro- or bromo- precursor for the radiosynthesis of the structurally modified analogs of FACH is subject of a patent application by Helmholtz-Zentrum Dresden-Rossendorf (DMPA registration No. 10 2019 112 040.3) with following contributing inventors: Rareş-Petru Moldovan, Masoud Sadeghzadeh, Barbara Wenzel, Mathias Kranz, Steffen Fischer, Rodrigo Teodoro, Friedrich-Alexander Ludwig, Magali Toussaint and Peter Brust.

Supplementary Materials: The following are available online. Figure S1: (**A**) Representative semi-preparative radio- and UV-HPLC chromatograms representing two peaks a/b which are supposed to reflect neutral and deprotonated form of the radiotracer ([^{18}F]**1a/b**) (conditions: Reprosil-Pur 120 CN, 10 µm, 250 × 20 mm, 45% MeOH/20 mM NH$_4$OAc (aq.), 7.0 mL/min); (**B**) Analytical radio- and UV-HPLC chromatograms detected two peaks a/b which assumed to be the neutral and deprotonated form of the final product ([^{18}F]**1a/b**) in the sample solution co-eluted with the non-radioactive reference 1 (conditions: Reprosil-Pur C18-AQ, 250 × 4.6 mm, gradient with an eluent mixture of ACN/20 mM NH$_4$OAc (aq.), 1.0 mL/min)., Figure S2: Analytical UV- and radio-HPLC chromatograms representing two peaks a/b which are supposed to reflect the neutral and deprotonated form of the radiotracer ([^{18}F]**1a/b**) in the mouse brain sample at 30 min p.i. measured under reversed-phase conditions (Reprosil-Pur C18-AQ, 250 × 4.6 mm, gradient with an eluent mixture of ACN/20 mM NH$_4$OAc (aq.), 370 nm, 1.0 mL/min).

Author Contributions: Conceptualization, M.S., B.W., W.D.-C., R.-P.M., L.R.D. and P.B.; Data curation, D.G., M.T., F.-A.L., and G.S.; Formal analysis, D.G., W.D.-C., F.L., S.J., and S.K.J.; Funding acquisition, M.S., L.R.D., and P.B.; Investigation, M.S., B.W., D.G., W.D.-C., M.T., S.J., S.K.J., G.S., and V.R.M.; Methodology, M.S., B.W., S.F., R.T., G.S., and V.R.M.; Project administration, M.S. and P.B.; Resources, W.D.-C., R.T., S.F., and V.R.M.; Software, D.G. and G.S.; Supervision, L.R.D. and P.B.; Validation, B.W., D.G., W.D.-C, R.-P.M., F.-A.L., R.T., S.K.J., V.R.M., L.R.D., and P.B.; Visualization, M.S.; Writing—original draft, M.S., B.W., G.S., and D.G.; Writing—review & editing, M.S., B.W., D.G., W.D.-C., M.T., R.-P.M., S.F., F.-A.L., R.T., S.J., S.K.J., G.S., V.R.M., L.R.D., and P.B. All authors have read and agreed to the published version of the manuscript.

Funding: Masoud Sadeghzadeh was kindly financially supported by the Alexander von Humboldt Foundation and L. R. Drewes was supported by the University of Minnesota Foundation. This research received no further external funding.

Acknowledgments: The Alexander von Humboldt Foundation and the University of Minnesota Duluth are acknowledged for financial supports. We are thankful to K. Franke and A. Mansel for providing [^{18}F]fluoride as well as Matthias Scheunemann and Susann Schröder for their scientific supports. We also thank the staff of the Institute of Analytical Chemistry, Department of Chemistry and Mineralogy of the University of Leipzig, for recording and processing the NMR and HR-MS spectra.

Conflicts of Interest: The authors declare no conflict of interest. The Humboldt Foundation had no role in the design of the study; in the collection, analyses, or interpretation of data; in the writing of the manuscript, or in the decision to publish the results.

References

1. Halestrap, A.P.; Meredith, D. The SLC16 gene family—From monocarboxylate transporters (MCTs) to aromatic amino acid transporters and beyond. *Pflüger, Archiv für die Gesammte Physiologie des Menschen und der Thiere* **2004**, *447*, 619–628. [CrossRef] [PubMed]
2. Jones, R.S.; Morris, A.M.E. Monocarboxylate Transporters: Therapeutic Targets and Prognostic Factors in Disease. *Clin. Pharmacol. Ther.* **2016**, *100*, 454–463. [CrossRef] [PubMed]
3. Halestrap, A.P.; Wilson, M.C. The monocarboxylate transporter family—Role and regulation. *IUBMB Life* **2011**, *64*, 109–119. [CrossRef]

4. Hong, C.S.; Graham, N.A.; Gu, W.; Camacho, C.E.; Mah, V.; Maresh, E.L.; Alavi, M.; Bagryanova, L.; Krotee, P.A.L.; Gardner, B.K.; et al. MCT1 Modulates Cancer Cell Pyruvate Export and Growth of Tumors that Co-express MCT1 and MCT4. *Cell Rep.* **2016**, *14*, 1590–1601. [CrossRef] [PubMed]
5. Park, S.J.; Smith, C.P.; Wilbur, R.R.; Cain, C.P.; Kallu, S.R.; Valasapalli, S.; Sahoo, A.; Guda, M.R.; Tsung, A.J.; Velpula, K.K. An overview of MCT1 and MCT4 in GMB: Small molecule transporters with large implications. *Am. J. Cancer Res.* **2018**, *8*, 1967–1976. [PubMed]
6. Cao, Y.-W.; Liu, Y.; Dong, Z.; Guo, L.; Kang, E.-H.; Wang, Y.-H.; Zhang, W.; Niu, H.-T. Monocarboxylate transporters MCT1 and MCT4 are independent prognostic biomarkers for the survival of patients with clear cell renal cell carcinoma and those receiving therapy targeting angiogenesis. *Urol. Oncol. Semin. Orig. Investig.* **2018**, *36*, 311.e15–311.e25. [CrossRef]
7. Marchiq, I.; Pouysségur, J. Hypoxia, cancer metabolism and the therapeutic benefit of targeting lactate/H(+) symporters. *J. Mol. Med.* **2015**, *94*, 155–171. [CrossRef]
8. Roy, D.; Sheng, G.Y.; Herve, S.; Carvalho, E.; Mahanty, A.; Yuan, S.; Sun, L. Interplay between cancer cell cycle and metabolism: Challenges, targets and therapeutic opportunities. *Biomed. Pharm.* **2017**, *89*, 288–296. [CrossRef]
9. Zhu, A.; Lee, D.; Shim, H. Metabolic positron emission tomography imaging in cancer detection and therapy response. *Semin. Oncol.* **2011**, *38*, 55–69. [CrossRef]
10. O'Neill, H.; Malik, V.; Johnston, C.; Reynolds, J.V.; O'Sullivan, J. Can the Efficacy of [^{18}F]FDG-PET/CT in Clinical Oncology Be Enhanced by Screening Biomolecular Profiles? *Pharmaceuticals* **2019**, *12*, 16. [CrossRef]
11. Endo, K.; Oriuchi, N.; Higuchi, T.; Iida, Y.; Hanaoka, H.; Miyakubo, M.; Ishikita, T.; Koyama, K. PET and PET/CT using ^{18}F-FDG in the diagnosis and management of cancer patients. *Int. J. Clin. Oncol.* **2006**, *11*, 286–296. [CrossRef] [PubMed]
12. Ganapathy-Kanniappan, S.; Geschwind, J.-F. Tumor glycolysis as a target for cancer therapy: Progress and prospects. *Mol. Cancer* **2013**, *12*, 152. [CrossRef] [PubMed]
13. Draoui, N.; Feron, O. Lactate shuttles at a glance: From physiological paradigms to anti-cancer treatments. *Dis. Model. Mech.* **2011**, *4*, 727–732. [CrossRef] [PubMed]
14. Ocaña, M.C.; Poveda, B.M.; Quesada, A.R.; Medina, M. Ángel Metabolism within the tumor microenvironment and its implication on cancer progression: An ongoing therapeutic target. *Med. Res. Rev.* **2018**, *39*, 70–113. [CrossRef] [PubMed]
15. Beinat, C.; Patel, C.; Haywood, T.; Murty, S.; Alam, I.; Xie, Y.; Gandhi, H.; Holley, D.; Gambhir, S. Evaluation of [^{18}F]DASA-23 for non-invasive measurement of aberrantly expressed pyruvate kinase M2 in glioblastoma: Preclinical and first in human studies. *J. Nucl. Med.* **2019**, *60*, 52.
16. Fu, Y.; Liu, S.; Yin, S.; Niu, W.; Xiong, W.; Tan, M.; Li, G.; Zhou, M. The reverse Warburg effect is likely to be an Achilles' heel of cancer that can be exploited for cancer therapy. *Oncotarget* **2017**, *8*, 57813–57825. [CrossRef]
17. Herrero, P.; Dence, C.S.; Coggan, A.R.; Kisrieva-Ware, Z.; Eisenbeis, P.; Gropler, R.J. L-3-^{11}C-Lactate as a PET Tracer of Myocardial Lactate Metabolism: A Feasibility Study. *J. Nucl. Med.* **2007**, *48*, 2046–2055. [CrossRef]
18. Yokoi, F.; Hara, T.; Iio, M.; Nonaka, I.; Satoyoshi, E. 1-[^{11}C]pyruvate turnover in brain and muscle of patients with mitochondrial encephalomyopathy. A study with positron emission tomography (PET). *J. Neurol. Sci.* **1990**, *99*, 339–348. [CrossRef]
19. Graham, K.; Müller, A.; Lehmann, L.; Koglin, N.; Dinkelborg, L.; Siebeneicher, H. [^{18}F]Fluoropyruvate: Radiosynthesis and initial biological evaluation. *J. Label. Compd. Radiopharm.* **2014**, *57*, 164–171. [CrossRef]
20. Van Hée, V.F.; Labar, D.; Dehon, G.; Grasso, D.; Grégoire, V.; Muccioli, G.G.; Frédérick, R.; Sonveaux, P. Radiosynthesis and validation of (±)-[^{18}F]-3-fluoro-2-hydroxypropionate ([18F]-FLac) as a PET tracer of lactate to monitor MCT1-dependent lactate uptake in tumors. *Oncotarget* **2017**, *8*, 24415. [CrossRef]
21. Tateishi, H.; Tsuji, A.B.; Kato, K.; Sudo, H.; Sugyo, A.; Hanakawa, T.; Zhang, M.-R.; Saga, T.; Arano, Y.; Higashi, T. Synthesis and evaluation of ^{11}C-labeled coumarin analog as an imaging probe for detecting monocarboxylate transporters expression. *Bioorganic Med. Chem. Lett.* **2017**, *27*, 4893–4897. [CrossRef] [PubMed]
22. Wang, X.; Levi, A.J.; Halestrap, A.P. Substrate and inhibitor specificities of the monocarboxylate transporters of single rat heart cells. *Am. J. Physiol. Circ. Physiol.* **1996**, *270*, H476–H484. [CrossRef] [PubMed]
23. Hildyard, J.C.W.; Halestrap, A.P. Identification of the mitochondrial pyruvate carrier in Saccharomyces cerevisiae. *Biochem. J.* **2003**, *374*, 607–611. [CrossRef] [PubMed]

24. Gurrapu, S.; Jonnalagadda, S.; Alam, M.A.; Nelson, G.L.; Sneve, M.G.; Drewes, L.R.; Mereddy, V.R. Monocarboxylate Transporter 1 Inhibitors as Potential Anticancer Agents. *ACS Med. Chem. Lett.* **2015**, *6*, 558–561. [CrossRef] [PubMed]
25. Jonnalagadda, S.; Jonnalagadda, S.K.; Ronayne, C.T.; Nelson, G.L.; Solano, L.N.; Rumbley, J.; Holy, J.; Mereddy, V.R.; Drewes, L.R. Novel N,N-dialkyl cyanocinnamic acids as monocarboxylate transporter 1 and 4 inhibitors. *Oncotarget* **2019**, *10*, 2355–2368. [CrossRef] [PubMed]
26. Sadeghzadeh, M.; Moldovan, R.-P.; Fischer, S.; Wenzel, B.; Ludwig, F.-A.; Teodoro, R.; Deuther-Conrad, W.; Jonnalagadda, S.; Jonnalagadda, S.K.; Gudelis, E.; et al. Development and radiosynthesis of the first ^{18}F-labeled inhibitor of monocarboxylate transporters (MCTs). *J. Label. Compd. Radiopharm.* **2019**, *62*, 411–424. [CrossRef]
27. Sadeghzadeh, M.; Moldovan, R.; Wenzel, B.; Kranz, M.; Deuther-Conrad, W.; Toussaint, M.; Fischer, S.; Ludwig, F.-A.; Teodoro, R.; Jonnalagadda, S.; et al. Development of the first F-18-labeled MCT1/MCT4 lactate transport inhibitor: Radiosynthesis and preliminary in vivo evaluation in mice. In Proceedings of the 23rd International Symposium on Radiopharmaceutical Sciences, Beijing, China, 26–31 May 2019; Volume 62, pp. S59–S60.
28. Sadeghzadeh, M.; Moldovan, R.-P.; Teodoro, R.; Brust, P.; Wenzel, B. One-step radiosynthesis of the MCTs imaging agent [^{18}F]FACH by aliphatic 18F-labelling of a methylsulfonate precursor containing an unprotected carboxylic acid group. *Sci. Rep.* **2019**, *9*, 1–8. [CrossRef]
29. Colen, C.B.; Seraji-Bozorgzad, N.; Marples, B.; Galloway, M.; Sloan, A.E.; Mathupala, S.P. Metabolic Remodeling Of Malignant Gliomas For Enhanced Sensitization During Radiotherapy. *Neurosurgery* **2006**, *59*, 1313–1324. [CrossRef]
30. Mathupala, S.P.; Colen, C.B.; Parajuli, P.; Sloan, A.E. Lactate and malignant tumors: A therapeutic target at the end stage of glycolysis. *J. Bioenerg. Biomembr.* **2007**, *39*, 73–77. [CrossRef]
31. Omori, K.; Tachikawa, M.; Hirose, S.; Taii, A.; Akanuma, S.-I.; Hosoya, K.-I.; Terasaki, T. Developmental changes in transporter and receptor protein expression levels at the rat blood-brain barrier based on quantitative targeted absolute proteomics. *Drug Metab. Pharmacokinet.* **2020**, *35*, 117–123. [CrossRef]
32. Colen, C.B.; Shen, Y.; Ghoddoussi, F.; Yu, P.; Francis, T.B.; Koch, B.J.; Monterey, M.D.; Galloway, M.; Sloan, A.E.; Mathupala, S.P. Metabolic Targeting of Lactate Efflux by Malignant Glioma Inhibits Invasiveness and Induces Necrosis: An In Vivo Study. *Neoplasia* **2011**, *13*, 620–632. [CrossRef] [PubMed]
33. Miranda-Gonçalves, V.; Honavar, I.; Pinheiro, C.; Martinho, O.; Pires, M.; Pinheiro, C.; Cordeiro, M.; Bebiano, G.; Costa, P.; Palmeirim, I.; et al. Monocarboxylate transporters (MCTs) in gliomas: Expression and exploitation as therapeutic targets. *Neuro Oncol.* **2012**, *15*, 172–188. [CrossRef] [PubMed]
34. Serafimova, I.M.; Pufall, M.A.; Krishnan, S.; Duda, K.; Cohen, M.S.; Maglathlin, R.L.; McFarland, J.M.; Miller, R.M.; Frödin, M.; Taunton, J. Reversible targeting of noncatalytic cysteines with chemically tuned electrophiles. *Nat. Methods* **2012**, *8*, 471–476. [CrossRef] [PubMed]
35. Halestrap, A.P. The mechanism of the inhibition of the mitochondrial pyruvate transportater by α-cyanocinnamate derivatives. *Biochem. J.* **1976**, *156*, 181–183. [CrossRef] [PubMed]
36. Koley, M.; Wimmer, L.; Schnürch, M.; Mihovilovic, M.D. Regioselective Syntheses of 2,3-Substituted Pyridines by Orthogonal Cross-Coupling Strategies. *Eur. J. Org. Chem.* **2011**, *2011*, 1972–1979. [CrossRef]
37. Begouin, A.; Hesse, S.; Queiroz, M.J.R.P.; Kirsch, G. Palladium-Catalyzed Buchwald–Hartwig Coupling of Deactivated Aminothiophenes with Substituted Halopyridines. *Eur. J. Org. Chem.* **2007**, *2007*, 1678–1682. [CrossRef]
38. Schumacher, R.A.; Hopper, A.T.; Tehim, A.; Hess, H.-E.; Unterbeck, A.; Kuester, E.; Brubaker, W.F.; Dunn, R.F. Phosphodiesterase 4 Inhibitors, including N-Substituted Aniline and Diphenylamine Analogs. U.S. Patents 7,405,230 B2, 29 July 2008.
39. Vilsmeier, A.; Haack, A. Über die einwirkung von halogenphosphor auf alkyl-formanilide. Eine neue methode zur darstellung sekundärer und tertiärer P-alkylamino-benzaldehyde. *Eur. J. Inorg. Chem.* **1927**, *60*, 4. [CrossRef]
40. Roux, F.; Couraud, P.-O. Rat brain endothelial cell lines for the study of blood-brain barrier permeability and transport functions. *Cell. Mol. Neurobiol.* **2005**, *25*, 41–57. [CrossRef]
41. Pike, V.W. PET radiotracers: Crossing the blood–brain barrier and surviving metabolism. *Trends Pharmacol. Sci.* **2009**, *30*, 431–440. [CrossRef]
42. US EPA. *Estimation Programs Interface Suite™ for Microsoft® Windows*; Version 4.1; EPA: Washington, DC, USA, 2012.

43. *ACD/Percepta*, Version 2726; ACD/Labs 2015 Release (Build 2726 27 November 2014); ACD/Labs: Toronto, ON, Canada. Available online: www.acdlabs.com (accessed on 4 May 2020).
44. Schüürmann, G. Ecotoxic Modes of Action of Chemical Substances. In *Ecotoxicology: Ecological Fundamentals, Chemical Exposure and Biological Effects*; Schüürmann, G., Markert, B., Eds.; John Wiley; Spektrum Akademischer Verlag: New York, NY, USA, 1998; pp. 665–749.
45. Wang, Q.; Lu, Y.; Yuan, M.; Darling, I.M.; Repasky, E.A.; Morris, A.M.E. Characterization of Monocarboxylate Transport in Human Kidney HK-2 Cells. *Mol. Pharm.* **2006**, *3*, 675–685. [CrossRef]
46. Becker, H.M.; Mohebbi, N.; Perna, A.; Ganapathy, V.; Capasso, G.; Wagner, C.A. Localization of members of MCT monocarboxylate transporter family *Slc16* in the kidney and regulation during metabolic acidosis. *Am. J. Physiol. Physiol.* **2010**, *299*, F141–F154. [CrossRef] [PubMed]
47. Lindemann, M.; Hinz, S.; Deuther-Conrad, W.; Namasivayam, V.; Dukic-Stefanovic, S.; Teodoro, R.; Toussaint, M.; Kranz, M.; Juhl, C.; Steinbach, J.; et al. Radiosynthesis and in vivo evaluation of a fluorine-18 labeled pyrazine based radioligand for PET imaging of the adenosine A 2B receptor. *Bioorganic Med. Chem.* **2018**, *26*, 4650–4663. [CrossRef] [PubMed]
48. Nakao, R.; Amini, N.; Halldin, C. Simultaneous Determination of Protein-Free and Total Positron Emission Tomography Radioligand Concentrations in Plasma Using High-Performance Frontal Analysis Followed by Mixed Micellar Liquid Chromatography: Application to [^{11}C]PBR28 in Human Plasma. *Anal. Chem.* **2013**, *85*, 8728–8734. [CrossRef] [PubMed]
49. Pierre, K.; Pellerin, L. Monocarboxylate transporters in the central nervous system: Distribution, regulation and function. *J. Neurochem.* **2005**, *94*, 1–14. [CrossRef] [PubMed]
50. Vijay, N.; Morris, A.M.E. Role of monocarboxylate transporters in drug delivery to the brain. *Curr. Pharm. Des.* **2014**, *20*, 1487–1498. [CrossRef] [PubMed]
51. Spector, R. Micronutrient Homeostasis in Mammalian Brain and Cerebrospinal Fluid. *J. Neurochem.* **1989**, *53*, 1667–1674. [CrossRef]
52. Friedland, R.P.; Mathis, C.A.; Budinger, T.F.; Moyer, B.R.; Rosen, M. Labeled choline and phosphorylcholine: Body distribution and brain autoradiography: Concise communication. *J. Nucl. Med.* **1983**, *24*, 812–815.
53. Rosen, M.A.; Jones, R.M.; Yano, Y.; Budinger, T.F. Carbon-11 choline: Synthesis, purification, and brain uptake inhibition by 2-dimethylaminoethanol. *J. Nucl. Med.* **1985**, *26*, 1424–1428.
54. Hara, T.; Kosaka, N.; Shinoura, N.; Kondo, T. PET imaging of brain tumor with [methyl-^{11}C]choline. *J. Nucl. Med.* **1997**, *38*, 842–847.
55. Dardel, N.T.; Gómez-Río, M.; Triviño-Ibáñez, E.; Llamas-Elvira, J.M. Clinical applications of PET using C-11/F-18-choline in brain tumours: A systematic review. *Clin. Transl. Imaging* **2017**, *5*, 101–119. [CrossRef]
56. Evangelista, L.; Briganti, A.; Fanti, S.; Joniau, S.; Reske, S.; Schiavina, R.; Stief, C.; Thalmann, G.N.; Picchio, M. New Clinical Indications for ^{18}F/^{11}C-choline, New Tracers for Positron Emission Tomography and a Promising Hybrid Device for Prostate Cancer Staging: A Systematic Review of the Literature. *Eur. Urol.* **2016**, *70*, 161–175. [CrossRef] [PubMed]
57. Bergersen, L.H. Is lactate food for neurons? Comparison of monocarboxylate transporter subtypes in brain and muscle. *Neuroscience* **2007**, *145*, 11–19. [CrossRef]
58. Perez-Escuredo, J.; Van Hée, V.; Sboarina, M.; Falces, J.; Payen, V.L.; Pellerin, L.; Sonveaux, P. Monocarboxylate transporters in the brain and in cancer. *Biochim. Biophys. Acta (BBA) Bioenerg.* **2016**, *1863*, 2481–2497. [CrossRef] [PubMed]
59. Miranda-Gonçalves, V.; Bezerra, F.; Costa-Almeida, R.; Freitas-Cunha, M.; Soares, R.; Martinho, O.; Reis, R.M.; Pinheiro, C.; Baltazar, F. Monocarboxylate transporter 1 is a key player in glioma-endothelial cell crosstalk. *Mol. Carcinog.* **2017**, *56*, 2630–2642. [CrossRef]
60. Teodoro, R.; Wenzel, B.; Oh-Nishi, A.; Fischer, S.; Peters, D.; Suhara, T.; Deuther-Conrad, W.; Brust, P. A high-yield automated radiosynthesis of the alpha-7 nicotinic receptor radioligand [^{18}F]NS10743. *Appl. Radiat. Isot.* **2015**, *95*, 76–84. [CrossRef] [PubMed]

Sample Availability: Samples of the compounds **1**, **2** and **15** are available from the authors.

© 2020 by the authors. Licensee MDPI, Basel, Switzerland. This article is an open access article distributed under the terms and conditions of the Creative Commons Attribution (CC BY) license (http://creativecommons.org/licenses/by/4.0/).

Article

Sigma-1 Receptor Positron Emission Tomography: A New Molecular Imaging Approach Using (S)-(−)-[^{18}F]Fluspidine in Glioblastoma

Magali Toussaint [1,*], Winnie Deuther-Conrad [1], Mathias Kranz [1,2,3], Steffen Fischer [1], Friedrich-Alexander Ludwig [1], Tareq A. Juratli [4], Marianne Patt [5], Bernhard Wünsch [6], Gabriele Schackert [4], Osama Sabri [5] and Peter Brust [1]

[1] Helmholtz-Zentrum Dresden-Rossendorf (HZDR), Institute of Radiopharmaceutical Cancer Research, Department of Neuroradiopharmaceuticals, Research site Leipzig, 04318 Leipzig, Germany; w.deuther-conrad@hzdr.de (W.D.-C.); mathias.kranz@uit.no (M.K.); s.fischer@hzdr.de (S.F.); f.ludwig@hzdr.de (F.-A.L.); p.brust@hzdr.de (P.B.)
[2] PET Imaging Center, University Hospital of North Norway (UNN), 9009 Tromsø, Norway
[3] Nuclear Medicine and Radiation Biology Research Group, The Arctic University of Norway, 9009 Tromsø, Norway
[4] Department of Neurosurgery, Technische Universität Dresden (TUD), University Hospital Carl Gustav Carus, 01307 Dresden, Germany; tareq.juratli@uniklinikum-dresden.de (T.A.J.); gabriele.schackert@uniklinikum-dresden.de (G.S.)
[5] Department of Nuclear Medicine, University Hospital Leipzig, 04318 Leipzig, Germany; marianne.patt@medizin.uni-leipzig.de (M.P.); osama.sabri@medizin.uni-leipzig.de (O.S.)
[6] Institute of Pharmaceutical and Medicinal Chemistry, University of Münster, 48149 Münster, Germany; wuensch@uni-muenster.de
* Correspondence: m.toussaint@hzdr.de; Tel.: +49-341-234-179-4616

Academic Editors: Anne Roivainen and Xiang-Guo Li
Received: 13 April 2020; Accepted: 2 May 2020; Published: 6 May 2020

Abstract: Glioblastoma multiforme (GBM) is the most devastating primary brain tumour characterised by infiltrative growth and resistance to therapies. According to recent research, the sigma-1 receptor (sig1R), an endoplasmic reticulum chaperone protein, is involved in signaling pathways assumed to control the proliferation of cancer cells and thus could serve as candidate for molecular characterisation of GBM. To test this hypothesis, we used the clinically applied sig1R-ligand (S)-(−)-[^{18}F]fluspidine in imaging studies in an orthotopic mouse model of GBM (U87-MG) as well as in human GBM tissue. A tumour-specific overexpression of sig1R in the U87-MG model was revealed in vitro by autoradiography. The binding parameters demonstrated target-selective binding according to identical K_D values in the tumour area and the contralateral side, but a higher density of sig1R in the tumour. Different kinetic profiles were observed in both areas, with a slower washout in the tumour tissue compared to the contralateral side. The translational relevance of sig1R imaging in oncology is reflected by the autoradiographic detection of tumour-specific expression of sig1R in samples obtained from patients with glioblastoma. Thus, the herein presented data support further research on sig1R in neuro-oncology.

Keywords: sigma-1 receptor availability; orthotopic xenograft of glioblastoma in mouse; small animal Positron Emission Tomography/Magnetic Resonance Imaging (PET/MRI); (S)-(−)-[^{18}F]fluspidine; imaging-based biomarker

1. Introduction

Glioblastoma multiforme (GBM) is the most common primary tumour of the central nervous system. Although the global incidence is rare with less than 10 per 100,000 people, the median survival rates for patients with GBM remain dramatically low despite complex surgical, pharmacological and radiation therapy approaches [1,2]. An important aspect contributing to this poor outcome is the genetic heterogeneity of GBM, which translates into heterogeneous expression patterns of potentially druggable targets [3]. Accordingly, the development of new targeted therapies as well as of biomarkers for predictions of treatment response would benefit from an improved understanding of how such spatiotemporal patterns evolve and change during pathogenesis [4–9]. Nuclear medicine imaging techniques offer a unique possibility to noninvasively assess the distribution and amount of certain biological targets and thus to contribute significantly to the drug-discovery process and later on to the evaluation of the treatment efficacy [10–12].

By application of suitable radiolabeled molecules, positron emission tomography (PET) in particular can assess such alterations with high sensitivity. Imaging agents for the investigation of the catabolic and anabolic metabolism can detect cancer-specific alterations in high-capacity processes such as glycolysis (by [^{18}F]FDG), amino acid transport (by [^{11}C]MET or [^{18}F]FET), and membrane turnover (by [^{18}F]FMC) [13,14]. They are currently utilized to improve the clinical management of brain cancer patients. Furthermore, the PET technology offers the principal possibility to investigate differences in the expression pattern and activity of diagnostically and therapeutically relevant proteins, such as receptors or enzymes, and to correlate them with tumour heterogeneity and aggressiveness. The current development of radiolabelled probes to image e.g., isocitrate dehydrogenase mutations (IDH1R132H) [15], or the glutamate carboxypeptidase II (prostate-specific membrane antigen, PSMA) [16], reflects the interest in preclinical and clinical research on detailed and targeted molecular characterisation of malignancies in the brain, which is a prerequisite to define the role of nuclear medicine imaging for the individualized treatment of patients with GBM [14,17].

Our research on the identification of new targets for brain cancer imaging focuses on the sigma-1 receptor (sig1R), an intracellular chaperone protein highly expressed in a variety of cancers including GBM [18,19]. Under physiological conditions, the sig1R is localized at the mitochondrion-associated endoplasmic reticulum membrane (MAM) and at the plasma membrane and is involved by interactions with other proteins in a number of pathways related to the metabolism and proliferation of cells. Accordingly, albeit at different levels, sig1R is expressed in all peripheral organs as well as in the central nervous system. Widely distributed in the brain [20], the sig1R is involved in memory, emotional, and sensory functions and changes in its expression are related to neurodegenerative diseases such as Huntington's disease and Alzheimer's disease, as well as in stroke, depression and pain disorders [20,21]. Most likely, due to the translocation of the intracellular receptor from MAM to the plasma membrane and the cell nucleus, which is triggered under pathological conditions [22–25], sig1R is functionally involved into a variety of cellular pathways related to stress response and survival [25–27]. In addition, the expression of sig1R seems to be upregulated by cancer-specific mechanisms, as indicated by the high levels of sig1R protein discovered in many cancer cell lines [19,26–28]. The antiproliferative effect of pharmacological inhibition of sig1R by putative antagonistic ligands on cancer cell lines further substantiates the potential role of sig1R in cancer biology [27,29–33]. Sig1R ligands influence apoptosis, migration, and cell cycle progression pathways through their interaction with voltage-dependent K^+ channels, volume-regulated Cl^- channels, or endoplasmic reticulum Ca^{2+} release [22,31,32,34]. Altogether, the available data present strong evidence of an important role of sig1R in tumour biology, and for that reason, PET noninvasive molecular imaging of sig1R is assumed to improve our understanding of the role of this particular protein in tumour pathophysiology and to promote the development of a sig1R-targeted-therapies [35].

Already different radiotracers have been developed to investigate the expression of sig1R by PET such as [^{18}F]FMSA4503 [36], [^{18}F]SFE or [^{18}F]FTC-146 [37,38]. However, only [^{11}C]SA4503 and (S)-(−)-[^{18}F]fluspidine were applied in research on the in vivo imaging of sig1R in brain tumours using heterotopic brain tumour models, as by the group of van Waarde, or orthotopic models, as by our group [39,40].

Encouraged by the approval of the in-house developed radiopharmaceutical (S)-(−)-[^{18}F]fluspidine for clinical trials (EudraCT Numbers: 2014-005427-27, 2016-001757-41), the promising results of a collaborative pilot study in an orthotopic mouse model of GBM [40], and the establishment of orthotopic brain tumour models in our group, we decided to investigate further the relevance of sig1R in brain cancer biology and to evaluate the potential of (S)-(−)-[^{18}F]fluspidine-PET to characterise brain tumours on a molecular level. We report herein on the assessment of GBM-specific expression of sig1R by a combination of in vitro and in vivo approaches using mice bearing intracranial U87-MG tumours as a preclinical orthotopic model of human GBM and the sig1R-specific radioligand (S)-(−)-[^{18}F]fluspidine. Initially, we validated the in vivo selectivity of (S)-(−)-[^{18}F]fluspidine for sig1R in a sig1R-knockout mouse model. Then, we validated by radioligand binding assays the suitability of the U87-MG cell line used for the orthotopic GBM model regarding the presence of the target, and confirmed by autoradiography the unimpaired overexpression under in vivo conditions. By immunohistochemistry as radioligand-independent method, we could confirm the expression and overexpression of sig1R protein in U87-MG cells in 2D culture as well as in the cellular environment of the mouse brain. Finally, we performed dynamic PET/MRI (magnetic resonance imaging) studies to assess the pharmacokinetics of (S)-(−)-[^{18}F]fluspidine in the orthotopic U87-MG mouse model of human GBM, and report on the very first detection of sig1R protein in human GBM tissue by means of in vitro autoradiography, validating the relevance of this target.

2. Results

2.1. Expression of sig1R in U87-MG Cells

2.1.1. Expression of sig1R in U87-MG Cells in 2D Cell Culture

We initially evaluated the expression of sig1R in U87-MG cells, a human primary glioblastoma cell line, grown in 2D cell culture by radioligand binding assays and immunohistochemistry to determine their suitability for the intended orthotopic mouse model of GBM. By a single saturation assay using (+)-[^{3}H]pentazocine, an established sig1R-specific radioligand, a B_{max} value of 129 fmol/mg protein and a K_D value of 2.4 nM was determined. Specific binding of (S)-(−)-fluspidine towards (+)-[^{3}H]pentazocine-labeled binding sites in U87-MG cells has been proven by displacement studies, and the affinity of (S)-(−)-[^{18}F]fluspidine to sig1R has been determined with a K_D of 16.7 nM.

To verify the identity of the specific binding site of the two radioligands by an independent method, we further performed immunohistochemistry using a sig1R-specific antibody. The thereby determined cytoplasmic staining of protein in isolated U87-MG cells, which corresponds with the labelling of sig1R in the positive control HEK-293 cells overexpressing human sig1R, is demonstrated in Figure 1A, B, respectively.

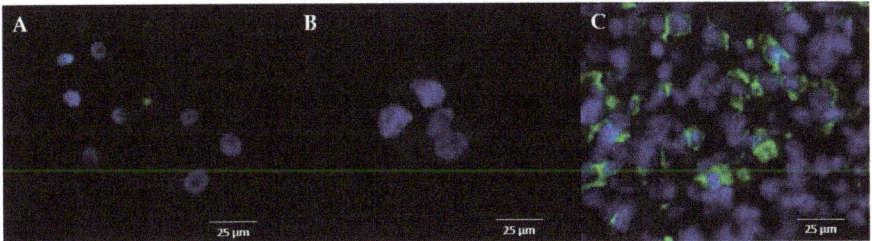

Figure 1. Immunofluorescent staining of sigma-1 receptors (sig1R). Representative image of the sig1R staining (**A**) in U87-MG cells grew in vitro, (**B**) in HEK-293 cells overexpressing human sigma-1 receptor (hsig1R) grew in vitro and (**C**) in a cryosection of U87-MG tumour cells orthotopically implanted in a mouse brain (scale bar: 25 µm, x40, green channel: sig1R staining, blue channel: nucleus staining).

2.1.2. Expression of sig1R in U87-MG Cells Grown In Vivo

Because transplantation of human cancer cells into mice might be associated with an altered expression profile due to the significant change in the microenvironment, we subsequently investigated orthotopically implanted U87-MG tumour cells by immunohistochemistry and radioligand binding studies.

The strong immunofluorescence signal determined in tumour cells in cryosections obtained from mouse brain at day 27 after intracerebral transplantation of U87-MG cells indicates the persistently high expression of sig1R in the orthotopic GBM (Figure 1C). Besides, the fluorescent staining is informative in that not all cells in the field of view express sig1R, indicating a heterogeneous expression profile within the tumour bulk (Figure 1C). Thus, the conservation of sig1R expression irrespective of the environment (culture medium or brain dynamic environment) was confirmed (Figure 1).

Complementary autoradiography performed with the sig1R-specific PET tracer (S)-(−)-[^{18}F]fluspidine confirmed the expression of sig1R in orthotopically growing U87-MG cells. The autoradiographic images presented in Figure 2 indicate a high density of binding sites in the tumour region (Figure 2B). The radioactive signal is nearly completely abolished by co-administration of the sig1R-specific ligand SA4503, a selective agonist commonly used as competitive agent (IC_{50} = 17.4 nM [41]) (Figure 2C). Interestingly, the macroscopic distribution pattern of (S)-(−)-[^{18}F]fluspidine in the orthotopic tumour reflects a heterogeneous accumulation of activity within the tumour, similar to what was observed by immunohistochemistry on cellular level. By saturation studies, performed as homogenous radioligand displacement experiments by co-incubation of (S)-(−)-[^{18}F]fluspidine with different concentrations of (S)-(−)-fluspidine, we determined the kinetic binding parameters of (S)-(−)-[^{18}F]fluspidine in the tumour (T) and an internal reference region, the contralateral striatum (CL). In both regions, (S)-(−)-[^{18}F]fluspidine bound specifically and with comparable affinities of $K_{D,T}$ = 17.5 ± 1.3 nM and $K_{D,CL}$ = 17.0 ± 4.8 nM. However, the sig1R density was ~1.7 times higher in the tumour area compared to the CL area, as reflected by values of $B_{max,T}$ = 704 ± 16 fmol/mg protein vs. $B_{max,CL}$ = 414 ± 36 fmol/mg protein.

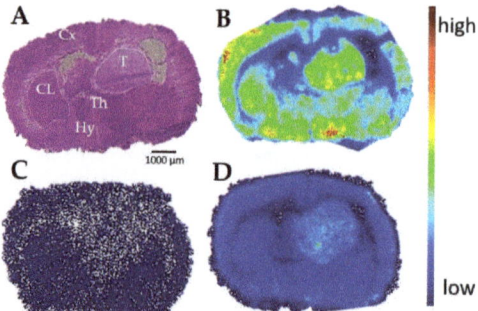

Figure 2. In vitro autoradiography of the mouse brain bearing an orthotopic U87-MG xenograft. Representative autoradiographic images of the coronal plane of mouse brain slices: (**A**) Hematoxylin-eosin staining; (**B**) in vitro distribution of activity after incubation with 0.1 MBq/mL (S)-(−)-[^{18}F]fluspidine, (**C**) co-incubation with 10 µM SA4503 to determine the nonspecific binding and (**D**) with 10 nM of (S)-(−)-fluspidine as competing agent. Cx: cortex; CL: contralateral striatum; Th: thalamus; Hy: hypothalamus; T: tumour. Width of a mouse brain ~1 cm.

2.2. Assessment of GBM-Specific (S)-(−)-[^{18}F]Fluspidine Kinetics In Vivo by Dynamic Small-Animal PET Imaging Studies

2.2.1. Additional Approval of Target Specificity of (S)-(−)-[^{18}F]Fluspidine PET

Due to access to a limited number of sig1R-knockout mice [42], we performed a single head-to-head dynamic PET study with (S)-(−)-[^{18}F]fluspidine in sig1R-knockout ($n = 1$) and control ($n = 3$) mice with the main focus on the tracer kinetics in the striatum. Mean injected activity was 3.5 MBq and 1.9 MBq with a molar activity of 24 GBq/µmol and 21 GBq/µmol at the time of injection, resulting in a mean chemical concentration of 4.7 nmol/kg and 3.2 nmol/kg in the control group and the sig1R-knockout mouse respectively. The corresponding time-activity curves (TACs) are presented in Figure 3. Both the sig1R-knockout and the control animals showed a rapid uptake of activity within the first minutes after i.v. injection of (S)-(−)-[^{18}F]fluspidine (initial peak between 1 and 2 min p.i. of 1.45 and 1.12 SUV, respectively) along with a much stronger and faster washout observed in the sig1R-knockout mouse in comparison to the control animals (SUV at 10, 20, and 40 min p.i. of 0.84, 0.64, and 0.25 vs. 1.07, 1.03, and 0.67, respectively) confirming the selectivity in vivo of (S)-(−)-[^{18}F]fluspidine for sig1R (Figure 3A). In order to correct for potential model-related differences in the brain perfusion, we calculated an SUV ratio (SUVR) from the TAC data obtained in striatum and blood at each time points. The corresponding SUVR curves, presented in Figure 3B, indicate for both animal models nearly stable SUVR values at 15 to 60 min p.i., albeit with clearly different values. While for the control animals SUVR values in the range of 0.8 to 1 (corresponding to an area under the curve (AUC) value of 48.6 ± 8.3) has been estimated, the notably lower value of about 0.4 (corresponding to an AUC value of 23.4) determined in the single sig1R-knockout mouse confirms the target specificity of (S)-(−)-[^{18}F]fluspidine.

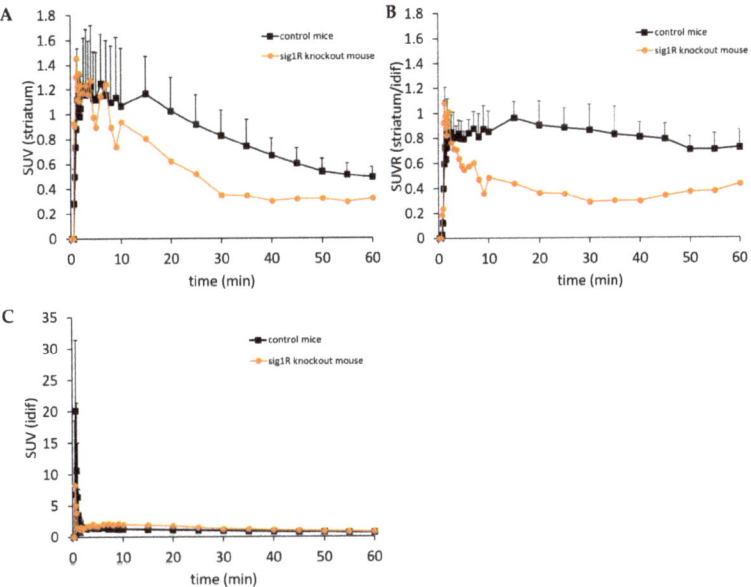

Figure 3. Positron Emission Tomography/Magnetic Resonance (PET/MR) imaging of control mice ($n = 3$) and of sig1R-knockout mouse ($n = 1$) after i.v. administration of (S)-(−)-[^{18}F]fluspidine. (**A**) Average striatal time-activity curves for control mice (black squares) and sig1R-knockout mouse (orange dots). (**B**) Average time-varying SUVRs of the striatum over the blood (defined from the image-derived input function (idif)) of control mice (black squares) and sig1R-knockout mouse (orange dots) (**C**) Average time-varying SUV of the blood (defined from the idif) of control mice (black squares) and sig1R-knockout mouse (orange dots).

2.2.2. GBM-Specific Pharmacokinetics of (S)-(−)-[^{18}F]Fluspidine in the Orthotopic U87-MG Mouse Model

Encouraged by the in vitro and in vivo findings on the specific expression of sig1R in orthotopically grown U87-MG tumours and the sig1R-specific binding of (S)-(−)-[^{18}F]fluspidine, we proceeded with PET studies with (S)-(−)-[^{18}F]fluspidine performed under baseline conditions in nude mice-bearing orthotopic U87-MG (n = 3). Mean injected activity was 9.1 MBq with a molar activity of 92.5 GBq/µmol at the time of injection, resulting in a mean chemical concentration of 4.2 nmol/kg. The tumour growth was assessed by MRI with a T2-weighted sequence, and a 60 min dynamic PET scan, followed by T1- and T2-weighted sequences, performed when the tumour size was 28 ± 8 mm^3 (i.e., 23 to 30 days after implantation). The regions-of-interest (ROIs) were delineated on the T2-weighted MR images and then applied on the PET data to generate the regional TACs.

As reflected by the TACs presented in Figure 4, although not statistically significant, the uptake of (S)-(−)-[^{18}F]fluspidine in the tumour is lower and more slowly than the activity uptake in the control region with maximal SUV values of 0.82 at 3 min p.i. and 1.24 at 1 min p.i., respectively. However, because the washout from the tumour was slower, the tumour and CL TACs intersected at about 30 min p.i., demonstrating with SUVs of 0.38 and 0.28 at 60 min p.i. a higher retention of activity in the tumour region compared to the CL, respectively. The retarded washout of the sig1R-specific (S)-(−)-[^{18}F]fluspidine from the orthotopic tumour is in accordance with the autoradiographic data, indicating a higher availability of sig1R in the U87-MG tumour tissue in comparison to CL.

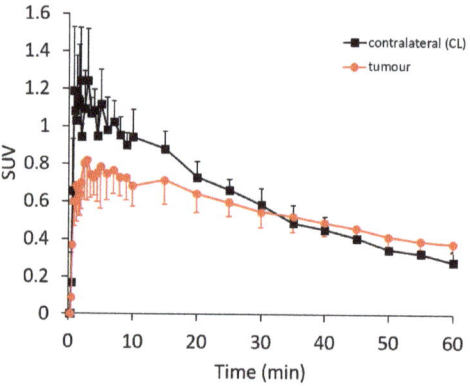

Figure 4. PET/MR imaging of sig1R in mice with orthotopic xenograft of human GBM cells (U87-MG). Average time-activity curves after i.v. administration of (S)-(−)-[^{18}F]fluspidine of the tumour (red dots) and the contralateral (black squares) regions of interest (n = 3). Statistical test: Student t-test, * $p < 0.05$.

The intratumoral heterogeneity of sig1R expression already discovered by the radioligand and antibody investigations in vitro was detectable also by the in vivo imaging study. The early PET images between 2 and 9 min after injection show an heterogeneous uptake of (S)-(−)-[^{18}F]fluspidine into the tumour (Figure 5D, upper panel). According to the histological analyses of the explanted tumour tissue, performed immediately after the PET scans, the tumour inner part is characterised by a lower cell density compared to the periphery along with extra-cellular oedema area highlighting presumably areas of necrosis (Figure 5A–C). Therefore, the heterogeneous uptake of (S)-(−)-[^{18}F]fluspidine may also (or additionally) be caused by reduced blood supply to the tumour centre. The PET image at later time points (45 to -60 min p.i.; Figure 5D, lower panel) pictures a more homogenous uptake of the tracer, along with a low slope, reflecting an accumulation.

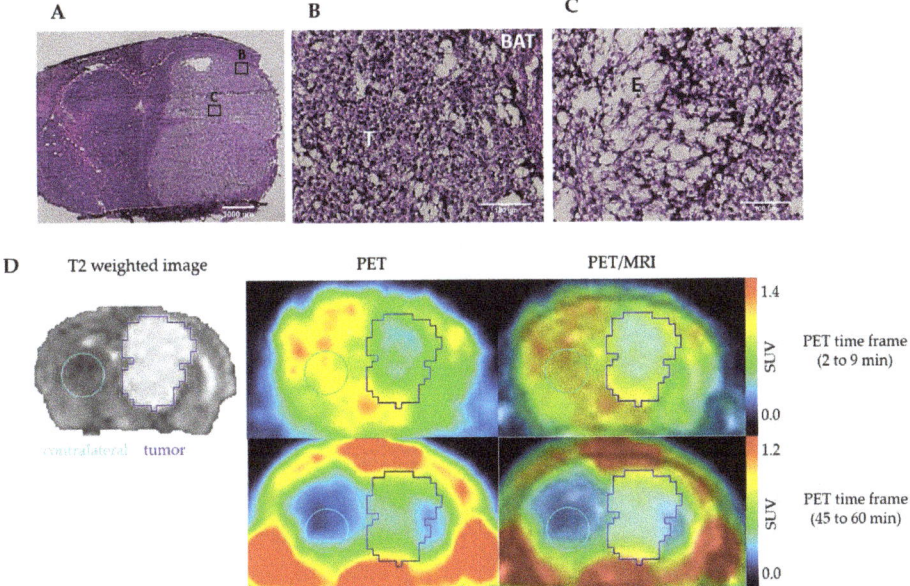

Figure 5. Hematoxylin-eosin staining of U87-MG tumour: (**A**) tumour bulk in the right striatum of a mouse brain (×2, scale bar: 1000 µm); (**B**) tumour periphery presents area of high density of cell nuclei; (**C**) tumour centre presents area of lower cell density accompanied by oedema. (×40, BAT: brain adjacent to tumour; T: tumour, E: oedema. Scale bar: 100 µm. (**D**) Representatives coronal PET/MR images of U87-MG tumour-bearing mouse after i.v. administration of (S)-(−)-[^{18}F]fluspidine. The upper panel exhibits the distribution of (S)-(−)-[^{18}F]fluspidine at early times p.i. (averaged time frames from 2 to 9 min), and the lower panel exhibits the distribution of (S)-(−)-[^{18}F]fluspidine at later times (averaged time frames from 45 to 60 min).

2.3. Presence of (S)-(−)-[^{18}F]Fluspidine Binding Sites in Human GBM Tissue

To initially assess the suitability of sig1R as specific target for molecular characterisation of human GBM, we performed in vitro autoradiography with (S)-(−)-[^{18}F]fluspidine using cryosections of tissue samples obtained from 3 patients diagnosed with Glioblastoma multiforme IV. Total and nonspecific binding of the PET ligand was determined by incubation with only (S)-(−)-[^{18}F]fluspidine or with co-administration of a high concentration of haloperidol to block the sig1R, followed by histological staining of the respective cryosections. As shown in Figure 6, the autoradiographic images indicate a heterogeneous pattern of binding sites of (S)-(−)-[^{18}F]fluspidine in all three GBM samples with the highest density in regions histologically characterised by a high density of cells which we assume might be related to the highly proliferating tumour cells. Accordingly, although it was not possible within this preliminary study to confirm by immunohistochemistry the distribution pattern of sig1R in the cryosections or to identify the type of cells possessing high specific binding of (S)-(−)-[^{18}F]fluspidine, these preliminary data motivate us to design a complementary study on the investigation of sig1R protein in a larger number of GBM samples by means of specific radioligands and antibodies.

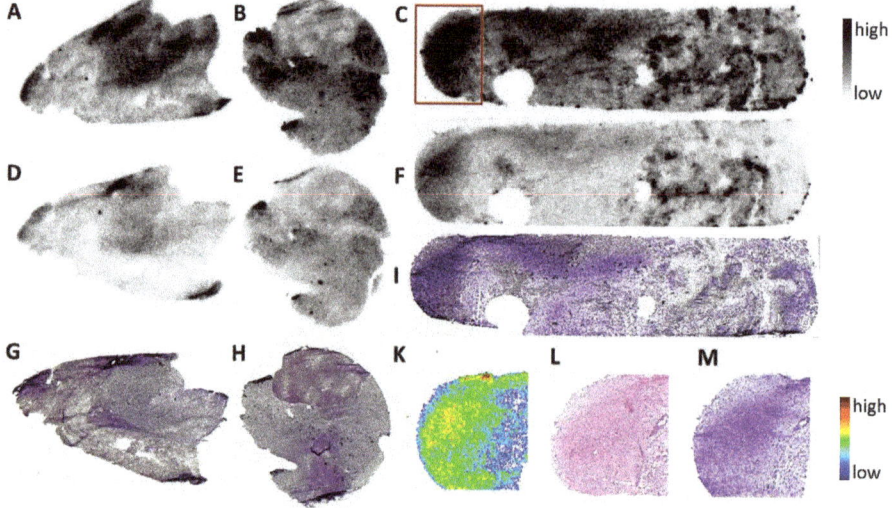

Figure 6. Sig1R autoradiography with the sig1R-specific PET ligand (S)-(−)-[^{18}F]fluspidine in human GBM in vitro. Binding of (S)-(−)-[^{18}F]fluspidine at 4.5 nM in cryosections (12 µm) of tumour tissue obtained from three patients (**A,B,C**) demonstrated heterogeneous distribution throughout the slices. By co-incubation with 1 µM haloperidol (**D,E,F**), a substantial reduction in activity accumulation was obtained. Histochemical analysis of corresponding sections was performed by Nissl staining (**G,H,I**). Analysis of one sample at higher magnification (red square in **C**) demonstrated correlation of the activity accumulation (**K**) with highly cell dense regions (H&E staining: (**L**); Nissl staining: (**M**). Length of the biopsies samples ~1 cm.

3. Discussion

In this study, we evaluated the availability of sig1R in an orthotopic mouse model of human GBM. High expression of sig1R in different cancer cell lines derived from prostate, breast, colon, melanoma, small and non-small cell lung cancer, brain tumours including GBM, neuroblastoma, and meningioma have already been reported [18,19,26,27,43–48]. The involvement of sig1R in many but selective protein interactions, the antiproliferative effect of putative antagonists, as well as their nonpleiotropic effects make sig1R a potent drug target prone to overcome adaptive drug resistance alone or in combination with other drugs [49,50]. It is also known that the upregulation of sig1R on both mRNA and protein level in the same cancer subtype differs from one cell line to another and from one patient to another, probably reflecting a context-dependent expression of sig1R [51–53]. Consequently, an improved understanding of how such patterns evolve and change during pathogenesis, by the use of noninvasive PET imaging, would promote the development of sig1R-based therapies. In this context, we chose to evaluate by PET the suitability of the clinically approved imaging agent (S)-(−)-[^{18}F]fluspidine for the analysis of the expression of sig1R in GBM.

We first investigated in vitro the level of expression of sig1R protein in U87-MG cells, a human GBM cell line widely applied for orthotopic brain cancer mouse models. As consistently reported, sig1R is located at the endoplasmic reticulum-mitochondria interface and redistributes ligand-mediated and under conditions of cellular stress dynamically to the plasma membrane and the nucleus envelope [54,55]. The examination of sig1R expression in U87-MG cells by using (+)-[^{3}H]pentazocine, a selective sig1R ligand widely applied in radioligand binding assays [56,57], demonstrated high affinity binding towards a single binding site expressed in U87-MG cells grown in 2D cell culture. The density of sig1R in this cell line, B_{max} = 129 fmol/mg protein, is in the range of values determined by

(+)-[^{3}H]pentazocine in other cancer cell lines such as 42 fmol/mg protein in the C6 murine glioblastoma cells, 76.5 fmol/mg protein in the NB41A3 neuroblastoma cells or 1115 fmol/mg protein in the U-138-MG cells [19]. We identified the U87-MG cells as suitable for the orthotopic GBM model applied in this study.

As the microenvironment is known to influence gene expression, e.g., the hypoxia-stimulated HIF-1α expression in glioma or the culture mode (2D vs. 3D)-dependent differential gene expression of colorectal cell lines [58–61], we further investigated the expression of sig1R in the intracerebral U87-MG tumour. The comparable immunofluorescence staining on 2D-cultivated U87-MG cells and on the orthotopically grown U87-MG tumour along with the similarity of the K_D values of the PET tracer (S)-(−)-[^{18}F]fluspidine, indicate the conservation of sig1R expression and conformation over the translation from in vitro culture to in vivo implantation. Furthermore, the cytoplasmic localization of the sig1R fluorescence signal which was observed in vitro and in the explanted brain tumours matches with the cellular localization of the receptor found in rat astrocytes and mouse neurons [62,63].

Subsequently, we quantified the number of sig1R expressed in the U87-MG tumours implanted in the right striatum as well as in the internal control region, the left striatum. The equivalent K_D values obtained for (S)-(−)-[^{18}F]fluspidine in both regions indicate that the PET radiotracer binds to the same target, i.e., the sig1R, in both compartments. The analysis of the binding parameter B_{max} excludes conformational differences between sig1R in cancer and normal cells, as discussed by Kim et al., as a possible reason for the higher accumulation of sig1R-targeting radioligands in tumour tissue, but clearly indicates an about 2-fold higher density of sig1R in the U87-MG tumour in comparison to the healthy brain [51]. Thus, the herein exploited orthotopic U87-MG GBM mouse model is appropriate for the following imaging studies. As an add-on to the extensive and validated data on the selectivity of the clinically applied PET radioligand (S)-(−)-[^{18}F]fluspidine obtained mainly by pharmacological intervention studies [64,65], we made use of access to a sig1R-knockout mouse model to measure the actual contribution of the off-target binding of the radiotracer to the uptake of activity in the brain in imaging studies in mouse [66]. In accordance with the fast washout kinetics observed in the knockout model, we supposed only a weak background signal in imaging studies with (S)-(−)-[^{18}F]fluspidine in the orthotopic brain cancer model and no relevant interaction with off-target binding sites in vivo.

Eventually, the results of the fundamental characterisation of the components of the experimental setting, i.e., the mouse model and the PET radioligand, with respect to availability of and selectivity to sig1R, prompted us to proceed with dynamic PET studies in the orthotopic U87-MG glioblastoma mouse model. Only few studies have explored the use of PET radiotracers for sig1R imaging of tumours, and even less have addressed brain tumours in particular [36,39,67–70] such as the investigation of sig1R in an ectopic glioma rat model as well as in spontaneous pituitary tumours in rats using [^{11}C]SA4503 by the group of van Waarde [39,69,70]. To the best of our knowledge, we are the first exploring the sig1R availability of human glioblastoma in an orthotopic tumour mouse model. The in vivo imaging studies revealed a tumour-to-background ratio (TBR) of only slightly higher than 1, detectable from the late PET images. Even though we observed a continuous washout of activity from the tumour, this process was slower than in contralateral tissue. Accordingly, the activity concentration in the tumour surpassed that in the contralateral striatum over time.

Despite this, the TBR value determined in our study is in fact notably lower than the values reported for the [^{11}C]SA4503 PET studies mentioned above. However, we assume that this discrepancy is related mainly to the characteristics of the background region, in particular the physiological expression of sig1R in the different grafting sites. An ectopic tumour obtained by e.g., implantation of C6 glioma in the shoulder in the soft tissue [71], close to the muscle, benefit of an ideal background tissue with low expression of sig1R [20], leading to a TBR values > 4. Such values are not comparable to orthotopically transplanted brain tumours due to the comparatively high expression of sig1R in the surrounding nondiseased brain, as indicated by e.g., in the herein performed PET studies with (S)-(−)-[^{18}F]fluspidine in healthy mice [72,73].

The reasons for the discrepant results obtained in the present paper regarding the in vitro and in vivo quantification of sig1R in the U87-MG tumours are not clear at the moment. We assume, that factors such as microenvironment, vascularisation, or interstitial fluid pressure affect the binding parameters of (S)-(−)-[^{18}F]fluspidine in vivo. The U87-MG tumour is known to be highly vascularised and presenting necrotic foci [62], suggesting a first uptake in the vascularised periphery and a later accumulation by diffusion in the core of the tumour tissue. However, a systematic investigation of these processes was beyond the scope of this study. Notwithstanding this limitation, a detailed investigation of the PET images of the intracranial U87-MG tumours revealed that the heterogeneous pattern of activity accumulation discovered already in vitro could be detectable by the in vivo imaging approach as well. Interestingly, a similar distribution of [^{11}C]SA4503 in the tumour outer rim was reported in the already mentioned PET study of the ectopic C6 glioma model as well as in a patient with non-small cell lung cancer in the tumour tissue [74,75]; noteworthy is the discrepancy between the distribution patterns of [^{11}C]SA4503 and [^{18}F]FDG [69,74]. Since [^{18}F]FDG PET images may be misleading due to an increased glucose metabolism in noncancerous but inflammatory tissues, the authors suggested the use of sig1R PET imaging to discriminate between tumour and inflammation [69].

A final aspect addressed in this study on the suitability of PET imaging of sig1R in glioblastoma was the investigation of the expression of sig1R in human GBM tissue. The accordingly performed receptor autoradiography with (S)-(−)-[^{18}F]fluspidine on cryosections of human glioblastoma obtained from three patients consistently showed a heterogeneous distribution of binding sites of the sig1R-targeting radioligand with high-density binding in cell-dense regions as suggested by the subsequent histological analysis. However, although sig1R appears to play a role in proliferation, this preliminary examination does not allow to speculate about a correlation between receptor expression and tumour proliferation but nevertheless suggests to design a respective large-scale study [28].

4. Materials and Methods

All experimental work including animals has been conducted in accordance with the national legislation on the use of animals for research (Tierschutzgesetz (TierSchG), Tierschutz-Versuchstierverordnung (TierSchVersV)) and has been approved by the responsible research ethics committee (TVV 30/17; TVV 18/18 Landesdirektion Sachsen).

4.1. Radiochemistry

Enantiomerically pure (S)-(−)-[^{18}F]fluspidine was prepared on a TRACERlab FXN synthesizer (GE Healthcare, Waukesha, WI, USA) as described in previous publications [65]. The radiochemical purity of (S)-(−)-[^{18}F]fluspidine was >99%, and the molar activity (A_m) at the end of the synthesis (EOS) was 89–180 GBq/µmol ($n = 2$).

4.2. Cell Culture

U87-MG cells (obtained from Jens Pietzsch/Birgit Belter, Department Radiopharmaceutical and Chemical Biology, Helmholtz-Zentrum Dresden-Rossendorf, Rossendorf, Germany) and human hsig1R-transfected Human Embryonic Kidney (HEK) cells (obtained from Olivier Soriani, Institut de Biologie Valrose—University Côte d'Azur, Sophia Antipolis, France) were maintained in monolayer culture (37 °C, 5% CO_2, 95% O_2) in Dulbecco's Modified Eagle Medium (DMEM, Gibco, Invitrogen, Dun Laoghaire, Ireland) supplemented with 10% heat inactivated fetal bovine serum (Gibco, Invitrogen, Dun Laoghaire, Ireland), 5% penicillin and streptomycin, 1.25% sodium pyruvate, 1% L-glutamine (Gibco, Invitrogen, Ireland) and 1 µg/mL puromycin (Gibco, Invitrogen, Dun Laoghaire, Ireland) only for the transfected cells.

4.3. In Vivo Competitive Radioligand Binding Assay

Cell membrane homogenates of U87-MG cells were obtained by gentle scraping the cells grown to confluency in one 175 cm^2 flask, followed by sedimentation of the cells suspended in cell culture medium by centrifugation at 800 rpm for 3 min at room temperature, re-suspension of the cells in 1 mL 50 mM TRIS-HCl, pH 7.4/4 °C and incubation on ice for 20 min, centrifugation of the suspension at 15,000 rpm for 15 min at 4 °C, and finally re-suspension of the pellet in 200 µL 50 mM TRIS-HCl, pH 7.4/4 °C and storage at −25 °C. The radioligand binding assay was performed by incubating the U87-MG cell membrane homogenate (226 µg protein/mL) with the Sig1R agonist (+)-[^{3}H] pentazocine (working concentration = 3.25 nM; A_m = 995 GBq/mmol; PerkinElmer LAS GmbH, Rodgau, Germany) in incubation buffer (50 mM TRIS-HCl, pH 7.4, 120 mM NaCl, 5 mM KCl, 2 mM $CaCl_2$, 1 mM $MgCl_2$) without (total binding, TB; n = 3) or with co-incubation of 1 µM haloperidol (nonspecific binding, NB; n = 3) at room temperature for 60 min. The incubation was terminated by filtration via a Whatman® glass microfibre filter (Grade GF/B, pre-incubated in freshly prepared polyethyleneimine (3%) at room temperature for 90 min), followed by quadruplicate washing with 50 mM TRIS-HCl, pH 7.4/4 °C using a semi-automated cell harvester (48-samples; Brandel, Gaithersburg, MD, USA). Filter-bound radioactivity was detected in terms of DPM/vial by liquid scintillation counting (Beckman LS 6500; Beckman Coulter Inc., Fullerton, CA, USA) of the isolated filters immersed for two hours in liquid scintillation cocktail (Ultima Gold; PerkinElmer LAS GmbH, Rodgau, Germany). Specific binding (SB) was calculated by SB (DPM/vial) = TB (DPM/vial) − NB (DPM/vial). The B_{max} and the K_D values were estimated by a nonlinear regression model (equation: one-site binding (hyperbola)) using GraphPad Prism, Version 4.1 (GraphPad Inc., La Jolla, CA, USA).

4.4. In Vitro Autoradiography on Human Glioblastoma Tissue

Cryosections of brain tumour tissue from three patients (Glioblastoma multiforme IV) were obtained using a microtome (MICROM HM560, Fisher Scientific GmbH, Schwerte, Germany), mounted on microscopy slides (SuperFrost, Thermo Scientific Menzel, Fisher Scientific GmbH, Schwerte, Germany), dried for ~2 h at room temperature, and stored at −25 °C until the autoradiography study. For the experiment, the slides were taken out from the freezer, the cryosections dried under a stream of cold air, and pre-incubated with incubation buffer (50 mM TRIS-HCl, pH 7.4, 120 mM NaCl, 5 mM KCl, 2 mM $CaCl_2$, 1 mM $MgCl_2$) at room temperature for 15 min. The pre-incubation solution was decanted, the slices dried again under a stream of cold air, and covered afterwards with the incubation solution ((S)-(−)-[^{18}F]fluspidine, 197 kBq/mL incubation buffer = 4.5 nM at the time of incubation, without (total binding) or with co-incubation with 1 µM haloperidol to assess nonspecific binding). Incubation at room temperature was terminated after 60 min, the slides were washed two times in 50 mM TRIS-HCl, pH 7.4 at 4 °C, on ice for two minutes each followed by dipping in ice-cold demineralized water for 5 s and rapid drying under a stream of cold air. Afterwards, the slides were exposed to a phosphor imager plate (BAS-IP TR 2025, FujiFilm Corporation, Tokyo, Japan) along with standards obtained by pipetting and drying 1 µL of each concentration of a serial dilution of the radioligand solution on to a microscopic slide. The exposed phosphor-imaging plates were scanned using a high resolution scanner (HD-CR 35 Bio; Dürr NDT GmbH & Co. KG, Bietigheim-Bissingen, Germany) at a laser spot size of 12.5 µm (pixel size: 12.5 µm^2) followed by two-dimensional analysis of the digitized images (AIDA 4.27; Elysia-raytest GmbH, Straubenhardt, Germany). The tracer distribution in the autoradiographic images obtained for total and nonspecific binding was compared by visual inspection and correlated with the histochemical staining (Nissl- and Hematoxiline-eosin staining) of the corresponding tissue sections.

4.5. In Vitro Autoradiography on Mice Brain-Bearing Glioblastoma

Cryosections of brains obtained from female athymic nude mice (Rj:NMRI-Foxn1 nu/nu) (10–12 weeks old, 25–38 g), were obtained as described above. The same protocol as in Section 4.3 was used. The incubation step was performed with 0.1–0.2 MBq/mL (S)-(−)-[^{18}F]fluspidine in buffer for 60 min at room temperature. Nonspecific binding was determined in the presence of 10 µM of SA4503 (Tocris, Bio-Techne GmbH, Wiesbaden-Nordenstadt, Germany) or 100 µM to 10 nM of (S)-(−)-fluspidine, respectively. Developed autoradiographs were analysed in a phosphor imager (HD-CR 35; Dürr NDT GmbH & Co. KG, Bietigheim-Bissingen, Germany). The quantification was performed by using 2D-densitometric analysis (AIDA 2.31 software; raytest Isotopenmessgeräte GmbH, Straubenhardt, Germany). The B_{max} and the K_D values were estimated by a linear regression model (equation: one-site binding (hyperbola)) using GraphPad Prism, Version 4.1 (GraphPad Inc., La Jolla, CA, USA).

4.6. Immunohistochemistry

Tissues were cryopreserved by incubation in 2-Methylbutane at −25 °C (Merck, Germany). The brains were cut into coronal sections 10 µm thickness with cryostat (MICROM HM560, Fisher Scientific GmbH, Schwerte, Germany) and kept at −25 °C. Immunostaining was performed after fixation in PFA 4% for 20 min at 4 °C of the slides. Detection of the sig1R protein was performed by overnight incubation at 4 °C of the slides with the primary mouse monoclonal antibody (1:500 in blocking buffer 5% normal goat serum, B-5: sc-137075, Santa Cruz Biotechnology, Inc., Dallas, TX, USA). After washing with a solution of 1% BSA in PBS, the slides were incubated for 1 h at room temperature with the secondary polyclonal goat anti-mouse antibody (1:200 in dilution buffer 1% BSA, Alexa Fluor® 488; ab150117, Abcam, Berlin, Germany). A Hoechst counterstaining, 10 min at room temperature, was performed to visualize the nuclei of the cells (1:1000 in PBS, Hoechst 33258, Life Technologies, Carlsbad, Ca, USA). After a step of washing and drying, slides were cover up with mounting medium. (Aquapolymount, Polysciences Europe GmbH, Hirschberg an der Bergstrasse, Germany). Visualization of the slides was performed by fluorescence microscopy (Leica, DMi8, software Leica LASX, Leica Mikrosysteme Vertrieb GmbH, Wetzlar, Germany).

4.7. Animals and Orthotopic Brain Tumour Model

Female athymic nude mice (Rj:NMRI-Foxn1 nu/nu) were chosen for this study (Janvier labs, France). The mice were used for tumour implantation at the age of 8 weeks (26–30 g). During microsurgery mice were anesthetized with a mixture of air and isoflurane concentrate (1.5–2% depending on the breathing) under sterile conditions. The mice were placed into a Stoelting stereotactic frame (just for mouseTM, Stoelting Europe, Dublin, Ireland). A midline incision was done and a burr hole was drilled 0.5 mm anterior and 2.5 mm lateral to the bregma. 5×10^4 U87-MG cells were suspended in 1 µL Hank's Buffered Salt Solution (HBSS, 1X) and were injected 3.0 mm into the brain parenchyma with a flow of 0.1 µL/min using a 10 µL Hamilton syringe. After injection, the burr hole was filled with bonewax (Ethicon, US, LLC), the scalp incision sutured (Vicryl 6.0, Ethicon, US, LLC) and the surface antiseptically cleaned. Animal sacrifice was performed by induction of anesthesia with a mixture of air and isoflurane concentrate followed by cervical dislocation.

4.8. Small Animal PET/MR Imaging

For the time of the experiments, female CD-1 mice ($n = 3$; age: 10 weeks; weight: 30–35 g) or *nude* mice ($n = 3$; age: 10 weeks; weight: 25–30 g) (Janvier Labs, Le Genest-Saint-Isle, France) and one CD-1 sig1R-knockout mouse ($n = 1$; age: 10 weeks; weight: 27 g) (Envigo RMS, SARL, Bresso, Italy) were kept in a dedicated climatic chamber with free access to water and food under a 12:12h dark:light cycle at a constant temperature of 24/26 °C. The animals were anaesthetized (Anaesthesia Unit U-410, AgnTho's, Lidingö, Sweden) with isoflurane (1.8%, 0.35 L/min) delivered in a 60% oxygen/40% air mixture (Gas Blender 100 Series, MCQ instruments, Rome, Italy) and maintained at 37 °C with a thermal bed system. (S)-(−)-[^{18}F]fluspidine was injected into the lateral tail vein (control group: 3.5 ± 1.9 MBq, A_m: 94 ± 7 GBq/µmol EOS; *h*sig1R-knockout mouse: 1.9 MBq; A_m: 89 GBq/µmol EOS; tumour group: 5.7 ± 3.7 MBq; A_m: 119 ± 41 GBq/µmol EOS) followed by a 60 min PET/MR scan (nanoScan®, Mediso, Hungary). Each PET image was corrected for random coincidences, dead time, scatter and attenuation (AC), based on a whole body (WB) MR scan. The list mode data were sorted into sonograms using a framing scheme of 12×10 s, 6×30 s, 5×300 s, 9×600 s. The reconstruction parameters for the list mode data are: 3D-ordered subset expectation maximization (OSEM), 4 iterations, 6 subsets, energy window: 400–600 keV, coincidence mode: 1–5, ring difference: 81. The mice were positioned prone in a special mouse bed (heated up to 37 °C), with the head fixed to a mouth piece for the anesthetic gas supply with isoflurane in 40% air and 60% oxygen. The animal head was positioned in the center of the field of view in order to benefit from the highest spatial resolution possible (spatial resolution center of the FOV: 900 µm). A dynamic PET scan of a duration of 60 min was performed followed by a T2 weighted sequence (Fast Spin Echo, TR/TE: 4377/88.5 ms, NEX: 4, FOV: 70 × 70 mm, matrix: 256 × 256, SI: 0.9 mm) and a T1 weighted sequence (Gradient Echo, TR/TE: 15/2.59 ms, NEX: 4, FOV: 60 × 60 mm, matrix: 160 × 160, slice thickness: 0.5 mm) for anatomical orientation and AC correction respectively. Image registration and evaluation of the region of interest (ROI) was done with PMOD (PMOD Technologies LLC, v. 3.9, Zurich, Switzerland). The respective brain regions were identified using the T2 weighted sequence and the tumour area and the contralateral area were delineated manually. The hypersignal due to the tumour in T2 weighted images was manually segmented and described as "tumour ROI", and due to the compression of the contralateral side a fixed circled shape ROI was used to delineate the striatum avoiding nearby structure (cortex, ventricles, hypothalamus). The image-derived input function (IDIF) was extracted from a voxel of interest (VOI) segmented on the inferior vena cava (IVC). The IVC was identified using the first time frames showing the first passage of (S)-(−)-[^{18}F]fluspidine bolus. An automatic algorithm from PMOD was used to identify the IVC signal avoiding heart and kidney area [76]. The activity data are expressed as mean standardized uptake value (SUV) of the overall ROI or as SUV ratio of the striatum ROI over the IDIF (SUVR). Data are presented as mean ± standard deviation (SD). Microsoft Excel was used to perform statistical tests. A parametric student *t*-test preceded by a Fischer test for variance were used to compare the groups with $p < 0.05$.

5. Conclusions

To conclude, we showed for the first time in an orthotopic GBM model, the U87-MG mouse model of glioblastoma and the suitability of a sig1R-targeting PET radioligand, (S)-(−)-[^{18}F]fluspidine, to investigate the tumour-specific expression pattern of sig1R by in vivo imaging. Whether the inferior outcome in vivo in comparison to in vitro is caused by the physiological expression of sig1R in the healthy brain or by certain pathophysiological characteristics of the orthotopic mouse model of GBM, remains to be elucidated. Nevertheless, this first evaluation of the sig1R availability in an orthotopic in vivo model of brain tumour contributes to a better understanding of this model and suggests an expression of sig1R in the tumour periphery as found in other studies, which may be related to proliferation and invasiveness. In conclusion, the data obtained in the U87-MG mouse model of GBM along with the detection of sig1R in human GBM tissue for the first time by a PET radioligand, indicate not only the relevance of this target but also the suitability of (S)-(−)-[^{18}F]fluspidine for sig1R-targeted cancer research and drug development.

Author Contributions: Conceptualization, P.B., W.D.-C., G.S.; methodology, M.T., M.K., W.D.-C., S.F., F.-A.L., M.P., O.S., T.A.J., G.S.; formal analysis, M.T., W.D.-C.; investigation, M.T., M.K., W.D.-C.; tracer synthesis and analysis, O.S., M.P.; writing—original draft preparation, M.T.; writing—review and editing, M.T., W.D.-C., M.K., M.P., O.S., P.B, F.A.L., S.F, B.W. All authors have read and agreed to the published version of the manuscript.

Funding: This research received no external funding.

Acknowledgments: The technical assistance of Tina Spalholz is acknowledged. The sig1R knockout mice were kindly provided by Envigo RMS SARL.

Conflicts of Interest: The authors declare no conflicts of interest.

References

1. Tykocki, T.; Eltayeb, M. Ten-year survival in glioblastoma. A systematic review. *J. Clin. Neurosci.* **2018**, *54*, 7–13. [CrossRef] [PubMed]
2. Witthayanuwat, S.; Pesee, M.; Supaadirek, C.; Supakalin, N.; Thamronganantasakul, K.; Krusun, S. Survival Analysis of Glioblastoma Multiforme. *Asian Pac. J. Cancer Prev.* **2018**, *19*, 2613–2617. [PubMed]
3. Aldape, K.; Brindle, K.M.; Chesler, L.; Chopra, R.; Gajjar, A.; Gilbert, M.R.; Gottardo, N.; Gutmann, D.H.; Hargrave, D.; Holland, E.C.; et al. Challenges to curing primary brain tumours. *Nat. Rev. Clin. Oncol.* **2019**, *16*, 509–520. [CrossRef] [PubMed]

4. Verger, A.; Langen, K.-J. PET Imaging in glioblastoma: Use in clinical practice. In *Glioblastoma*; De Vleeschouwer, S., Ed.; Codon Publications: Brisbane, QLD, Australia, 2017; pp. 155–172.
5. Wardak, M.; Schiepers, C.; Cloughesy, T.F.; Dahlbom, M.; Phelps, M.E.; Huang, S.-C. ^{18}F-FLT and ^{18}F-FDOPA PET kinetics in recurrent brain tumors. *Eur. J. Nucl. Med. Mol. Imaging* **2014**, *41*, 1199–1209. [CrossRef]
6. Fukuma, R.; Yanagisawa, T.; Kinoshita, M.; Shinozaki, T.; Arita, H.; Kawaguchi, A.; Takahashi, M.; Narita, Y.; Terakawa, Y.; Tsuyuguchi, N.; et al. Prediction of IDH and TERT promoter mutations in low-grade glioma from magnetic resonance images using a convolutional neural network. *Sci. Rep.* **2019**, *9*, 20311. [CrossRef]
7. Sampson, J.H.; Maus, M.V.; June, C.H. Immunotherapy for Brain Tumors. *J. Clin. Oncol.* **2017**, *35*, 2450–2456. [CrossRef]
8. Guidotti, G.; Brambilla, L.; Rossi, D. Exploring Novel Molecular Targets for the Treatment of High-Grade Astrocytomas Using Peptide Therapeutics: An Overview. *Cells* **2020**, *9*, 490. [CrossRef]
9. Golub, D.; Iyengar, N.; Dogra, S.; Wong, T.; Bready, D.; Tang, K.; Modrek, A.S.; Placantonakis, D.G. Mutant Isocitrate Dehydrogenase Inhibitors as Targeted Cancer Therapeutics. *Front. Oncol.* **2019**, *9*, 417. [CrossRef]
10. Lewis, D.Y.; Soloviev, D.; Brindle, K.M. Imaging tumor metabolism using positron emission tomography. *Cancer J.* **2015**, *21*, 129–136. [CrossRef]
11. Zhu, A.; Lee, D.; Shim, H. Metabolic PET imaging in cancer detection and therapy response. *Semin. Oncol.* **2011**, *38*, 55–69. [CrossRef]
12. Bhattacharyya, S. Application of positron emission tomography in drug development. *Biochem. Pharmacol.* **2012**, *1*, e128. [CrossRef] [PubMed]
13. Treglia, G.; Muoio, B.; Trevisi, G.; Mattoli, M.V.; Albano, D.; Bertagna, F.; Giovanella, L. Diagnostic Performance and prognostic value of PET/CT with different tracers for brain tumors: A Systematic review of published meta-analyses. *Int. J. Mol. Sci.* **2019**, *20*, 4669. [CrossRef] [PubMed]
14. Drake, L.R.; Hillmer, A.T.; Cai, Z. Approaches to PET imaging of glioblastoma. *Molecules* **2020**, *25*, 568. [CrossRef] [PubMed]
15. Chitneni, S.K.; Yan, H.; Zalutsky, M.R. Synthesis and evaluation of a ^{18}F-Labeled triazinediamine analogue for imaging mutant idh1 expression in gliomas by PET. *ACS Med. Chem. Lett.* **2018**, *9*, 606–611. [CrossRef] [PubMed]
16. Salas Fragomeni, R.A.; Menke, J.R.; Holdhoff, M.; Ferrigno, C.; Laterra, J.J.; Solnes, L.B.; Javadi, M.S.; Szabo, Z.; Pomper, M.G.; Rowe, S.P. Prostate-specific membrane antigen–targeted imaging With [^{18}F]DCFPyL in high-grade gliomas. *Clin. Nucl. Med.* **2017**, *42*, e433–e435. [CrossRef]
17. Werner, J.-M.; Lohmann, P.; Fink, G.R.; Langen, K.-J.; Galldiks, N. Current landscape and emerging fields of pet imaging in patients with brain tumors. *Molecules* **2020**, *25*, 1471. [CrossRef]
18. Thomas, G.E.; Szucs, M.; Mamone, J.Y.; Bem, W.T.; Rush, M.D.; Johnson, F.E.; Coscia, C.J. Sigma and opioid receptors in human brain tumors. *Life Sci.* **1990**, *46*, 1279–1286. [CrossRef]
19. Vilner, B.J.; John, C.S.; Bowen, W.D. Sigma-1 and sigma-2 receptors are expressed in a wide variety of human and rodent tumor cell lines. *Cancer Res.* **1995**, *55*, 408–413.
20. The Human Protein Atlas—SigmaR1—Organ Expression. Available online: https://www.proteinatlas.org/ENSG00000147955-SIGMAR1/tissue (accessed on 19 March 2020).
21. Cobos, E.; Entrena, J.; Nieto, F.; Cendan, C.; Pozo, E. Pharmacology and therapeutic potential of sigma1 receptor ligands. *Curr. Neuropharmacol.* **2008**, *6*, 344–366. [CrossRef]
22. Crottès, D.; Guizouarn, H.; Martin, P.; Borgese, F.; Soriani, O. The sigma-1 receptor: A regulator of cancer cell electrical plasticity? *Front. Physiol.* **2013**, *4*, 175. [CrossRef]
23. Palmer, C.P.; Mahen, R.; Schnell, E.; Djamgoz, M.B.A.; Aydar, E. Sigma-1 receptors bind cholesterol and remodel lipid rafts in breast cancer cell lines. *Cancer Res.* **2007**, *67*, 11166–11175. [CrossRef] [PubMed]
24. Hayashi, T.; Su, T.-P. Cholesterol at the endoplasmic reticulum: Roles of the sigma-1 receptor chaperone and implications thereof in human diseases. In *Cholesterol Binding and Cholesterol Transport Proteins*; Harris, J.R., Ed.; Subcellular Biochemistry; Springer: Dordrecht, The Netherlands, 2010; Volume 51, pp. 381–398.
25. Tsai, S.-Y.A.; Chuang, J.-Y.; Tsai, M.-S.; Wang, X.; Xi, Z.-X.; Hung, J.-J.; Chang, W.-C.; Bonci, A.; Su, T.-P. Sigma-1 receptor mediates cocaine-induced transcriptional regulation by recruiting chromatin-remodeling factors at the nuclear envelope. *Proc. Natl. Acad. Sci. USA* **2015**, *112*, E6562–E6570. [CrossRef] [PubMed]
26. Bem, W.T.; Thomas, G.E.; Mamone, J.Y.; Homan, S.M.; Levy, B.K.; Johnson, F.E.; Coscia, C.J. Overexpression of sigma receptors in nonneural human tumors. *Cancer Res.* **1991**, *51*, 6558–6562.

27. Spruce, B.A.; Campbell, L.A.; McTavish, N.; Cooper, M.A.; Appleyard, M.V.L.; O'Neill, M.; Howie, J.; Samson, J.; Watt, S.; Murray, K.; et al. Small molecule antagonists of the sigma-1 receptor cause selective release of the death program in tumor and self-reliant cells and inhibit tumor growth in vitro and in vivo. *Cancer Res.* **2004**, *64*, 4875–4886. [CrossRef]
28. Aydar, E.; Onganer, P.; Perrett, R.; Djamgoz, M.B.; Palmer, C.P. The expression and functional characterization of sigma-1 receptors in breast cancer cell lines. *Cancer Lett.* **2006**, *242*, 245–257. [CrossRef] [PubMed]
29. Brent, P.J.; Pang, G.; Little, G.; Dosen, P.J.; Van Helden, D.F. The sigma receptor ligand, reduced haloperidol, induces apoptosis and increases intracellular-free calcium levels [Ca^{2+}]i in colon and mammary adenocarcinoma cells. *Biochem. Biophys. Res. Commun.* **1996**, *219*, 219–226. [CrossRef]
30. Vilner, B.J.; Bowen, W.D. Sigma receptor-active neuroleptics are cytotoxic to C6 glioma cells in culture. *Eur. J. Pharmacol.* **1993**, *244*, 199–201. [CrossRef]
31. Megalizzi, V.; Mathieu, V.; Mijatovic, T.; Gailly, P.; Debeir, O.; De Neve, N.; Van Damme, M.; Bontempi, G.; Haibe-Kains, B.; Decaestecker, C.; et al. 4-IBP, a sigma1 receptor agonist, decreases the migration of human cancer cells, including glioblastoma cells, in vitro and sensitizes them in vitro and in vivo to cytotoxic insults of proapoptotic and proautophagic drugs. *Neoplasia* **2007**, *9*, 358–369. [CrossRef]
32. Megalizzi, V.; Decaestecker, C.; Debeir, O.; Spiegl-Kreinecker, S.; Berger, W.; Lefranc, F.; Kast, R.E.; Kiss, R. Screening of anti-glioma effects induced by sigma-1 receptor ligands: Potential new use for old anti-psychiatric medicines. *Eur. J. Cancer* **2009**, *45*, 2893–2905. [CrossRef] [PubMed]
33. Colabufo, N.A.; Berardi, F.; Contino, M.; Niso, M.; Abate, C.; Perrone, R.; Tortorella, V. Antiproliferative and cytotoxic effects of some sigma2 agonists and sigma1 antagonists in tumour cell lines. *Naunyn. Schmiedebergs Arch. Pharmacol.* **2004**, *370*, 106–113. [CrossRef]
34. Soriani, O.; Rapetti-Mauss, R. Sigma receptors: Their role in disease and as therapeutic targets, advances in experimental medicine and biology. In *Advances in Experimental Medicine and Biology*; Smith, S.B., Su, T.-P., Eds.; Springer International Publishing AG: Cham, Switzerland, 2017; Volume 964, pp. 63–77.
35. Van Waarde, A.; Rybczynska, A.A.; Ramakrishnan, N.; Ishiwata, K.; Elsinga, P.H.; Dierckx, R.A.J.O. Sigma receptors in oncology: Therapeutic and diagnostic applications of sigma ligands. *Curr. Pharm. Des.* **2010**, *16*, 3519–3537. [CrossRef] [PubMed]
36. Kawamura, K.; Tsukada, H.; Shiba, K.; Tsuji, C.; Harada, N.; Kimura, Y.; Ishiwata, K. Synthesis and evaluation of fluorine-18-labeled SA4503 as a selective sigma1 receptor ligand for positron emission tomography. *Nucl. Med. Biol.* **2007**, *34*, 571–577. [CrossRef] [PubMed]
37. Waterhouse, R.N.; Chang, R.C.; Zhao, J.; Carambot, P.E. In vivo evaluation in rats of [^{18}F]1-(2-fluoroethyl)-4-[(4-cyanophenoxy)methyl]piperidine as a potential radiotracer for PET assessment of CNS sigma-1 receptors. *Nucl. Med. Biol.* **2006**, *33*, 211–215. [CrossRef] [PubMed]
38. Cipriano, P.W.; Lee, S.-W.; Yoon, D.; Shen, B.; Tawfik, V.L.; Curtin, C.M.; Dragoo, J.L.; James, M.L.; McCurdy, C.R.; Chin, F.T.-N.; et al. Successful treatment of chronic knee pain following localization by a sigma-1 receptor radioligand and PET/MRI: A case report. *J. Pain Res.* **2018**, *11*, 2353–2357. [CrossRef] [PubMed]
39. Rybczynska, A.A.; Elsinga, P.H.; Sijbesma, J.W.; Ishiwata, K.; de Jong, J.R.; de Vries, E.F.; Dierckx, R.A.; van Waarde, A. Steroid hormones affect binding of the sigma ligand [^{11}C]SA4503 in tumour cells and tumour-bearing rats. *Eur. J. Nucl. Med. Mol. Imaging* **2009**, *36*, 1167–1175. [CrossRef] [PubMed]
40. Kranz, M.; Bergmann, R.; Kniess, T.; Belter, B.; Neuber, C.; Cai, Z.; Deng, G.; Fischer, S.; Zhou, J.; Huang, Y.; et al. Bridging from Brain to Tumor Imaging: (S)-(−)- and (R)-(+)-[^{18}F]Fluspidine for Investigation of Sigma-1 Receptors in Tumor-Bearing Mice. *Molecules* **2018**, *23*, 702. [CrossRef]
41. Lever, J.R.; Gustafson, J.L.; Xu, R.; Allmon, R.L.; Lever, S.Z. σ1 and σ2 receptor binding affinity and selectivity of SA4503 and fluoroethyl SA4503. *Synapse* **2006**, *59*, 350–358. [CrossRef]
42. Langa, F.; Codony, X.; Tovar, V.; Lavado, A.; Gimenez, E.; Cozar, P.; Cantero, M.; Dordal, A.; Hernandez, E.; Perez, R.; et al. Generation and phenotypic analysis of sigma receptor type I (sigma1) knockout mice. *Eur. J. Neurosci.* **2003**, *18*, 2188–2196. [CrossRef]
43. Crawford, K.W.; Bowen, W.D. Sigma-2 receptor agonists activate a novel apoptotic pathway and potentiate antineoplastic drugs in breast tumor cell lines. *Cancer Res.* **2002**, *62*, 313–322.
44. Schrock, J.M.; Spino, C.M.; Longen, C.G.; Stabler, S.M.; Marino, J.C.; Pasternak, G.W.; Kim, F.J. Sequential cytoprotective responses to Sigma1 ligand-induced endoplasmic reticulum stress. *Mol. Pharmacol.* **2013**, *84*, 751–762. [CrossRef]

45. Maneckjee, R.; Minna, J.D. Biologically active MK-801 and SKF-10,047 binding sites distinct from those in rat brain are expressed on human lung cancer cells. *Mol. Biol. Cell* **1992**, *3*, 613–619. [CrossRef] [PubMed]
46. John, C.S.; Bowen, W.D.; Varma, V.M.; McAfee, J.G.; Moody, T.W. Sigma receptors are expressed in human non-small cell lung carcinoma. *Life Sci.* **1995**, *56*, 2385–2392. [CrossRef]
47. Moody, T.W.; Leyton, J.; John, C. Sigma ligands inhibit the growth of small cell lung cancer cells. *Life Sci.* **2000**, *66*, 1979–1986. [CrossRef]
48. Ryan-Moro, J.; Chien, C.C.; Standifer, K.M.; Pasternak, G.W. Sigma binding in a human neuroblastoma cell line. *Neurochem. Res.* **1996**, *21*, 1309–1314. [CrossRef] [PubMed]
49. Bozic, I.; Allen, B.; Nowak, M.A. Dynamics of targeted cancer therapy. *Trends Mol. Med.* **2012**, *18*, 311–316. [CrossRef] [PubMed]
50. Schwartz, H.; Scroggins, B.; Zuehlke, A.; Kijima, T.; Beebe, K.; Mishra, A.; Neckers, L.; Prince, T. Combined HSP90 and kinase inhibitor therapy: Insights from The Cancer Genome Atlas. *Cell Stress Chaperones* **2015**, *20*, 729–741. [CrossRef]
51. Kim, F.J.; Maher, C.M. Sigma1 Pharmacology in the Context of Cancer. In *Sigma Proteins: Evolution of the Concept of Sigma Receptors*; Kim, F.J., Pasternak, G.W., Eds.; Handbook of Experimental Pharmacology; Springer International: Cham, Switzerland, 2017; Volume 244, pp. 237–308.
52. Cancer Cell Line Encyclopedia. Available online: https://portals.broadinstitute.org/ccle/page?gene=SIGMAR1 (accessed on 8 March 2020).
53. The Human Protein Atlas—sigmaR1—IHC. Available online: https://www.proteinatlas.org/ENSG00000147955-SIGMAR1/pathology/glioma#ihc (accessed on 21 March 2020).
54. Chu, U.B.; Ruoho, A.E. Biochemical pharmacology of the sigma-1 receptor. *Mol. Pharmacol.* **2016**, *89*, 142–153. [CrossRef]
55. Su, T.-P.; Su, T.-C.; Nakamura, Y.; Tsai, S.-Y. The sigma-1 receptor as a pluripotent modulator in living systems. *Trends Pharmacol. Sci.* **2016**, *37*, 262–278. [CrossRef] [PubMed]
56. DeHaven-Hudkins, D.L.; Fleissner, L.C.; Ford-Rice, F.Y. Characterization of the binding of [^{3}H](+)-pentazocine to σ recognition sites in guinea pig brain. *Eur. J. Pharmacol.* **1992**, *227*, 371–378. [CrossRef]
57. Chu, U.B.; Ruoho, A.E. Sigma Receptor Binding Assays. *Curr. Protoc. Pharmacol.* **2015**, *71*, 1–34. [CrossRef]
58. Renfrow, J.J.; Soike, M.H.; Debinski, W.; Ramkissoon, S.H.; Mott, R.T.; Frenkel, M.B.; Sarkaria, J.N.; Lesser, G.J.; Strowd, R.E. Hypoxia-inducible factor 2α: A novel target in gliomas. *Future Med. Chem.* **2018**, *10*, 2227–2236. [CrossRef] [PubMed]
59. Cooper, L.A.D.; Gutman, D.A.; Chisolm, C.; Appin, C.; Kong, J.; Rong, Y.; Kurc, T.; Van Meir, E.G.; Saltz, J.H.; Moreno, C.S.; et al. The Tumor Microenvironment Strongly Impacts Master Transcriptional Regulators and Gene Expression Class of Glioblastoma. *Am. J. Pathol.* **2012**, *180*, 2108–2119. [CrossRef] [PubMed]
60. Jaal, J.; Kase, M.; Minajeva, A.; Saretok, M.; Adamson, A.; Junninen, J.; Metsaots, T.; Jõgi, T.; Joonsalu, M.; Vardja, M.; et al. VEGFR-2 Expression in Glioblastoma Multiforme Depends on Inflammatory Tumor Microenvironment. *Int. J. Inflam.* **2015**, 1–7. [CrossRef]
61. Luca, A.C.; Mersch, S.; Deenen, R.; Schmidt, S.; Messner, I.; Schäfer, K.-L.; Baldus, S.E.; Huckenbeck, W.; Piekorz, R.P.; Knoefel, W.T.; et al. Impact of the 3D microenvironment on phenotype, gene expression, and EGFR inhibition of colorectal cancer cell lines. *PLoS ONE* **2013**, *8*, e59689. [CrossRef] [PubMed]
62. Montilla-Garcia, A.; Perazzoli, G.; Tejada, M.A.; Gonzalez-Cano, R.; Sanchez-Fernandez, C.; Cobos, E.J.; Baeyens, J.M. Modality-specific peripheral antinociceptive effects of μ-opioid agonists on heat and mechanical stimuli: Contribution of sigma-1 receptors. *Neuropharmacology* **2018**, *135*, 328–342. [CrossRef]
63. Zhang, Y.; Lv, X.; Bai, Y.; Zhu, X.; Wu, X.; Chao, J.; Duan, M.; Buch, S.; Chen, L.; Yao, H. Involvement of sigma-1 receptor in astrocyte activation induced by methamphetamine via up-regulation of its own expression. *J. Neuroinflammation* **2015**, *12*, 29. [CrossRef]
64. Wiese, C.; Maestrup, E.G.; Schepmann, D.; Vela, J.M.; Holenz, J.; Buschmann, H.; Wünsch, B. Pharmacological and metabolic characterisation of the potent σ1 receptor ligand 1′-benzyl-3-methoxy-3H-spiro[[2]benzofuran-1,4′-piperidine]. *J. Pharm. Pharmacol.* **2009**, *61*, 631–640. [CrossRef]
65. Brust, P.; Deuther-Conrad, W.; Becker, G.; Patt, M.; Donat, C.K.; Stittsworth, S.; Fischer, S.; Hiller, A.; Wenzel, B.; Dukic-Stefanovic, S.; et al. Distinctive In Vivo Kinetics of the New σ1 Receptor Ligands (R)-(+)- and (S)-(−)-[^{18}F]Fluspidine in Porcine Brain. *J. Nucl. Med.* **2014**, *55*, 1730–1736. [CrossRef]

66. Shen, B.; James, M.L.; Andrews, L.; Lau, C.; Chen, S.; Palner, M.; Miao, Z.; Arksey, N.C.; Shuhendler, A.J.; Scatliffe, S.; et al. Further validation to support clinical translation of [^{18}F]FTC-146 for imaging sigma-1 receptors. *EJNMMI Res.* **2015**, *5*, 49. [CrossRef]
67. Ogawa, K.; Kanbara, H.; Kiyono, Y.; Kitamura, Y.; Kiwada, T.; Kozaka, T.; Kitamura, M.; Mori, T.; Shiba, K.; Odani, A. Development and evaluation of a radiobromine-labeled sigma ligand for tumor imaging. *Nucl. Med. Biol.* **2013**, *40*, 445–450. [CrossRef]
68. Xie, F.; Bergmann, R.; Kniess, T.; Deuther-Conrad, W.; Mamat, C.; Neuber, C.; Liu, B.; Steinbach, J.; Brust, P.; Pietzsch, J.; et al. ^{18}F-labeled 1,4-dioxa-8-azaspiro [4.5]decane derivative: Synthesis and biological evaluation of a sigma1 receptor radioligand with low lipophilicity as potent tumor imaging agent. *J. Med. Chem.* **2015**, *58*, 5395–5407. [CrossRef] [PubMed]
69. Van Waarde, A.; Jager, P.L.; Ishiwata, K.; Dierckx, R.A.; Elsinga, P.H. Comparison of sigma-ligands and metabolic PET tracers for differentiating tumor from inflammation. *J. Nucl. Med.* **2006**, *47*, 150–154. [PubMed]
70. Ramakrishnan, N.K.; Rybczynska, A.A.; Visser, A.K.D.; Marosi, K.; Nyakas, C.J.; Kwizera, C.; Sijbesma, J.W.A.; Elsinga, P.H.; Ishiwata, K.; Pruim, J.; et al. Small-animal PET with a sigma-ligand, [^{11}C]SA4503, detects spontaneous pituitary tumors in aged rats. *J. Nucl. Med.* **2013**, *54*, 1377–1383. [CrossRef] [PubMed]
71. Nonnekens, J.; Schottelius, M. "Luke! Luke! Don't! It's a trap!"—spotlight on bias in animal experiments in nuclear oncology. *Eur. J. Nucl Med. Mol. Imaging* **2020**, *47*, 1024–1026. [CrossRef] [PubMed]
72. Fischer, S.; Wiese, C.; Maestrup, E.G.; Hiller, A.; Deuther-Conrad, W.; Scheunemann, M.; Schepmann, D.; Steinbach, J.; Wunsch, B.; Brust, P. Molecular imaging of sigma receptors: Synthesis and evaluation of the potent sigma1 selective radioligand [^{18}F]fluspidine. *Eur. J. Nucl. Med. Mol. Imaging* **2011**, *38*, 540–551. [CrossRef] [PubMed]
73. Brust, P.; Deuther-Conrad, W.; Lehmkuhl, K.; Jia, H.; Wunsch, B. Molecular imaging of sigma1 receptors in vivo: Current status and perspectives. *Curr. Med. Chem.* **2014**, *21*, 35–69. [CrossRef]
74. Van Waarde, A.; Rybczynska, A.A.; Ramakrishnan, N.K.; Ishiwata, K.; Elsinga, P.H.; Dierckx, R.A.J.O. Potential applications for sigma receptor ligands in cancer diagnosis and therapy. *Biochim. Biophys. Acta* **2015**, *1848*, 2703–2714. [CrossRef]
75. Elsinga, P.; Pruim, J.; Ishiwata, K.; Dierckx, R.; Groen, H. PET-imaging of sigma receptors in non-small cell lung cancer patients. *J. Nucl. Med.* **2006**, *47*, 477P.
76. Lanz, B.; Poitry-Yamate, C.; Gruetter, R. Image-Derived Input Function from the Vena Cava for ^{18}F-FDG PET Studies in Rats and Mice. *J. Nucl. Med.* **2014**, *55*, 1380–1388. [CrossRef]

Sample Availability: Samples of the compounds are not available from the authors.

© 2020 by the authors. Licensee MDPI, Basel, Switzerland. This article is an open access article distributed under the terms and conditions of the Creative Commons Attribution (CC BY) license (http://creativecommons.org/licenses/by/4.0/).

Article

Preclinical Incorporation Dosimetry of [^{18}F]FACH—A Novel ^{18}F-Labeled MCT1/MCT4 Lactate Transporter Inhibitor for Imaging Cancer Metabolism with PET

Bernhard Sattler [1,*,†], Mathias Kranz [2,3,4,†], Barbara Wenzel [2], Nalin T. Jain [2], Rareş-Petru Moldovan [2], Magali Toussaint [2], Winnie Deuther-Conrad [2], Friedrich-Alexander Ludwig [2], Rodrigo Teodoro [2], Tatjana Sattler [5], Masoud Sadeghzadeh [2,‡], Osama Sabri [1,‡] and Peter Brust [2,‡]

1 Department of Nuclear Medicine, University Hospital Leipzig, 04103 Leipzig, Germany; osama.sabri@medizin.uni-leipzig.de
2 Helmholtz-Zentrum Dresden-Rossendorf, Institute of Radiopharmaceutical Cancer Research, Department of Neuroradiopharmaceuticals, 04318 Leipzig, Germany; mathias.kranz@unn.no (M.K.); m.sadeghzadeh@hzdr.de (M.S.); nalintjain@gmail.com (N.T.J.); b.wenzel@hzdr.de (B.W.); r.moldovan@hzdr.de (R.-P.M.); m.toussaint@hzdr.de (M.T.); w.deuther-conrad@hzdr.de (W.D.-C.); f.ludwig@hzdr.de (F.-A.L.); r.teodoro@hzdr.de (R.T.); p.brust@hzdr.de (P.B.)
3 Tromsø PET Center, University Hospital of North Norway, 9009 Tromsø, Norway
4 Nuclear Medicine and Radiation Biology Research Group, The Arctic University of Norway, 9009 Tromsø, Norway
5 Department of Claw Animals, University of Leipzig, 04103 Leipzig, Germany; tasat@vetmed.uni-leipzig.de
* Correspondence: bernhard.sattler@medizin.uni-leipzig.de; Tel.: +49-(0)341-97-18034
† These authors contributed equally to this work.
‡ These authors contributed equally to this work.

Academic Editors: Anne Roivainen and Xiang-Guo Li
Received: 28 March 2020; Accepted: 20 April 2020; Published: 26 April 2020

Abstract: Overexpression of monocarboxylate transporters (MCTs) has been shown for a variety of human cancers (e.g., colon, brain, breast, and kidney) and inhibition resulted in intracellular lactate accumulation, acidosis, and cell death. Thus, MCTs are promising targets to investigate tumor cancer metabolism with positron emission tomography (PET). Here, the organ doses (ODs) and the effective dose (ED) of the first ^{18}F-labeled MCT1/MCT4 inhibitor were estimated in juvenile pigs. Whole-body dosimetry was performed in three piglets (age: ~6 weeks, weight: ~13–15 kg). The animals were anesthetized and subjected to sequential hybrid Positron Emission Tomography and Computed Tomography (PET/CT) up to 5 h after an intravenous (iv) injection of 156 ± 54 MBq [^{18}F]FACH. All relevant organs were defined by volumes of interest. Exponential curves were fitted to the time–activity data. Time and mass scales were adapted to the human order of magnitude and the ODs calculated using the ICRP 89 adult male phantom with OLINDA 2.1. The ED was calculated using tissue weighting factors as published in Publication 103 of the International Commission of Radiation Protection (ICRP103). The highest organ dose was received by the urinary bladder (62.6 ± 28.9 µSv/MBq), followed by the gall bladder (50.4 ± 37.5 µSv/MBq) and the pancreas (30.5 ± 27.3 µSv/MBq). The highest contribution to the ED was by the urinary bladder (2.5 ± 1.1 µSv/MBq), followed by the red marrow (1.7 ± 0.3 µSv/MBq) and the stomach (1.3 ± 0.4 µSv/MBq). According to this preclinical analysis, the ED to humans is 12.4 µSv/MBq when applying the ICRP103 tissue weighting factors. Taking into account that preclinical dosimetry underestimates the dose to humans by up to 40%, the conversion factor applied for estimation of the ED to humans would rise to 20.6 µSv/MBq. In this case, the ED to humans upon an iv application of ~300 MBq [^{18}F]FACH would be about 6.2 mSv. This risk assessment encourages the translation of [^{18}F]FACH into clinical study phases and the further investigation of its potential as a clinical tool for cancer imaging with PET.

Keywords: preclinical radiopharmaceutical dosimetry; image-based internal dosimetry; OLINDA; MCT1/MCT4 lactate transporter inhibitor; [^{18}F]FACH; radiation safety

1. Introduction

Aerobic glycolysis is a common feature of cancer physiology. Even under adequate oxygenation, cancer cells generate energy mainly via the cytosolic conversion of glucose into pyruvate and not via the downstream mitochondrial respiratory chain, known as the Warburg effect [1]. Due to the conversion of pyruvate to lactate, this phenomenon results in intracellularly high concentrations of lactate along with a decrease in the pH. Consequently, to avoid apoptosis caused by an acidic environment in the cytoplasm, cancer cells facilitate the proton-coupled efflux of pyruvate and lactate [2,3] by the transmembrane monocarboxylate transporters (MCTs). Upregulation of several MCTs was observed in different cancer types [4–8] and pharmacological inhibition of MCT1/MCT4 impairs cancer cell proliferation and tumor growth [9]. Increased expression of MCT1 and MCT4 in brain malignancies is assumed to be linked to the pathogenesis of glioblastoma in particular [8]. Accordingly, MCT1 and MCT4 are interesting targets for inhibitor-based molecular imaging and pharmacological treatment of this aggressive form of glioma.

Recently, [^{18}F]FLac, [^{18}F]FP, and an ^{11}C-labeled coumarin analog were developed as radiotracers to monitor MCT1 with positron emission tomography (PET). However, defluorination, nonspecific binding, and insufficient affinity hamper the applicability of these tracers [10,11]. Recently, with [^{18}F]FACH, a fluorinated analog of α-cyano-4-hydroxycinnamic acid (α-CHC) [6] ((E)-2-cyano-3-{4-[(3-[^{18}F]fluoropropyl)(propyl)amino]-2-methoxyphenyl}acrylic acid), a novel radiolabeled MCT1/4 inhibitor was developed by our group. The inhibitory potency of FACH (IC$_{50, MCT1}$ = 11 nM; IC$_{50, MCT4}$ = 6.5 nM) indicates the suitability of [^{18}F]FACH for in vivo molecular imaging of MCT1/MCT4 with PET [12,13].

For the clinical translation of newly developed radiopharmaceuticals, a safety and tolerability assessment is mandatory. One integral part of the assessment is the preclinical radiation dose assessment to estimate the radiation exposure caused by systemic, i.e., intravenous (iv), application of the radiotracer. For the estimation of organ doses (ODs) and the effective dose (ED), the pharmacokinetics and tissue distribution of the tracer have to be determined. The activity concentration in the different organs and tissues is required to determine time-integrated activity concentrations (TIAC) and finally calculate the region-specific numbers of disintegration (NOD). As recently shown, our approach to perform respective PET imaging studies combined with CT or Magnet Resonance Imaging (MRI) in a small number of piglets is suitable to generate pharmacokinetic data for dosimetric calculations [14–17]. With this fully imaging-based approach [18], the total number of animals used for dosimetry studies can be reduced to a minimum and the acquired data can be re-used for further research questions. It is important to note that when using animals for dose estimation in humans, an interspecies scaling has to be applied [19]. However, as also shown by our previous studies [14–16], preclinical dosimetry underestimates the ED in humans by up to 40% independent of the species and size of the test animal, and the radiation dose assessment has to be adapted accordingly.

Herein, we present the results of a preclinical dosimetry study of the newly developed MCT1/MCT4-specific radioligand [^{18}F]FACH performed in juvenile pigs with a clinical PET/CT system and report on the estimation of the ED in humans as required for approval of application of the new radiopharmaceutical in clinical studies.

2. Results

The radiation safety of the newly developed MCT1/MCT4-specific radioligand [^{18}F]FACH was preclinically investigated. Following intravenous injection of 156 ± 54 MBq (0.63 ± 0.49 µg) of the radiopharmaceutical, no adverse effects were observed based on vital sign monitoring, and three piglets were subjected to sequential PET/CT up to 252 min post injection (p.i.). Afterwards, volumes of interest (VOIs) of the various organs were manually defined using the respective whole-body CT dataset for anatomical orientation. The estimated TIACs were extrapolated to the human entity (Equation (2)), followed by estimation of the organ doses with OLINDA 2.1 and calculation of the effective dose using tissue weighting factors as published in Publication 103 of the International Commission of Radiation Protection (ICRP103) [20]. Finally, it can be concluded that [^{18}F]FACH is safe with respect to the radiation risk that is caused by its systemic application for PET studies. This supports and encourages the translation of this newly developed radiopharmaceutical to clinical study phases.

The TIAC courses in organs allow for determination of the fraction of administered activity, also referred to as percent of injected dose (%ID), in the particular region, followed by calculation of the NOD by integration over time applied at mono-, bi-, or tri-exponentially fitted TIACs. Figure 1 shows six examples of the exponential fits of the fractions of activity (%ID) in different organs over time (all scaled to human dimensions). The fits of all organs of the three investigated animals as well as the exact exponential equations of the fits can be found in the Supplementary Materials (Figures S1–S3).

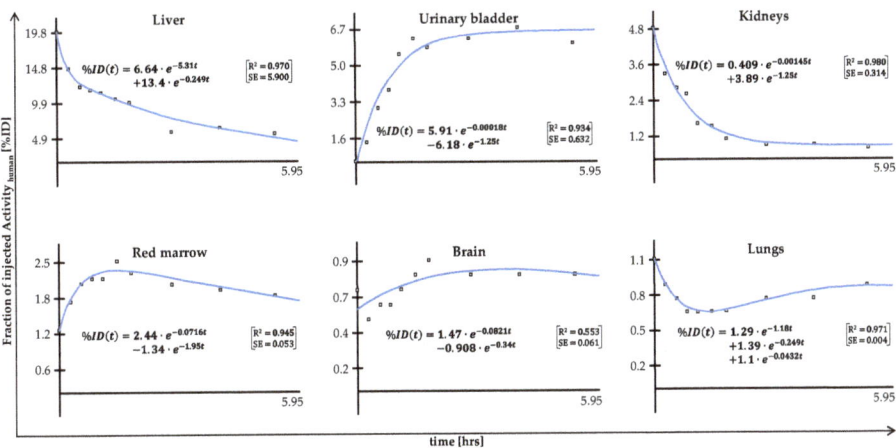

Figure 1. Examples of the mono-, bi-, or tri-exponential fits, including exponential fit equations and parameters of fit goodness (R-squared and squared error) of the human scale time–activity data using the EXM-Module of OLINDA. All fits are presented in the Supplementary Materials (Figures S1–S3).

In Figure 2, the distribution of activity in one representative animal at different times p.i. of [^{18}F]FACH is presented, showing a high uptake in the liver, pancreas, small intestines, and spleen. The rapid uptake in the liver and later also in the gall bladder indicates hepatobiliary excretion. Early accumulation in the bone structures might be related to defluorination. At later time points, activity also accumulated in the gall bladder and the urinary bladder.

An example of manually delineated VOIs using the PET or CT information to extract the accumulated amount of activity in the respective organ is presented in Figure 3. All biodistribution data, including TIACs for all organs, are available in the Supplementary Materials (Tables S1–S4; Figure S1).

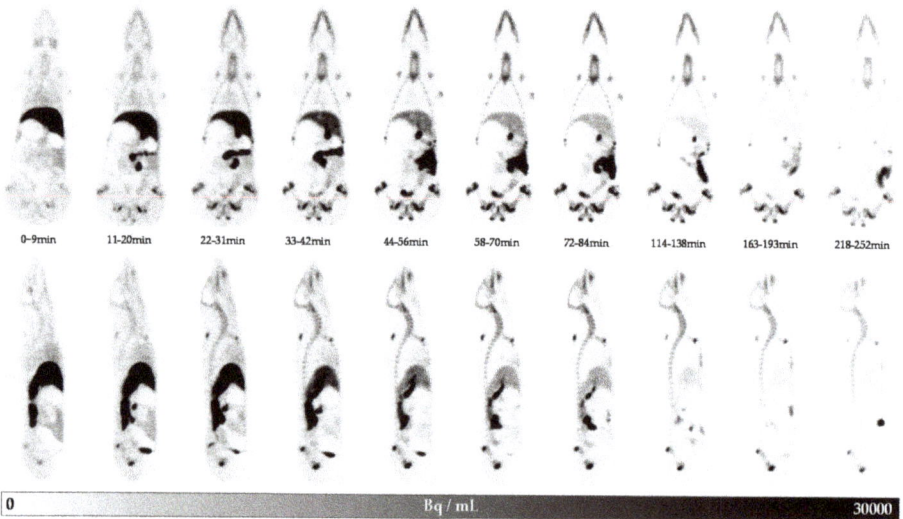

Figure 2. Whole body dynamic positron emission tomography (PET) images in coronal (upper row) and sagittal (lower row) views of pig 1 after intravenous (iv) application of 191 MBq [^{18}F]FACH.

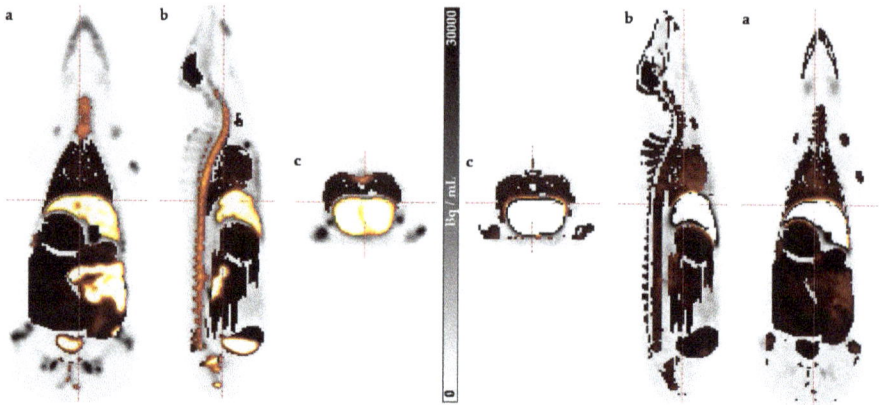

Figure 3. PET image with volumes of interest (VOIs) superimposed in (**a**) coronal, (**b**) sagittal, and (**c**) transversal views. The left panel, pig 1, obtained at 42 min post injection (p.i.), shows the organs except for the peripheral bone, whereas the right panel, pig 2, obtained at 56 min. p.i., shows all delineated organs. (The gray scale in the middle refers to the original quantitative PET data, not to the colored VOI structures).

Based on the biokinetic data extracted from the VOI analysis in piglets and the following extrapolation to the human entity (Equations (3) and (4)), the ODs and the ED were estimated for the adult male model. The resulting mean values obtained for 24 organs are presented in Table 1.

Table 1. Complete results of the dose assessment of the three investigated piglets using OLINDA 2.1 (mean value ± standard deviation (SD)).

Target Organ	OD in mSv/MBq	ED Contr. in ± mSv/MBq	ED Contr. in mSv/MBq	SD in ± mSv/MBq
Adrenals	1.42×10^{-2}	5.69×10^{-4}	1.31×10^{-4}	4.73×10^{-6}
Brain	4.69×10^{-3}	8.09×10^{-4}	4.69×10^{-5}	8.09×10^{-6}
Esophagus	9.03×10^{-3}	1.21×10^{-3}	3.61×10^{-4}	4.86×10^{-5}
Eyes	6.91×10^{-3}	1.68×10^{-3}	0.00	0.00
Gall Bladder Wall	5.04×10^{-2}	3.75×10^{-2}	4.65×10^{-4}	3.47×10^{-4}
Left Colon	1.37×10^{-2}	2.91×10^{-3}	6.63×10^{-4}	1.41×10^{-4}
Small Intestine	1.84×10^{-2}	8.44×10^{-3}	1.71×10^{-4}	7.81×10^{-5}
Stomach Wall	1.09×10^{-2}	3.07×10^{-3}	1.31×10^{-3}	3.64×10^{-4}
Right Colon	2.40×10^{-2}	7.40×10^{-3}	1.16×10^{-3}	3.62×10^{-4}
Rectum	1.38×10^{-2}	1.15×10^{-3}	3.18×10^{-4}	2.60×10^{-5}
Heart Wall	9.46×10^{-3}	7.51×10^{-4}	8.72×10^{-5}	6.72×10^{-6}
Kidneys	2.62×10^{-2}	2.19×10^{-3}	2.42×10^{-4}	2.00×10^{-5}
Liver	2.33×10^{-2}	1.36×10^{-2}	9.34×10^{-4}	5.46×10^{-4}
Lungs	7.88×10^{-3}	2.27×10^{-4}	9.46×10^{-4}	2.72×10^{-5}
Pancreas	3.05×10^{-2}	2.73×10^{-2}	2.81×10^{-4}	2.52×10^{-4}
Prostate	1.30×10^{-2}	2.85×10^{-3}	6.01×10^{-5}	1.32×10^{-5}
Salivary Glands	7.90×10^{-3}	2.11×10^{-3}	7.90×10^{-5}	2.11×10^{-5}
Red Marrow	1.39×10^{-2}	2.19×10^{-3}	1.67×10^{-3}	2.63×10^{-4}
Osteogenic Cells	2.12×10^{-2}	1.88×10^{-3}	2.12×10^{-4}	1.88×10^{-5}
Spleen	9.08×10^{-3}	1.14×10^{-3}	8.37×10^{-5}	1.04×10^{-5}
Testes	8.45×10^{-3}	2.21×10^{-3}	3.38×10^{-4}	8.87×10^{-5}
Thymus	8.33×10^{-3}	1.73×10^{-3}	7.69×10^{-5}	1.59×10^{-5}
Thyroid	7.08×10^{-3}	8.70×10^{-4}	2.83×10^{-4}	3.49×10^{-5}
Urinary Bladder Wall	6.26×10^{-2}	2.89×10^{-2}	2.50×10^{-3}	1.15×10^{-3}
Total Body	9.76×10^{-3}	1.53×10^{-3}	0.00	0.00
ED			12.4×10^{-2}	

OD = organ equivalent dose; ED contr. = effective dose contribution.

Here, the effective dose (E) represents the tissue-weighted sum of the equivalent doses in all specified tissues and organs of the body (Equation (1)), where H_T is the equivalent dose in the respective tissue or organ, T, and w_T is the tissue weighting factor [20].

$$E = \sum_T w_T \times H_T \left[\frac{mSv}{MBq}\right] \qquad (1)$$

The highest OD was received by the urinary bladder (62.6 ± 28.9 µSv/MBq), followed by the gall bladder (50.4 ± 37.5 µSv/MBq), the pancreas (30.5 ± 27.3 µSv/MBq), the kidneys (26.2 ± 2.2 µSv/MBq), the right colon (24.0 ± 7.4 µSv/MBq), and the liver (23.3 ± 13.6 µSv/MBq). When involving the tissue weighting factor w_T, which weights the organ equivalent dose in a tissue or organ to represent the relative contribution of that tissue or organ to the total health risk resulting from uniform irradiation of the body, this ranking changes: the highest contribution to the ED is by the urinary bladder (2.5 ± 1.1 µSv/MBq), followed by the red marrow (1.7 ± 0.3 µSv/MBq), the stomach (1.3 ± 0.4 µSv/MBq), the right colon (1.2 ± 0.4 µSv/MBq), the lungs (0.9 ± 0.05 µSv/MBq), and the liver (0.1 ± 0.05 µSv/MBq). The ED was found to be 12.4 µSv/MBq. According to this preclinically obtained data, a standard injection in humans of 300 MBq [18F]FACH for PET in three-dimensional (3D) mode would result in an ED of about 3.7 mSv. As known from other studies [21], preclinical incorporation dose estimates underrate the ED to humans by up to 40% [14–16]. Taking this into account, the conversion factor to estimate the ED to humans undergoing a clinical PET investigation using [18F]FACH would rise to 20.6 µSv/MBq and finally result in an ED of about 6.2 mSv/300 MBq of administered [18F]FACH.

In comparison to ED values determined for other clinically applied diagnostic radiotracers [21], this risk assessment encourages the translation of [^{18}F]FACH into clinical study phases and the further investigation of its potential as a clinical tool for PET imaging of MCT1/MCT4 in different pathologies, including oncological diseases.

3. Discussion

Due to metabolic reprogramming, highly proliferating cancer cells utilize large amounts of glucose and convert the glucose carbon mainly to lactate to support their anabolic requirements [22–24]. Given the availability of suitable radiolabeled agents, non-invasive molecular imaging by PET enables the investigation of cancer-related alterations in the metabolism of cells. For example, the enhanced uptake of glucose in cancer cells can be monitored by [^{18}F]fluoro-deoxyglucose ([^{18}F]FDG), a substrate of the glucose transporters (GLUT) overexpressed in cancer cells that has been clinically used for more than three decades for diagnosis, staging, and treatment monitoring in oncology [25,26]. Monocarboxylate transporters (MCTs), which mediate the proton-coupled transport of small carboxylic acids such as lactate, the end product of aerobic glycolysis, are also known to be overexpressed in different cancers [27,28]. Accordingly, ^{11}C- or ^{18}F-labeled substrates of MCTs such as [^{18}F]lactate and [^{11}C]pyruvate have already been developed and evaluated regarding their potential to image the expression of MCTs in tumors using PET [10,29,30]. Besides, highly affine and small molecule inhibitors of MCTs designed for targeted cancer therapies also bear the potential for the development of respective imaging agents. In this context, our group has recently reported on the development and evaluation of [^{18}F]FACH as a new MCT-targeting imaging agent possessing high inhibitory potency towards MCT1 and MCT4 [12,13]. For further assessment of the suitability of this radiopharmaceutical in clinical settings, a preclinical dosimetry study is necessary to estimate the doses delivered to humans to ensure the safe usage of [^{18}F]FACH.

According to the data obtained in this study, the overall dose estimate and vital signs monitoring confirm the radiation safety and tolerability of [^{18}F]FACH and support the translation of [^{18}F]FACH to further clinical study phases.

As found for other low-molecular-weight ^{18}F-labeled tracers, which were investigated by our group in recent years (Table 2), the organs involved in the renal as well as the hepatobiliary excretion received the highest ODs. Due to the high initial uptake of activity in the liver, although decreasing over time, the surrounding tissues, such as lung, pancreas, and kidney tissues, are exposed to comparatively high doses and thus belong to the organs contributing mainly to the ED. Furthermore, the increasing concentration in urine results in the high OD and subsequently ED of the urinary bladder wall. Notably, besides being excretory organs, the liver and kidneys are also target organs due to a high expression of MCT1/MCT4 [31–35].

Table 2. Dosimetry results of different PET radiopharmaceuticals preclinically estimated using piglet biokinetic data.

Tracer	Target/Organ	Preclinical (µSv/MBq)	Reference
[^{18}F]FACH	MCT1/4/brain tumor	12.4	this study
[^{18}F]DBT10	α7 receptor/brain tumor	13.7	[22]
(S)-(−)-[^{18}F]fluspidine (R)-(+)-[^{18}F]fluspidine	σ receptor/brain	12.9 14.0	[15]
(−)-[^{18}F]flubatine (+)-[^{18}F]flubatine	α4β2 receptor/brain	14.7 14.3	[16] [14]

A limitation of this preclinical study is that the piglets did not void during the entire imaging phase, which results in a simplification of the data analysis. As the voiding bladder model implemented in the OLINDA software cannot be applied, the dose estimation for the wall of the urinary bladder is purely imaging based and uses data based on a continuously filling bladder over the imaging time (see Figure 1). Therefore, to reduce the dose to this dose-limiting organ in humans, participants of clinical studies should be instructed to void the bladder before and immediately after each imaging session.

Another limitation of this study is related to the increasing uptake of activity in the bones. Anatomically, high amounts of the red marrow are found not only in the backbone, sternum, and pelvis but also in the epiphyseal plates of the peripheral bone of juvenile piglets, because bone growth is not yet complete. Hence, target-specific accumulation of activity due to physiological expression of MCTs in erythrocytes produced by the red bone marrow [36,37] may contribute to comparatively high ODs and finally explain the high ED contribution of bone marrow. In addition, accumulation of [^{18}F]fluoride is indicated by the radio-chromatographic analysis of plasma samples obtained during the PET imaging study (data are not shown).

Furthermore, a limitation of preclinical dosimetry studies is that the animals usually have to be anesthetized. Depending on the target of the tracer under investigation, anesthesia will have an influence on biodistribution and biokinetics. In this study, however, the anesthesia should have had a rather secondary influence. First and foremost, the anesthesia as described does not affect MCT1/MCT4 transporter inhibitors. It does act in the central nervous system (mainly the brain). With our protocol, we maintained a rather shallow anesthesia so that the animals respired spontaneously. Thus, the alterations of the biodistribution and biokinetics of [^{18}F]FACH by this anesthesia should have had a rather small influence on the dosimetry result. We expect a slight general deceleration of the metabolism, particularly that of the gastro-intestinal tract, just as while sleeping very deeply. However, animal anesthesia is generally necessary to render preclinical incorporation dosimetry possible. Although it is a well-known limitation of preclinical dosimetry studies, it does not invalidate their results. Moreover, alongside the other limitations of preclinical dosimetry described, it will be part of the comparison once clinical data have been acquired and be involved in the (up)scaling of preclinical ED results for the assessment of the ED to humans by a particular radiotracer.

4. Materials and Methods

4.1. Synthesis of [^{18}F]FACH

The radiosynthesis of [^{18}F]FACH has recently been published [38]. Briefly, [^{18}F]FACH is produced via a one-step radiosynthesis approach by using a mesylate precursor bearing an unprotected carboxylic acid function and the Kryptofix 2.2.2/K$_2$CO$_3$/[^{18}F]F-complex system. Isolation of [^{18}F]FACH was performed by semi-preparative HPLC (Reprosil-Pur C18-AQ column, 250 × 10 mm, 10 μm, 50% CH$_3$CN/20 mM NH$_4$HCO$_2$ (aq.), pH = 4–4.5, flow 3.5 mL/min). The tracer (chemical structure shown in Figure 4) was finally purified via solid-phase extraction (Sep-Pak® C18 light cartridge) and formulated in 10% EtOH/saline solution with molar activities in the range of 50–120 GBq/μmol (n = 8, at the end of synthesis) using starting activities of 1–3 GBq.

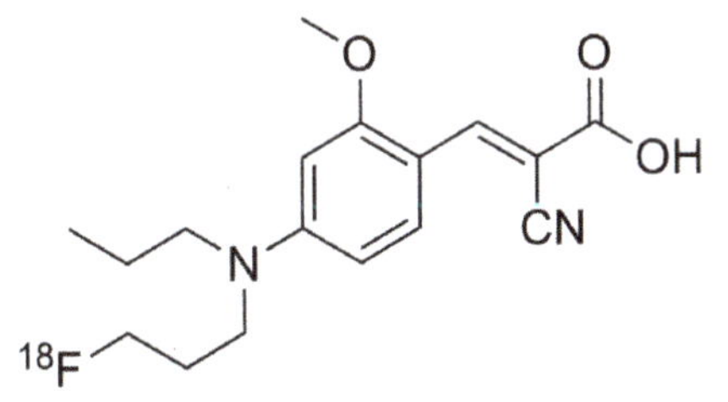

Figure 4. Chemical structure of [^{18}F]FACH.

4.2. Preclinical Dosimetry Studies—In Vivo PET/CT Imaging in Pigs

All animal experiments were approved by the responsible institutional and federal state authorities (Landesdirektion Leipzig; TVV 18/18, Reference Number DD24.1-5131/446/19).

Three piglets (age: ~6 weeks, weight: ~13–15 kg) were fasted on the day of imaging and received an intramuscular. injection of 1 mL azaperone and 4 mL ketamine to introduce anesthesia. After 15 min, 2 mL of ketamine and 1 mL of midazolam (5 mg/mL) were iv injected (ear vein, V. auricularis), followed by 5 mL of G40, 3 mL of ketamine, and 1.5 mL of midazolam in 50 mL of NaCl 0.9% with an infusion pump at a flow rate of 37.5 mL/h to maintain the narcosis throughout the entire investigation time. During narcosis, the animals maintained spontaneous respiration and no mandatory ventilation was applied. The subjects were sequentially imaged after an iv injection (contralateral ear vein) of 156 ± 54 MBq [^{18}F]FACH (0.63 ± 0.49 µg) in a PET/CT system (Biograph16, SIEMENS, Erlangen, Germany). The piglets were positioned prone with legs alongside the body on a custom-made plastic trough including a piglet head-holder (Figure 5). The PET acquisition was divided into a sequential (4 × 9 min, 3 × 12 min) and a static part (1 × 24 min, 1 × 30 min, 1 × 36 min), each of which was preceded by a low-dose CT to acquire structural data for attenuation correction (AC) and anatomical orientation (Figure 6). Post mortem, the urine was collected by bladder punctuation, weighted, and divided into three 1 mL samples for activity measurements in a gamma-counter (Packard Cobra II 5003 Auto Gamma Counting System, GMI, Ramsey, MN, USA).

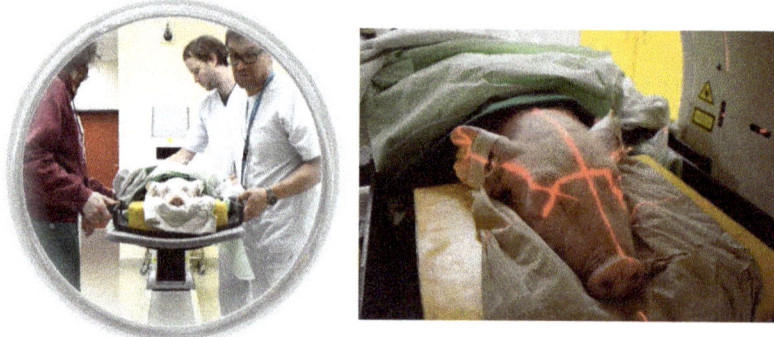

Figure 5. The animals were placed in a plastic trough and a special head rest to guarantee reproducible positioning and avoid the movement of artefacts.

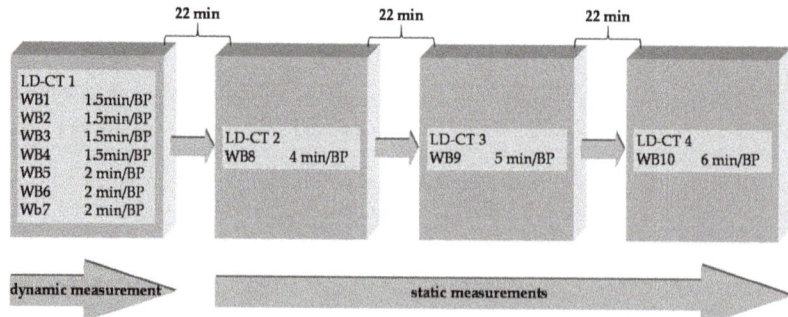

Figure 6. PET/CT imaging protocol comprising a dynamic and static part with increasing duration per bed position (BP) to compensate for decay and, thus, decreasing count statistics, preceded by a low-dose CT (LD-CT) for attenuation correction and anatomical orientation, respectively. Positioning of the animal in the PET field of view.

PET data reconstruction was done using low-dose CT attenuation correction (AC) and an iterative OSEM algorithm with 4 iterations and 8 subsets. As the PET/CT system is in daily clinical use, it is periodically subjected to detector normalization and activity calibration. Furthermore, all peripheral activity-measuring devices to be used for the investigation (dose calibrator, gamma counter) are cross-calibrated in terms of timing and a radioactivity adjustment.

4.3. Image Analysis

The image data were analyzed with ROVER (ABX, Radeberg, Germany; v. 3.0.46h). The organs were identified and manually delineated using 3D volumes of interest (VOI). The CT data were used for anatomical orientation and for image registration with the PET data. Relevant source organs like brain, gall bladder, large intestine, small intestine, stomach, heart, kidneys, liver, lungs, pancreas, red marrow (backbone, pelvis, sternum), spleen, thyroid, testes, skeleton (bone), and the urinary bladder were delineated and the time–activity data transformed into percentage of injected dose (%ID$_{organ}$) with Equation (2).

$$\%ID_{organ_t} = \frac{A_{organ_t} \times c_{scan_t}}{A_{0t}} \, [\%] \quad (2)$$

where A_{organ_t} is the activity in the organ at the time t; c_{scan_t} is a calibration factor representing the theoretical body activity (derived from a whole body mask of a volume that equals the body dimentions of the animal) decay corrected to t and divided by the imaged body activity in each image frame at the time t, and A_{0t} is the injected activity decay corrected to t_0.

4.4. Incorporation Dosimetry

The dose calculation has been described by our group in detail before [14–16]. Briefly, due to differences in weight, size, and metabolic rates between the animal species and the human volunteers, it is necessary to map the preclinical extracted biodistribution data to human circumstances. Thus, the animal biokinetic data (time scale and %ID values) were adapted to the human circumstances according to [22] to fit the human weight, size, and metabolic rates (Equations (3) and (4)).

$$t_{human} = t_{animal} \left[\frac{m_{human}}{m_{animal}} \right]^{0.25} \quad (3)$$

$$[\%ID]/organ_{human} = [\%ID]/g_{animal} \times \frac{m_{TB_{animal}}[g]}{m_{TB_{human}}[g]} \times m_{organ_{human}}[g] \quad (4)$$

The TIACs were estimated by exponential fitting and the dosimetry estimation was performed using OLINDA/EXM software (v. 2.1). Finally, organ doses (OD) were estimated and multiplied by tissue weighting factors as published in ICRP103 [20].

5. Conclusions

By extrapolation of the preclinical dosimetry data obtained in piglets, the effective dose as a measure of the overall radiation risk upon iv application of about 300 MBq [^{18}F]FACH to humans is estimated to be 3.7 mSv. However, as frequently observed for other PET tracers, the preclinical dosimetry underestimates the ED to humans by up to 40%. Accordingly, we conclude that the ED to humans by [^{18}F]FACH can be expected to be about 20.6 µSv/MBq. With a systemic application of 300 MBq, PET imaging with [^{18}F]FACH would yield an effective dose of about 6.2 mSv to human subjects. This value is well within the range of other ^{18}F-labeled radiopharmaceuticals. Despite the study-specific overestimation of the contributions of bone and the urinary bladder to the ED, this risk assessment encourages the transference of [^{18}F]FACH from preclinical to clinical study phases to further assess the suitability of this new radiopharmaceutical for PET imaging of oncological diseases.

Supplementary Materials: The following data are available online: Tables S1–S3: %ID values of the three piglets after i.v. injection of [^{18}F]FACH. Table S4: Detailed results of the dose calculation for the three animals: organ equivalent doses and effective dose contributions involving the tissue risk factor wT of ICRP103. Figures S1–S3: All time-integrated activity curves with fitting functions and fit goodness parameters (R-squared and error) of all organs and systems of organs that could be identified as taking up the tracer for all included animals.

Author Contributions: B.S., M.K., M.S., P.B., W.D.-C., and O.S. conceived and designed the experiments; M.K., W.D.-C., M.T., N.T.J., C.D., T.S., and B.S. performed the experiments; B.S., M.K., N.T.J., and O.S. analyzed the data; B.S., M.K., M.S., T.S., W.D.-C., O.S., and P.B. drafted the manuscript; M.S., R.-P.M., and P.B. designed the molecule and performed the organic chemistry; M.S., B.W., and R.T. performed the radiosynthesis; B.W., M.S., R.T., and F.-A.L. performed the metabolite analysis; T.S. handled the animals, conducted and maintained their anesthesia, and took all the blood and urine samples. All authors have read and agreed to the published version of the manuscript.

Funding: This research was funded by ALEXANDER VON HUMBOLDT FOUNDATION, grant number 7211172142 and the APC was funded by STRAHLENSCHUTZSEMINAR IN THüRINGEN E.V., grant number F2015_06.

Acknowledgments: We are very thankful to K. Franke, A. Mansel, and S. Fischer for providing [^{18}F]fluoride.

Conflicts of Interest: The authors declare no conflicts of interest.

References

1. Warburg, O. On the origin of cancer cells. *Science* **1956**, *123*, 309–314. [CrossRef] [PubMed]
2. Ponisovskiy, M.R. Warburg effect mechanism as the target for theoretical substantiation of a new potential cancer treatment. *Crit. Rev. Eukaryot. Gene Expr.* **2011**, *21*, 13–28. [CrossRef] [PubMed]
3. Koppenol, W.H.; Bounds, P.L.; Dang, C.V. Otto Warburg's contributions to current concepts of cancer metabolism. *Nat. Rev. Cancer* **2011**, *11*, 325. [CrossRef] [PubMed]
4. Halestrap, A.P. The SLC16 gene family–structure, role and regulation in health and disease. *Mol. Asp. Med.* **2013**, *34*, 337–349. [CrossRef]
5. Pinheiro, C.; Longatto-Filho, A.; Azevedo-Silva, J.; Casal, M.; Schmitt, F.C.; Baltazar, F. Role of monocarboxylate transporters in human cancers: State of the art. *J. Bioenerg. Biomembr.* **2012**, *44*, 127–139. [CrossRef]
6. Gurrapu, S.; Jonnalagadda, S.K.; Alam, M.A.; Nelson, G.L.; Sneve, M.G.; Drewes, L.R.; Mereddy, V.R. Monocarboxylate transporter 1 inhibitors as potential anticancer agents. *ACS Med. Chem. Lett.* **2015**, *6*, 558–561. [CrossRef]
7. Fang, J.; Quinones, Q.J.; Holman, T.L.; Morowitz, M.J.; Wang, Q.; Zhao, H.; Sivo, F.; Maris, J.M.; Wahl, M.L. The H+-linked monocarboxylate transporter (MCT1/SLC16A1): A potential therapeutic target for high-risk neuroblastoma. *Mol. Pharmacol.* **2006**, *70*, 2108–2115. [CrossRef]
8. Park, S.J.; Smith, C.P.; Wilbur, R.R.; Cain, C.P.; Kallu, S.R.; Valasapalli, S.; Sahoo, A.; Guda, M.R.; Tsung, A.J.; Velpula, K.K. An overview of MCT1 and MCT4 in GBM: Small molecule transporters with large implications. *Am. J. Cancer Res.* **2018**, *8*, 1967.
9. Payen, V.L.; Mina, E.; van Hée, V.F.; Porporato, P.E.; Sonveaux, P. Monocarboxylate transporters in cancer. *Mol. Metab.* **2020**, *33*, 48–66. [CrossRef]
10. van Hée, V.F.; Labar, D.; Dehon, G.; Grasso, D.; Grégoire, V.; Muccioli, G.G.; Frédérick, R.; Sonveaux, P. Radiosynthesis and validation of (±)-[^{18}F]-3-fluoro-2-hydroxypropionate ([^{18}F]-FLac) as a PET tracer of lactate to monitor MCT1-dependent lactate uptake in tumors. *Oncotarget* **2017**, *8*, 24415. [CrossRef]
11. Tateishi, H.; Tsuji, A.B.; Kato, K.; Sudo, H.; Sugyo, A.; Hanakawa, T.; Zhang, M.-R.; Saga, T.; Arano, Y.; Higashi, T. Synthesis and evaluation of ^{11}C-labeled coumarin analog as an imaging probe for detecting monocarboxylate transporters expression. *Bioorganic Med. Chem. Lett.* **2017**, *27*, 4893–4897. [CrossRef] [PubMed]
12. Sadeghzadeh, M.; Moldovan, R.-P.; Fischer, S.; Wenzel, B.; Ludwig, F.-A.; Teodoro, R.; Deuther-Conrad, W.; Jonnalagadda, S.; Jonnalagadda, S.K.; Gudelis, E.; et al. Development and radiosynthesis of the first ^{18}F-labeled inhibitor of monocarboxylate transporters (MCTs). *J. Label. Comp. Radiopharm.* **2019**, *62*, 411–424. [CrossRef] [PubMed]

13. Sadeghzadeh, M.; Moldovan, R.-P.; Wenzel, B.; Kranz, M.; Deuther-Conrad, W.; Toussaint, M.; Fischer, S.; Ludwig, F.-A.; Teodoro, R.; Jonnalagadda, S.K.; et al. Development of the first ^{18}F-labeled MCT1/MCT4 lactate transport inhibitor: Radiosynthesis and preliminary in vivo evaluation in mice. *J. Label. Comp. Radiopharm.* **2019**, *62*, S59–S60. [CrossRef] [PubMed]
14. Kranz, M.; Sattler, B.; Tiepolt, S.; Wilke, S.; Deuther-Conrad, W.; Donat, C.K.; Fischer, S.; Patt, M.; Schildan, A.; Patt, J. Radiation dosimetry of the α 4 β 2 nicotinic receptor ligand (+)-[^{18}F] flubatine, comparing preclinical PET/MRI and PET/CT to first-in-human PET/CT results. *EJNMMI Phys.* **2016**, *3*, 25. [CrossRef] [PubMed]
15. Kranz, M.; Sattler, B.; Wüst, N.; Deuther-Conrad, W.; Patt, M.; Meyer, P.; Fischer, S.; Donat, C.; Wünsch, B.; Hesse, S. Evaluation of the enantiomer specific biokinetics and radiation doses of [^{18}F] fluspidine—A new tracer in clinical translation for imaging of σ1 receptors. *Molecules* **2016**, *21*, 1164. [CrossRef] [PubMed]
16. Sattler, B.; Kranz, M.; Starke, A.; Wilke, S.; Donat, C.K.; Deuther-Conrad, W.; Patt, M.; Schildan, A.; Patt, J.; Smits, R. Internal Dose assessment of (−)-^{18}F-flubatine, comparing animal model datasets of mice and piglets with first-in-human results. *J. Nucl. Med.* **2014**, *55*, 1885–1892. [CrossRef]
17. Stabin, M.G.; Siegel, J.A. RADAR dose estimate report: A compendium of radiopharmaceutical dose estimates based on OLINDA/EXM version 2.0. *J. Nucl. Med.* **2018**, *59*, 154–160. [CrossRef]
18. McParland, B.J. *Nuclear Medicine Radiation Dosimetry: Advanced Theoretical Principles*; Springer: London, UK, 2010; ISBN 978-184-882-1262.
19. Stabin, M.G. *Fundamentals of Nuclear Medicine Dosimetry*; Springer: New York, NY, USA, 2008; ISBN 978-038-774-5794.
20. Valentin, J. *The 2007 Recommendations of the International Commission on Radiological Protection*; Elsevier: Oxford, UK, 2007; Volume 37, pp. 1–133. [CrossRef]
21. Zanotti-Fregonara, P.; Lammertsma, A.A.; Innis, R.B. Suggested pathway to assess radiation safety of 18 F-labeled PET tracers for first-in-human studies. *Eur. J. Nucl. Med. Mol. Imaging* **2013**, *40*, 1781–1783. [CrossRef]
22. Kranz, M.; Sattler, B.; Deuther-Conrad, W.; Teodoro, R.; Donat, C.; Wenzel, B.; Scheunemann, M.; Patt, M.; Sabri, O.; Brust, P. Preclinical dose assessment and biodistribution of [^{18}F]DBT10, a new α7 nicotinic acetylcholine receptor (α7-nAChR) imaging ligand. *J. Nucl. Med.* **2014**, *55*, 1143.
23. Tennant, D.A.; Durán, R.V.; Gottlieb, E. Targeting metabolic transformation for cancer therapy. *Nat. Rev. Cancer* **2010**, *10*, 267–277. [CrossRef]
24. Ward, P.S.; Thompson, C.B. Metabolic reprogramming: A cancer hallmark even warburg did not anticipate. *Cancer Cell* **2012**, *21*, 297–308. [CrossRef] [PubMed]
25. O'Neill, H.; Malik, V.; Johnston, C.; Reynolds, J.V.; O'Sullivan, J. Can the efficacy of ^{18}FFDG-PET/CT in clinical oncology be enhanced by screening biomolecular profiles? *Pharmaceuticals (Basel)* **2019**, *12*, 16. [CrossRef] [PubMed]
26. Endo, K.; Oriuchi, N.; Higuchi, T.; Iida, Y.; Hanaoka, H.; Miyakubo, M.; Ishikita, T.; Koyama, K. PET and PET/CT using ^{18}F-FDG in the diagnosis and management of cancer patients. *Int. J. Clin. Oncol.* **2006**, *11*, 286–296. [CrossRef] [PubMed]
27. Payen, V.L.; Hsu, M.Y.; Rädecke, K.S.; Wyart, E.; Vazeille, T.; Bouzin, C.; Porporato, P.E.; Sonveaux, P. Monocarboxylate transporter MCT1 promotes tumor metastasis independently of its activity as a lactate transporter. *Cancer Res.* **2017**, *77*, 5591–5601. [CrossRef]
28. Jones, R.S.; Morris, M.E. Monocarboxylate transporters: Therapeutic targets and prognostic factors in disease. *Clin. Pharmacol. Ther.* **2016**, *100*, 454–463. [CrossRef]
29. Herrero, P.; Dence, C.S.; Coggan, A.R.; Kisrieva-Ware, Z.; Eisenbeis, P.; Gropler, R.J. L-3-^{11}C-lactate as a PET tracer of myocardial lactate metabolism: A feasibility study. *J. Nucl. Med.* **2007**, *48*, 2046–2055. [CrossRef] [PubMed]
30. Yokoi, F.; Hara, T.; Iio, M.; Nonaka, I.; Satoyoshi, E. 1-^{11}Cpyruvate turnover in brain and muscle of patients with mitochondrial encephalomyopathy. A study with positron emission tomography (PET). *J. Neurol. Sci.* **1990**, *99*, 339–348. [CrossRef]
31. Koho, N.; Maijala, V.; Norberg, H.; Nieminen, M.; Pösö, A.R. Expression of MCT1, MCT2 and MCT4 in the rumen, small intestine and liver of reindeer (Rangifer tarandus tarandus L.). *Comp. Biochem. Physiol. Part A Mol. Integr. Physiol.* **2005**, *141*, 29–34. [CrossRef]

32. de Araujo, G.G.; Gobatto, C.A.; de Barros Manchado-Gobatto, F.; Teixeira, L.F.; Dos Reis, I.G.; Caperuto, L.C.; Papoti, M.; Bordin, S.; Cavaglieri, C.R.; Verlengia, R. MCT1 and MCT4 kinetic of mRNA expression in different tissues after aerobic exercise at maximal lactate steady state workload. *Physiol. Res.* **2015**, *64*, 513–522. [CrossRef]
33. Sepponen, K.; Ruusunen, M.; Pakkanen, J.A.; Pösö, A.R. Expression of CD147 and monocarboxylate transporters MCT1, MCT2 and MCT4 in porcine small intestine and colon. *Vet. J.* **2007**, *174*, 122–128. [CrossRef]
34. Becker, H.M.; Mohebbi, N.; Perna, A.; Ganapathy, V.; Capasso, G.; Wagner, C.A. Localization of members of MCT monocarboxylate transporter family Slc16 in the kidney and regulation during metabolic acidosis. *Am. J. Physiol. Renal Physiol.* **2010**, *299*, F141–F154. [CrossRef] [PubMed]
35. Wang, Q.; Lu, Y.; Yuan, M.; Darling, I.M.; Repasky, E.A.; Morris, M.E. Characterization of monocarboxylate transport in human kidney HK-2 cells. *Mol. Pharm.* **2006**, *3*, 675–685. [CrossRef] [PubMed]
36. Deuticke, B. Monocarboxylate transport in erythrocytes. *J. Membr. Biol.* **1982**, *70*, 89–103. [CrossRef] [PubMed]
37. Poole, R.C.; Cranmer, S.L.; Halestrap, A.P.; Levi, A.J. Substrate and inhibitor specificity of monocarboxylate transport into heart cells and erythrocytes. Further evidence for the existence of two distinct carriers. *Biochem. J.* **1990**, *269*, 827–829. [CrossRef] [PubMed]
38. Sadeghzadeh, M.; Moldovan, R.-P.; Teodoro, R.; Brust, P.; Wenzel, B. One-step radiosynthesis of the MCTs imaging agent [^{18}F]FACH by aliphatic ^{18}F-labelling of a methylsulfonate precursor containing an unprotected carboxylic acid group. *Sci. Rep.* **2019**, *9*, 18890. [CrossRef] [PubMed]

Sample Availability: Samples of the compounds are available from the authors.

© 2020 by the authors. Licensee MDPI, Basel, Switzerland. This article is an open access article distributed under the terms and conditions of the Creative Commons Attribution (CC BY) license (http://creativecommons.org/licenses/by/4.0/).

Article

Improved Detection of Molecular Markers of Atherosclerotic Plaques Using Sub-Millimeter PET Imaging

Jessica Bridoux [1], Sara Neyt [2], Pieterjan Debie [1], Benedicte Descamps [3], Nick Devoogdt [1], Frederik Cleeren [4], Guy Bormans [4], Alexis Broisat [5], Vicky Caveliers [1,6], Catarina Xavier [1], Christian Vanhove [3] and Sophie Hernot [1,*]

[1] Laboratory of In Vivo Cellular and Molecular Imaging (ICMI, BEFY-MIMA), Vrije Universiteit Brussel, Laarbeeklaan 103, 1090 Brussels, Belgium; jessica.bridoux@vub.be (J.B.); Pieterjan.debie@vub.be (P.D.); ndevoogd@vub.be (N.D.); vicky.caveliers@uzbrussel.be (V.C.); Catarina.xavier@vub.be (C.X.)
[2] Preclinical imaging, MOLECUBES NV, 9000 Ghent, Belgium; sara.neyt@molecubes.com
[3] IBiTech-MEDISIP, Ghent University, 9000 Ghent, Belgium; Benedicte.descamps@ugent.be (B.D.); Christian.Vanhove@ugent.be (C.V.)
[4] Radiopharmaceutical Research, KU Leuven, 3000 Leuven, Belgium; frederik.cleeren@kuleuven.be (F.C.); guy.bormans@kuleuven.be (G.B.)
[5] Radiopharmaceutiques Biocliniques, INSERM 1039, Université de Grenoble, 38400 Grenoble, France; alexis.broisat@inserm.fr
[6] Nuclear Medicine department, UZ Brussel, 1090 Brussels, Belgium
* Correspondence: sophie.hernot@vub.be; Tel.: +32-2-477-49-91

Academic Editors: Anne Roivainen and Xiang-Guo Li
Received: 30 March 2020; Accepted: 13 April 2020; Published: 16 April 2020

Abstract: Since atherosclerotic plaques are small and sparse, their non-invasive detection via PET imaging requires both highly specific radiotracers as well as imaging systems with high sensitivity and resolution. This study aimed to assess the targeting and biodistribution of a novel fluorine-18 anti-VCAM-1 Nanobody (Nb), and to investigate whether sub-millimetre resolution PET imaging could improve detectability of plaques in mice. The anti-VCAM-1 Nb functionalised with the novel restrained complexing agent (RESCA) chelator was labelled with [^{18}F]AlF with a high radiochemical yield (>75%) and radiochemical purity (>99%). Subsequently, [^{18}F]AlF(RESCA)-cAbVCAM1-5 was injected in ApoE$^{-/-}$ mice, or co-injected with excess of unlabelled Nb (control group). Mice were imaged sequentially using a cross-over design on two different commercially available PET/CT systems and finally sacrificed for ex vivo analysis. Both the PET/CT images and ex vivo data showed specific uptake of [^{18}F]AlF(RESCA)-cAbVCAM1-5 in atherosclerotic lesions. Non-specific bone uptake was also noticeable, most probably due to in vivo defluorination. Image analysis yielded higher target-to-heart and target-to-brain ratios with the β-CUBE (MOLECUBES) PET scanner, demonstrating that preclinical detection of atherosclerotic lesions could be improved using the latest PET technology.

Keywords: vulnerable plaque; molecular imaging; PET imaging; nanobody; single-domain antibody; sub-millimetre resolution; AlF-radiolabelling

---

1. Introduction

Atherosclerosis is the progressive narrowing of arteries caused by the accumulation of lipids and fibrous elements in the artery walls. Although lesion growth will lead to progressive blood vessel occlusion, a large proportion of patients show no sign of disease until the sudden rupture of so-called vulnerable plaques. Rupture is usually associated with thrombosis, causing myocardial infarctions, strokes or peripheral vascular disease [1]. Together, those severe clinical complications claim over 15 million lives every year, making cardiovascular diseases the leading cause of death worldwide [2]. The joint ESC Guidelines suggested that early diagnosis of high-risk patients could be equally effective as preventing new cases, leading to potential cost-savings, and consequently encouraged research regarding risk assessment by non- or minimally invasive imaging techniques [3]. Techniques such as multi-detector CT [4,5], intravascular ultrasound [6,7], MRI [8,9], or Optical Coherence Tomography [10] are extensively being investigated for the assessment of morphological or structural characteristics of atherosclerotic lesions. Among the molecular imaging modalities, which can reveal specific biological aspects of atherosclerotic plaques, positron emission tomography/computed tomography (PET/CT) is one of the preferred techniques in clinic [11]. Although PET/CT is a sensitive and quantitative technique, most of the commercially available pre-clinical PET scanners do not meet the necessary sensitivity and spatial resolution to fully support clinical translation of new promising tracers [12]. Recently a novel PET scanner (β-CUBE, Molecubes, Ghent, Belgium) became commercially available that uses monolithic scintillation detectors to obtain sub-mm spatial resolution in combination with high sensitivity [13], which might improve plaque detection in mice.

In order to visualise the recruitment of inflammatory cells in atherosclerotic plaques with PET imaging, we used a Nanobody (Nb)-based tracer (cAbVCAM1-5) targeting the vascular cell adhesion molecule-1 (VCAM-1) [14,15]. Nbs are small antigen-binding fragments derived from heavy-chain-only antibodies and proved to have ideal characteristics for PET imaging [16,17]. Furthermore, the biological half-life of Nbs matches the half-life of fluorine-18 (^{18}F) (109.8 min), the most commonly used positron-emitting isotope because of its favourable nuclear decay characteristics. ^{18}F-labelling of heat-sensitive biomolecules is commonly performed via prosthetic groups. However, this time-consuming process often has low efficiency. Herein, we overcame the previous issues by functionalising the Nb with the novel restrained complexing agent (RESCA) developed by Cleeren et al. [18], allowing fast and simple ^{18}F-labelling via the Al^{18}F-method [19,20].

In this study, the cAbVCAM1-5 Nb, was labelled via Al^{18}F(RESCA) chemistry, and evaluated as a tracer to image atherosclerosis plaques in Apolipoprotein E-deficient (ApoE$^{-/-}$) mice. In addition, we investigated whether the imaging could be improved using the latest β-CUBE PET technology.

2. Results

2.1. Conjugation with RESCA and Radiolabelling of the Nb

The produced cAbVCAM1-5 Nb was randomly modified through conjugation of tetrafluorophenyl TFP-RESCA (Figure 1) on its lysines for subsequent Al^{18}F-labelling. Electrospray ionisation and quadrupole time-of-flight mass spectrometry (ESI-Q-ToF-MS) analysis revealed successful conjugation of the cAbVCAM1-5 Nb with RESCA. For cAbVCAM1-5-(RESCA)n, a mass of 14,658 + $n \times$ 419.5 Da was expected. Measured mass was obtained for $n = 1$ (15,076 ± 2) Da, $n = 2$ (15,495 ± 2) Da, $n = 3$ (15,913 ± 2) Da and $n = 4$ (16,331 ± 2) Da (Figure S1).

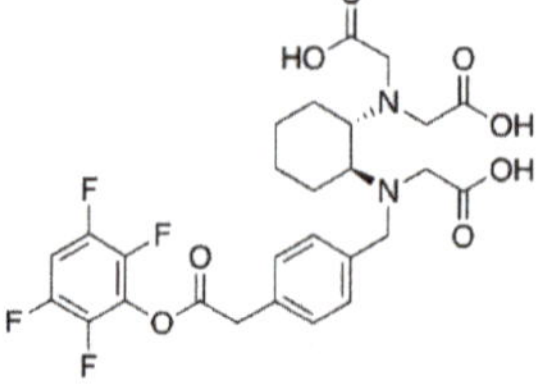

Figure 1. Structure of tetrafluorophenyl restrained complexing agent (TFP-RESCA).

Next, cAbVCAM1-5 randomly conjugated with RESCA was radiolabelled at room temperature (RT) with [^{18}F]AlF with a 78 ± 2% radiochemical yield (RCY). Separation of Nb from free [^{18}F]AlF was performed through a desalting PD10 column which was eluted in 500 µL fractions. The two fractions containing most of the activity were combined and filtered, allowing to obtain a radiochemical purity (RCP) of 99% (Figure 2) and an apparent molar activity of 24.5 ± 3.1 GBq/µmol. The radiolabelling and purification procedures were completed in less than an hour. [^{18}F]AlF(RESCA)-cAbVCAM1-5 Nb remained stable with a RCP of 96% (Figure S2A) over 3 h 30 min in injection buffer at RT, as well as in human serum at 37 °C over 1 h 30. At 2 h 30 min up to 6% defluorination was observed in human serum (Figure S2B).

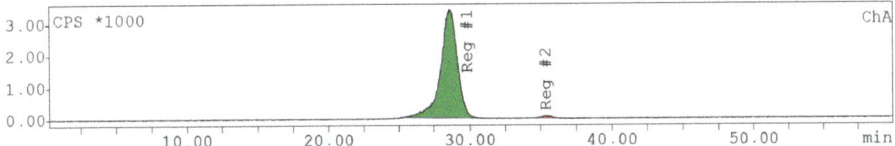

Figure 2. Size Exclusion Chromatography (SEC) profile of [^{18}F]AlF(RESCA)-cAbVCAM1-5 Nb before injection. Retention time (Rt) of [^{18}F]AlF(RESCA)-cAbVCAM1-5 = 28.7 min (99%), free [^{18}F]AlF and [^{18}F]F-Rt = 35.3 min (1%).

2.2. Imaging with the β-CUBE and LabPET8 Systems

In vivo PET imaging showed excretion of the tracer via the kidneys and bladder. The cohort injected with the [^{18}F]AlF(RESCA)-cAbVCAM1-5 Nb showed substantial signal in bone structures (Figure 3A, upper row). This signal was also observed in the control group (Figure 3A, lower row), where the [^{18}F]AlF(RESCA)-cAbVCAM1-5 Nb was co-injected with excess of unlabelled cAbVCAM1-5 Nb, indicating the non-specific character of the uptake.

Accumulation of [^{18}F]AlF(RESCA)-cAbVCAM1-5 Nb in the aortic arch of ApoE$^{-/-}$ mice was observed, which is the predominant site for atherosclerotic lesion formation in this model (Figure 3A, upper row). No signal was observed in the aortic arch of the control group (Figure 3A, lower row).

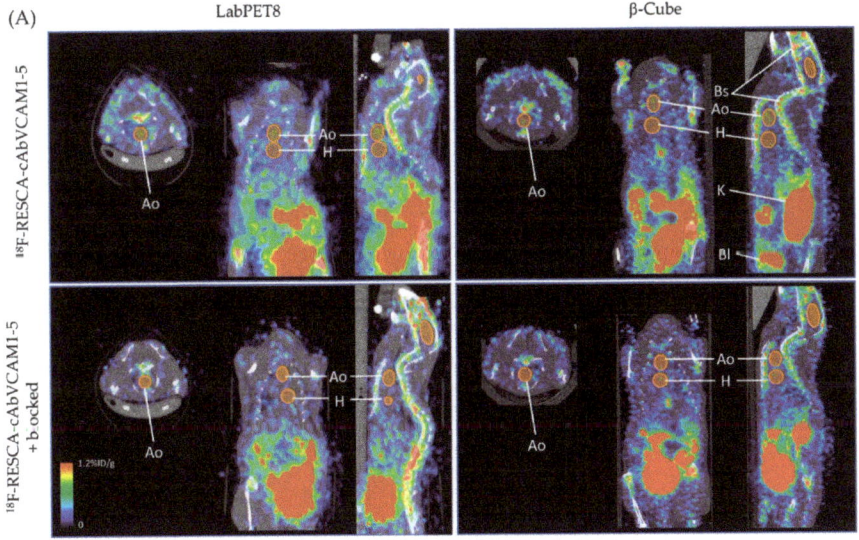

Figure 3. Cont.

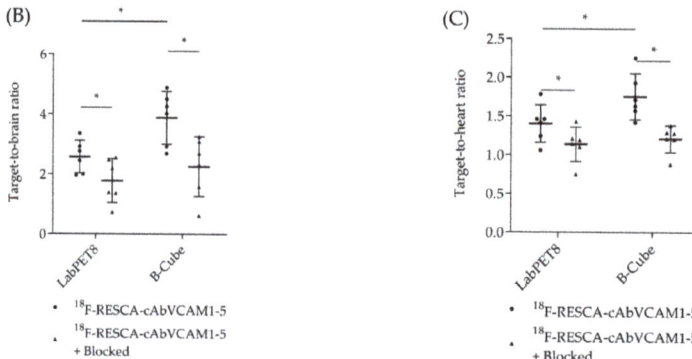

Figure 3. (**A**) Representative PET/CT images of the same mouse obtained with the LabPET8 (left) or β-CUBE (right) imaging system, demonstrating specific targeting of atherosclerotic lesions in the aortic arch (Ao) of ApoE$^{-/-}$ mice injected with [^{18}F]AlF(RESCA)-cAbVCAM1-5 Nb (upper row), while no uptake is seen at the level of the aortic arch of ApoE$^{-/-}$ mice co-injected with a 90-fold excess of unlabelled cAbVCAM1-5 Nb (blocking condition as control, unlabelled excess injected 15 min before injection of radiolabelled Nb) (lower row). Kidneys (K), bladder (Bl) and bone structures (Bs) are also visible on the images. Target-to-brain (T/B) (**B**) and target-to-heart (T/H) (**C**) ratios were calculated to compare the image quality between two commercially available preclinical PET scanners (β-CUBE and LabET8). The number of asterisks in the figures indicates the statistical significance (* $P < 0.05$).

When comparing the imaging data obtained with two distinct preclinical PET devices in a crossover study, better image quality was achieved with the β-CUBE than with the LabPET8 (Figure 3A). In vivo image contrast was evaluated by calculating target-to-brain (T/B) and target-to-heart (T/H) ratios. In both cases, significantly higher values were obtained with the β-CUBE than with the LabPET8 (T/B: 3.88 ± 0.88 vs. 2.57 ± 0.54, $p < 0.05$; T/H: 1.75 ± 0.30 vs. 1.40 ± 0.24, $p < 0.05$; respectively).

2.3. Ex Vivo Biodistribution and Atherosclerotic Plaque Targeting of [^{18}F]AlF(RESCA)-cAbVCAM1-5

The biodistribution of [^{18}F]AlF(RESCA)-cAbVCAM1-5 is summarised in Figure 4A and Table S1. Uptake in various organs and tissues is expressed as injected activity per gram (%IA/g). Constitutively VCAM-1 expressing organs such as the spleen (1.01 ± 0.34 %IA/g), lymph nodes (0.55 ± 0.15 %IA/g) and thymus (0.32 ± 0.09 %IA/g) showed specific uptake. These values were significantly lower when an excess of unlabelled Nb was co-injected (respectively 0.34 ± 0.14 %IA/g, 0.33 ± 0.22 %IA/g and 0.22 ± 0.06 %IA/g). In corroboration with the imaging data, high bone uptake was observed, which could not be reduced by competition (1.13 ± 0.33 vs. 0.96 ± 0.33 for the control). Other organs and tissues, except the kidneys (14.00 ± 3.75 %ID/g), showed no uptake of the tracer. Analysis of the dissected aortas and gamma counting confirmed the specific lesion targeting with [^{18}F]AlF(RESCA)-cAbVCAM1-5 Nb as seen on the PET/CT images. Uptake in the aortas of ApoE$^{-/-}$ mice was 2.15 ± 0.06 times higher ($p < 0.03$) as compared to the control group (Figure 4B). This was further confirmed by autoradiography of the dissected aortas even though some background in the blocked group is visible due to non-specific binding (Figure 4C).

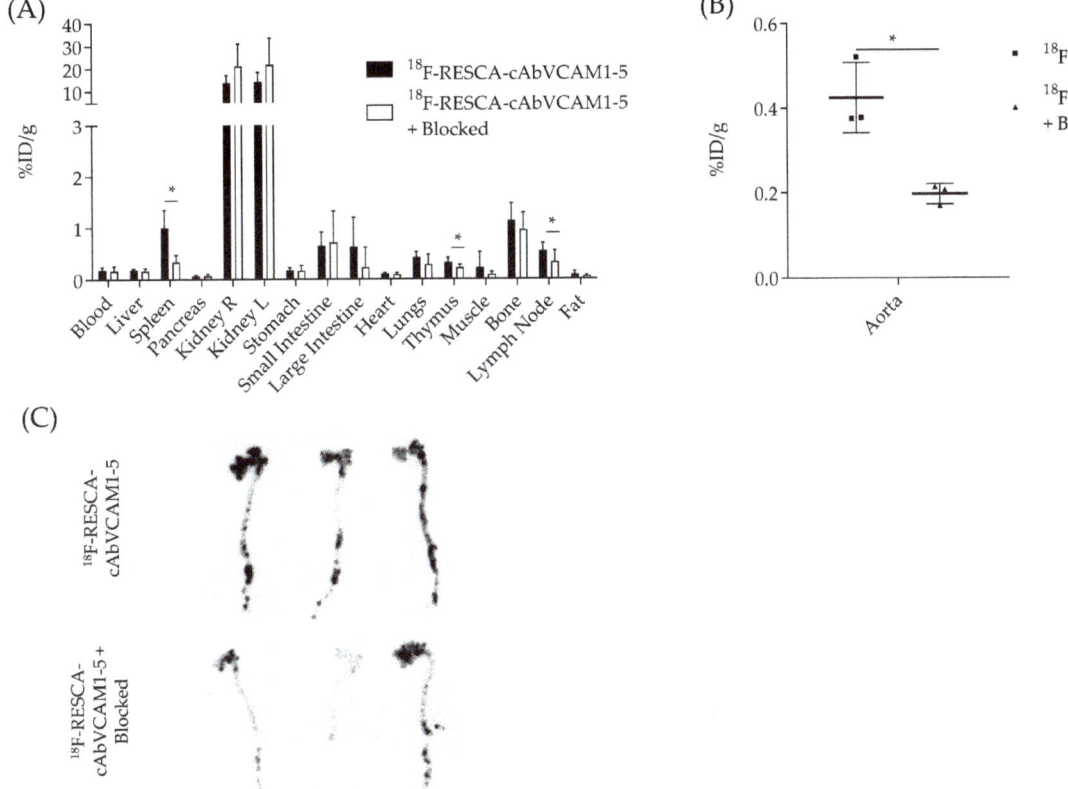

Figure 4. (**A**) Ex vivo biodistribution profile of [^{18}F]AlF(RESCA)-cAbVCAM1-5 Nb in ApoE$^{-/-}$ mice and ApoE$^{-/-}$ mice co-injected with a 90-fold excess of unlabelled Nb (blocking condition). (**B**) Ex vivo analysis of excised atherosclerotic aorta, showing significantly higher uptake (2.15 ± 0.06 times; $p < 0.03$) of the [^{18}F]AlF(RESCA)-cAbVCAM1-5 Nb compared to the control group (blocking condition). (**C**) Confirmation of the uptake by ex vivo autoradiography. The number of asterisks in the figures indicates the statistical significance (* $P < 0.05$).

3. Discussion

In the present study, as a proof of concept, a preclinical PET scanner capable of achieving sub-mm spatial resolution was used to image VCAM-1 expression in atherosclerotic lesions of ApoE$^{-/-}$ mice using an Al^{18}F(RESCA)-labelled Nb.

The lead compound cAbVCAM1-5 has previously been reported to target both the murine and human VCAM-1 receptor with nanomolar affinity, and was originally validated as technetium-99m (^{99m}Tc)-labelled tracer for SPECT imaging by Broisat et al. [14]. Although new generation clinical SPECT cameras are emerging [21], PET imaging remains the preferred imaging modality to quantitatively image molecular markers in the clinic. In addition, short-lived isotopes commonly used for PET imaging, such as fluorine-18 (^{18}F) or gallium-68 (^{68}Ga), match the short biological half-life of Nbs, allowing to decrease the radiation burden for the patients. To this end, the cAbVCAM1-5 Nb was radiolabelled with ^{18}F via the N-succinimidyl 4-[^{18}F]-fluorobenzoate ester ([^{18}F]SFB) prosthetic group [22]. Despite excellent in vivo results such as lower kidney retention and lower specific organ uptake, the production of [^{18}F]SFB remains a time-consuming multi-step procedure. In an attempt to facilitate its clinical translation, analogues of the cAbVCAM1-5 Nb labelled with radiometals via chelation were studied by Bala et al. [23] ^{68}Ga, however, has some disadvantages such as a very short half-life of 68 min, limiting the number of patients per synthesis, and requires a ^{68}Ge/^{68}Ga generator which is not available at every hospital. More recently, a new bifunctional chelator, RESCA-tetrafluorophenyl ester (TFP-RESCA) has been developed by Cleeren et al. [18,19]. Contrarily to most fluorination methods, the RESCA chelator, allows fast radiolabelling of biomolecules with [^{18}F]AlF at RT in aqueous solution.

We applied this strategy by coupling the RESCA chelator to the cAbVCAM1-5 Nb via its lysine residues, and an excellent radiolabelling yield with [^{18}F]AlF was obtained (>75% RCY in less than 60 min) (in comparison, radiolabelling of Nbs using [^{18}F]SFB have global yields ranging between 5 and 15% for a procedure time of 180 min) [24].

PET imaging of atherosclerotic lesions in the aortic arch of mice could be performed 2.5 h after intravenous (IV) administration of the radiolabelled Nb. Ex vivo analysis confirmed the imaging data, showing a significantly higher uptake of the tracer in excised aortas (0.42 ± 0.08 %IA/g) compared to the blocking condition (0.20 ± 0.02 %IA/g). This uptake, however, is lower than the previously reported uptake with [^{18}F]FB-labelled cAbVCAM1-5 Nb (1.18 ± 0.36 %IA/g) [22].

The non-invasive detection of small atherosclerotic lesions could be improved using a preclinical PET scanner of the latest generation (such as the β-CUBE), yielding significantly higher plaque-to-brain and plaque-to-heart ratios. This confirms the importance of PET scanners with sub-mm spatial resolution and high sensitivity to evaluate novel tracers for atherosclerotic plaque detection and characterisation.

Uptake of the radiolabelled cAbVCAM1-5 Nb in constitutively VCAM-1 expressing organs such as the spleen and the thymus was expected since this was already observed in previous studies [14,22,23]. This is attributed to specific targeting because uptake could be prevented by an excess of the cold analogue. Contrarily, bone uptake was noticeably high for both experimental groups. Although some specific uptake due to the expression of VCAM-1 in the bone marrow could be expected [25], the signal was observed in both groups, indicating the non-specific character of the uptake. The PET/CT images accurately co-localised the signal with bone structures (e.g., skull, limb bones, and vertebral column and sternum) and is most likely a result of uptake of degradation products derived from the tracer. The nature of this degradation product could either be formation of an active metabolite, or most likely in vivo decomplexation of [^{18}F]AlF, or defluorination, considering that radiometals and free fluorine tend to accumulate in the bone structures [26]. A hypothesis could be that after glomerular filtration and reabsorption in the proximal tubuli, the [^{18}F]AlF-RESCA complex is degraded in the lysosomes where the Nb is internalised. This results in the release of [^{18}F]AlF or [^{18}F]F$^-$ that returns into blood circulation and accumulates in the bone structures. As this instability appears to be lower in the case of other proteins [20,27], it would be interesting to further investigate the reasons behind the degradation in the case of Nbs.

In the context of atherosclerosis imaging, this bone uptake is particularly undesirable for two reasons. First of all, imaging must be performed at a time point that ensures the lowest achievable blood background signal, while keeping a significant signal in the VCAM-1-low-expressing targeted plaque. Even though Nbs are cleared very quickly from the blood, previous studies showed that imaging 2.5 to 3 h after tracer injection for the cAbVCAM1-5 Nb resulted in an optimal target-to-blood ratio [14] which unfortunately correlates with the formation of radio-metabolites observed during in vivo biodistribution studies. Secondly, [^{18}F]NaF is already being investigated in the clinic to image active calcification in atherosclerosis [28]. Thus, if free [^{18}F]F$^-$ or [^{18}F]AlF is entering the blood, it would become unclear which biological process is being imaged.

4. Material and Methods

All reagents and solvents were purchased from Sigma–Aldrich (Overijse, Belgium) or VWR (Oud-Heverlee, Belgium). The RESCA bifunctional derivative was synthetised as described in patent WO/2016/065435 [18]. All buffers used for derivatisation and labelling of the Nb were prepared in metal-free water (Fluka, Honeywell, Brussels, Belgium), chelexed (chelex, 100 sodium form (Sigma Aldrich, Overijse, Belgium), 2 g/L, overnight shaking at RT) and filtrated with a 0.2 µm PES membrane filter (VWR). ^{18}F was produced on site using a cyclotron (Cyclone KIUBE, IBA, Ottignies-Louvain-la-Neuve, Belgium) by irradiation of H2^{18}O with 18-MeV protons. Radioactivity was measured using an ionisation chamber-based dose calibrator (Veenstra Instruments, Joure, The Netherlands).

4.1. Nb Production

The cAbVCAM1-5 Nb was generated by immunisation of a dromedary in the context of a previous study, was produced in E. coli as a C-terminal His-tagged (six histidine residues) protein and was purified by immobilised metal affinity chromatography and size exclusion chromatography according to standard procedures as described elsewhere [14].

4.2. Random-Conjugation of the RESCA Chelator

The cAbVCAM1-5 Nb was modified by incubation with a 12-fold molar excess of TFP-RESCA (135 mM stock in DMSO) in 0.05 M sodium carbonate buffer (120 µM of Nb, pH 8.5–8.7) for 2 h at RT. The modified Nb was subsequently purified by SEC on a Superdex Peptide 10/300 GL column (GE Healthcare, Belgium) and eluted in 0.1 M NH$_4$OAc pH 7 on a medium-pressure chromatography system (Bio-Rad NGC, Belgium). The average number of chelators per Nb was determined by ESI-Q-ToF-MS (GIGA Proteomics, Liège, Belgium).

4.3. [^{18}F]NaF Production

The 2-mL solution of [^{18}F]F$^-$ in enriched water was manually applied to a preconditioned Sep-Pak Light QMA cartridge (WAT023525, Waters, Zellik, Belgium). The QMA was rinsed with 3 mL of water and [^{18}F]NaF was eluted with 300 µL of 0.9% NaCl solution (typically 5 GBq is eluted). The eluate was diluted with water to obtain a 10 MBq/µL solution.

4.4. Al^{18}F-Labelling of the Nb-RESCA

100 µL (1 GBq) of [^{18}F]NaF solution was added to 10 µL (20 nmol) of AlCl$_3$ solution (2 mM AlCl$_3$ trace-metal in 0.1 M NaOAc buffer pH 4.1) and left at RT for 5 min. The RESCA-conjugated cAbVCAM1-5 Nb (43 nmol, stock solution of 86.2 µM) in 0.1 M NH$_4$OAc pH 4.5 was added to the reactor containing the previously prepared [^{18}F]AlF solution and left at RT for 12 min.

4.5. Purification and Quality Control

The radiolabelled Nb was purified using a disposable PD-10 column (GE Healthcare, Machelen, Belgium) pre-conditioned with injection buffer (0.9% NaCl + 5 mg/mL vit. C, pH 6), and passed through a 0.22 µm PVDF membrane filter (Millex GV, Millipore, Darmstadt, Germany) before further use. RCP was assessed through instant thin layer chromatography (iTLC) on silica gel impregnated glass fiber sheets (Agilent Technologies, Machelen, Belgium) with 0.9% NaCl as mobile phase. iTLCs were analysed with a radio-TLC detector (RITA, Elysia Raytest, Angleur Belgium). RCP before in vivo injection was >99%. DC-RCY was calculated based on the activity obtained after PD-10 to the amount of starting activity used for Al18F-production, decay-corrected for the same time point.

4.6. In Vitro Stability Studies

At different time points, aliquots of filtered [^{18}F]AlF(RESCA)-cAbVCAM1-5 Nb were analysed for stability in injection buffer or in serum. For the latter, 400 µL of filtered [^{18}F]AlF(RESCA)-cAbVCAM1-5 Nb in injection buffer was added to 500 µL of human serum (Innovative research, Peary Court, FL, USA) and incubated at 37 °C. Analyses were performed via SEC on a HPLC system (Hitachi, Zaventem, Belgium) equipped with a radio-detector (GABI, Elysia Raytest, Angleur Belgium) and on a Superdex

75 10/300 GL column (GE Healthcare, Machelen, Belgium) equilibrated with Phosphate Buffer Saline (PBS) ([^{18}F]AlF(RESCA)-cAbVCAM1-5 Rt = 28.6 min, free [^{18}F]AlF and [^{18}F]F-Rt = 35.4 min).

4.7. Animal Model and Experimental Setup

All animal experiments were performed in accordance to the European guidelines for animal experimentation under the license LA1230272 and approved by the local Ethical Committee of the Vrije Universiteit Brussel (14-272-7). ApoE$^{-/-}$ mice were obtained from Charles River (L'Abresle, France). ApoE$^{-/-}$ mice were fed a high-fat Western diet with 1.25% cholesterol (D12108C, Research Diets, New Brunswick, NJ, USA) for 25–30 weeks to induce atherosclerotic lesions. VCAM-1 expression was assessed for this model in a previous study [14]. ApoE$^{-/-}$ mice ($N = 6$) were injected IV with (14.52 ± 8.98) MBq (12 μg) of [^{18}F]AlF(RESCA)-cAbVCAM1-5 Nb. The control group ($N = 6$) was injected with an excess of unlabelled cAbVCAM1-5 Nb (1.1 mg, 90-fold excess) followed by the injection of (14.34 ± 8.28) MBq (12 μg) of [^{18}F]AlF-RESCA-cAbVCAM1-5 Nb 15 min after the first injection.

4.8. In Vivo PET/CT Imaging and Image Processing

Two hours and 30 min after tracer injection (as determined to be the optimal time point for ideal T/B ratio in a previous study) [14], mice ($N = 6$/group) were imaged sequentially on two different PET/CT systems using a cross-over design: a β-CUBE (Molecubes, Ghent, Belgium) providing sub-mm (0.83 mm) spatial resolution and a LabPET8 (TriFoil Imaging, Chatsworth, CA, USA) with 1.2 mm spatial resolution. Both PET scans were acquired in list-mode with a total acquisition time of 30 min on the β-CUBE and 30 min on the LabPET8. PET data were reconstructed iteratively (OSEM for β-CUBE; MLEM for LabPET8) with a total of 50 iterations into a voxel size of 0.4 and 0.5 mm for the β-CUBE and LabPET8, respectively. Each PET scan was followed by a CT scan acquired for co-registration purposes using the CT-device of the same manufacturer (X-CUBE for β-CUBE; XO-CT for LabPET8). Volumes-of-interest were drawn at the level of the aortic arch, brain and heart, and T/B and T/H ratios were calculated.

4.9. Ex Vivo Analysis

Following in vivo PET/CT imaging, mice were euthanised to collect organs and tissues of interest. Aortas from the aortic root to the iliac bifurcation were collected as well. All samples were weighed and counted for radioactivity against a standard of known activity. Uptake was expressed as percentage of injected activity per gram (%IA/g), corrected for decay and extra-venous injection. Ex vivo autoradiography images were obtained after overnight exposure of the aortas to a dedicated phosphorscreen (Typhoon FLA 7000, GE Healthcare). Images were analyzed with ImageQuant.

4.10. Data Analysis and Statistics

Data are expressed as mean ± standard deviation. Comparisons between groups were performed using unpaired Student's t-test; for comparisons between scanners a paired Student's t-test was used. A p-value ≤ 0.05 was considered significant. Statistical analysis was performed using Prism 5 (Graph Pad Software) or SPSS Statistics software (version 24.0.0, IBM Company, Brussels, Belgium).

5. Conclusions

The cAbVCAM1-5 Nb could be easily radiolabelled with [^{18}F]AlF through chelation with RESCA. The potential of ^{18}F-labelled cAbVCAM1-5 Nb to target the atherosclerotic lesions and to provide good target-to-background ratios was demonstrated using the new β-CUBE imaging system. However, in vivo degradation leads to bone uptake of fluorine-18, which could interfere with the interpretation of imaging results. The excellent and uniform spatial resolution of the β-CUBE resulted in improved image quality and allowed better quantification as compared with an imaging system with lower resolution.

Supplementary Materials: The following are available online, Figure S1: Mass determination analysis of the modified Nb, Figure S2: SEC profile of [^{18}F]AlF(RESCA)-cAbVCAM1-5 Nb in human serum, Table S1: Biodistribution.

Author Contributions: Writing—original draft preparation, J.B.; writing—review and editing, S.H., C.V.; image acquisition & system handling, B.D; Image analysis, J.B., B.D., S.N., C.V., S.H.; Nanobody development, N.D., A.B.; radiochemistry, J.B., F.C.; data acquisition, J.B., P.D.; supervision, G.B., V.C., C.X., N.D., S.H.; All authors have read and agreed to the published version of the manuscript.

Funding: This project has received funding from FWO project N°G005815N and G0D8817N. This work was funded by a grant from the Scientific Fund W. Gepts UZ Brussel. J. Bridoux is funded by the EU H2020 MSCA ITN PET3D. Frederik Cleeren is a Postdoctoral Fellow of FWO (12R3119N).

Acknowledgments: The authors thank Cindy Peleman for technical assistance.

Conflicts of Interest: S. Neyt is an employee of MOLECUBES NV. A. Broisat and N. Devoogdt have patent on VCAM Nanobodies (PCT/EP2012/066348), granted amongst other countries in US and Europe (US9771423B2 and EP2748196B8)

References

1. Lusis, A.J. Atherosclerosis. *Nature* **2018**, *407*, 233–241. [CrossRef] [PubMed]
2. World Health Organization. The Top 10 Causes of Death. May 2018. Available online: https://www.who.int/news-room/fact-sheets/detail/the-top-10-causes-of-death (accessed on 15 April 2020).
3. Piepoli, M.F.; Hoes, A.W.; Agewall, S.; Albus, C.; Brotons, C.; Catapano, A.L.; Cooney, M.-T.; Corrà, U.; Cosyns, B.; Deaton, C.; et al. 2016 European Guidelines on cardiovascular disease prevention in clinical practice, Joint ESC Guidelines. *Eur. Heart J.* **2016**, *37*, 2315–2381. [CrossRef] [PubMed]
4. Van Gils, M.J.; Bodde, M.C.; Cremens, L.G.; Dippel, D.W.; van der Lugt, A. Determinants of calcification growth in atherosclerotic carotid arteries; a serial multi-detector CT angiography study. *Atherosclerosis* **2013**, *227*, 95–99. [CrossRef]
5. Guaricci, A.I.; De Santis, D.; Carbone, M.; Muscogiuri, G.; Guglielmo, M.; Baggiano, A.; Serviddio, G. Coronary Atherosclerosis Assessment by Coronary CT Angiography in Asymptomatic Diabetic Population: A Critical Systematic Review of the Literature and Future Perspectives. *Biomed. Res. Int.* **2018**, *2018*. [CrossRef] [PubMed]
6. Rafailidis, V.; Charitanti, A.; Tegos, T.; Destanis, E.; Chryssogonidis, L. Contrast-enhanced ultrasound of the carotid system: A review of the current literature. *J. Ultrasound* **2017**, *20*, 97–109. [CrossRef] [PubMed]
7. Katagiri, Y.; Tenekecioglu, E.; Serruys, P.W.; Collet, C.; Katsikis, A.; Asano, T.; Miyasaki, Y.; Piek, J.J.; Wykrzykowska, J.J.; Bourantas, C.; et al. What does the future hold for novel intravascular imaging devices: A focus on morphological and physiological assessment of plaque. *Expert Rev. Med. Devices* **2017**, *14*, 985–999. [CrossRef]
8. Raggi, P.; Baldassarre, D.; Day, S.; de Groot, E.; Fayad, Z.A. Non-invasive imaging of atherosclerosis regression with magnetic resonance to guide drug development. *Atherosclerosis* **2016**, *251*, 476–482. [CrossRef]
9. Eid, M.; De Cecco, C.N.; Schoepf, U.J.; Mangold, S.; Tesche, C.; Varga-Szemes, A.; Suranyi, P.; Stalcup, S.; Ball, B.D.; Caruso, D. The Role of MRI and CT in the Diagnosis of Atherosclerosis in an Aging Population. *Curr. Radiol. Rep.* **2016**, *4*, 12. [CrossRef]
10. Matthews, S.D.; Frishman, W.H. A Review of the Clinical Utility of Intravascular Ultrasound and Optical Coherence Tomography in the Assessment and Treatment of Coronary Artery Disease. *Cardiol. Rev.* **2017**, *25*, 68–76. [CrossRef]
11. Evans, N.R.; Tarkin, J.M.; Chowdhury, M.M.; Warburton, E.A.; Rudd, J.H. PET Imaging of Atherosclerotic Disease: Advancing Plaque Assessment from Anatomy to Pathophysiology. *Curr. Atheroscler. Rep.* **2016**, *18*, 30. [CrossRef]
12. Kuntner, C.; Stout, D. Quantitative preclinical PET imaging: Opportunities and challenges. *Front. Phys.* **2014**, *2*, 1–12. [CrossRef]
13. Krishnamoorthy, S.; Blankemeyer, E.; Mollet, P.; Surtis, S.; Van Holen, R.; Karp, J.S. Performance evaluation of the MOLECUBES b-CUBE—a high special resolution and high sensitivity small animal PET scanner utilizing monolithic LYSO scintillation detectors. *Phys. Med. Biol.* **2018**, *63*, 155013. [CrossRef] [PubMed]

14. Broisat, A.; Hernot, S.; Toczek, J.; De Vos, J.; Riou, L.M.; Martin, S.; Ahmadi, M.; Thielens, N.; Wernery, U.; Caveliers, V.; et al. Nanobodies targeting mouse/human VCAM1 for the nuclear imaging of atherosclerotic lesions. *Circ. Res.* **2012**, *110*, 921–937. [CrossRef] [PubMed]
15. Senders, M.L.; Hernot, S.; Carlucci, G.; van de Voort, J.C.; Fay, F.; Calcagno, C.; Tang, J.; Alaarg, A.; Zhao, Y.; Ishino, S.; et al. Nanobody-Facilitated Multiparametric PET/MRI Phenotyping of Atherosclerosis. *JACC Cardiovasc. Imaging* **2019**, *12*, 2015–2026. [CrossRef]
16. Chakravarty, R.; Goel, S.; Cai, W. Nanobody: The «Magic Bullet» for Molecular Imaging? *Theranostics* **2014**, *4*, 386–398. [CrossRef]
17. Debie, P.; Devoogdt, N.; Hernot, S. Targeted Nanobody-Based Molecular Tracers for Nuclear Imaging and Image-Guided Surgery. *Antibodies* **2019**, *8*, 12. [CrossRef]
18. Musthakahmed, A.M.S.; Billaud, E.; Bormans, G.M.; Cleeren, F.; Lecina, J.; Verbruggen, A. Methods for low temperature fluorine-18 radiolabelling of biomolecules (WO/216/065435). Available online: patents.google.com (accessed on 15 April 2020).
19. Cleeren, F.; Lecina, J.; Bridoux, J.; Devoogdt, N.; Tshibangu, T.; Xavier, C.; Bormans, G. Direct fluorine-18 labeling of heat-sensitive biomolecules for positron emission tomography imaging using the Al18F-RESCA method. *Nat. Protoc.* **2018**, *13*, 2330–2347. [CrossRef]
20. Cleeren, F.; Lecina, J.; Ahamed, M.; Raes, G.; Devoogdt, N.; Caveliers, V.; McQuade, P.; Rubins, D.J.; Li, W.; Verbruggen, A.; et al. Al18F-Labelling of Heat-Sensitive Biomolecules for Positron Emission Tomography Imaging. *Theranostics* **2017**, *14*, 2924–2939. [CrossRef]
21. Seo, Y.; Aparici, C.M.; Hasegawa, B.H. Technological Development and Advances in SPECT/CT. *Semin. Nucl. Med.* **2008**, *38*, 177–198. [CrossRef]
22. Bala, G.; Blykers, A.; Xavier, C.; Descamps, B.; Broisat, A.; Ghezzi, C.; Fagret, D.; Van Camp, G.; Caveliers, V.; Vanhove, C.; et al. Targeting of vascular cell adhesion molecule-1 by 18F-labelled nanobodies for PET/CT imaging of inflamed atherosclerotic plaques. *Eur. Heart J. Cardiovasc. Imaging* **2016**, *17*, 1001–1008. [CrossRef]
23. Bala, G.; Crauwels, M.; Blykers, A.; Remory, I.; Marschall, A.L.J.; Dübel, S.; Dumas, L.; Broisat, A.; Martin, C.; Ballet, S.; et al. Radiometal-labeled anti-VCAM-1 nanobodies as molecular tracers for atherosclerosis – impact of radiochemistry on pharmacokinetics. *Biol. Chem.* **2019**, *400*, 323–332. [CrossRef] [PubMed]
24. Xavier, C.; Blykers, A.; Vaneycken, I.; D'Huyvetter, M.; Heemskerk, J.; Lahoutte, T.; Devoogdt, N.; Caveliers, V. (^{18})F-nanobody for PET imaging of HER2 overexpressing tumors. *Nucl. Med. Biol.* **2016**, *43*, 247–252. [CrossRef] [PubMed]
25. Schaumann, D.H.; Tuischer, J.; Ebell, W.; Manz, R.A.; Lauster, R. VCAM-1-positive stromal cells from human bone marrow producing cytokines for B lineage progenitors and for plasma cells: SDF-1, flt3L, and BAFF. *Mol. Immunol.* **2007**, *44*, 1606–1612. [CrossRef] [PubMed]
26. Ogawa, K.; Saji, H. Advances in Drug Design of Radiometal-Based Imaging Agents for Bone Disorders. *Int. J. Mol. Imaging.* **2011**, *2011*, 537697. [CrossRef] [PubMed]
27. van der Veen, E.L.; Suurs, F.V.; Cleeren, F.; Bormans, G.; Elsinga, P.H.; Hospers, G.A.P.; Lub-de Hooge, M.N.; de Vries, E.F.J.; Antunes, I. Development and evaluation of interleukin-2 derived radiotracers for PET imaging of T-cells in mice. *JNM* **2020**. [CrossRef] [PubMed]
28. Blake, G.M.; Puri, T.; Siddique, M.; Frost, M.L.; Moore, A.E.B.; Fogelman, I. Site specific mesurements of bone formation using [^{18}F] sodium fluoride PET/CT. *Quant. Imaging Med. Surg.* **2018**, *8*, 47–59. [CrossRef]

Sample Availability: Authors are happy to provide Nanobody samples in case of a collaboration.

© 2020 by the authors. Licensee MDPI, Basel, Switzerland. This article is an open access article distributed under the terms and conditions of the Creative Commons Attribution (CC BY) license (http://creativecommons.org/licenses/by/4.0/).

Article

Thiol-Reactive PODS-Bearing Bifunctional Chelators for the Development of EGFR-Targeting [^{18}F]AlF-Affibody Conjugates

Chiara Da Pieve [1,*], Ata Makarem [2], Stephen Turnock [3], Justyna Maczynska [3], Graham Smith [1] and Gabriela Kramer-Marek [3,*]

[1] PET Radiochemistry, Division of Radiotherapy and Imaging, the Institute of Cancer Research, 123 Old Brompton Road, London SW7 3RP, UK; Graham.Smith@icr.ac.uk
[2] Division of Radiopharmaceutical Chemistry, German Cancer Research Center (DKFZ), Im Neuenheimer Feld 280, 69120 Heidelberg, Germany; a.makarem@dkfz-heidelberg.de
[3] Preclinical Molecular Imaging, Division of Radiotherapy and Imaging, the Institute of Cancer Research, 123 Old Brompton Road, London SW7 3RP, UK; Stephen.Turnock@icr.ac.uk (S.T.); justyna.maczynska@icr.ac.uk (J.M.)
* Correspondence: chiara.daPieve@icr.ac.uk (C.D.P.); Gabriela.Kramer-Marek@icr.ac.uk (G.K.-M.); Tel.: +44-(0)-2034-376-376 (C.D.P.); +44-(0)-0208-722-4412 (G.K.-M.)

Academic Editors: Anne Roivainen and Xiang-Guo Li
Received: 9 March 2020; Accepted: 27 March 2020; Published: 29 March 2020

Abstract: Site-selective bioconjugation of cysteine-containing peptides and proteins is currently achieved via a maleimide–thiol reaction (Michael addition). When maleimide-functionalized chelators are used and the resulting bioconjugates are subsequently radiolabeled, instability has been observed both during radiosynthesis and post-injection in vivo, reducing radiochemical yield and negatively impacting performance. Recently, a phenyloxadiazolyl methylsulfone derivative (PODS) was proposed as an alternative to maleimide for the site-selective conjugation and radiolabeling of proteins, demonstrating improved in vitro stability and in vivo performance. Therefore, we have synthesized two novel PODS-bearing bifunctional chelators (NOTA-PODS and NODAGA-PODS) and attached them to the EGFR-targeting affibody molecule Z$_{EGFR:03115}$. After radiolabeling with the aluminum fluoride complex ([^{18}F]AlF), both conjugates showed good stability in murine serum. When injected in high EGFR-expressing tumor-bearing mice, [^{18}F]AlF-NOTA-PODS-Z$_{EGFR:03115}$ and [^{18}F]AlF-NODAGA-PODS-Z$_{EGFR:03115}$ showed similar pharmacokinetics and a specific tumor uptake of 14.1 ± 5.3% and 16.7 ± 4.5% ID/g at 1 h post-injection, respectively. The current results are encouraging for using PODS as an alternative to maleimide-based thiol-selective bioconjugation reactions.

Keywords: [^{18}F]AlF; NOTA; NODAGA; PODS; thiol-reactive; linker; affibody molecule; bioconjugation; EGFR; tumor imaging

1. Introduction

Reactive sulfhydryl groups of cysteine residues are attractive sites for the chemical attachment of dyes, chelators, or drugs to biomolecules. Targeting cysteine residues on biomolecules has several key benefits. Firstly, the presence of the thiol group, a highly reactive nucleophile, allows for a fast and selective reaction at physiological pH. The natural low abundance of accessible and reduced cysteine residues prevents the formation of heterogeneous mixtures of the bioconjugates. Moreover, a customized site-specific incorporation of cysteine residues into a biomolecule can be easily achieved [1]. Different classes of thiol-targeting electrophilic compounds have been used, with maleimides being the most common choice [2–4]. However, maleimide conjugates have shown instability mostly as a consequence of thiol exchange reactions in vivo (e.g., retro-Michael addition) [5,6]. When maleimide-based bioconjugates are used as imaging agents, the above mentioned succinimidyl thioether linkage instability can lead to reduced accumulation in target tissues due to the release of the radioactive payload from the conjugate. Additionally, the thiol exchange reaction with endogenous thiol-containing biomolecules (e.g., albumin, cysteine and glutathione) can result in a higher background-to-noise ratio and a consequently reduced imaging contrast [7]. Thiol-reactive reagents have been investigated aiming at the formation of selective, fast, high-yielding and, most importantly, stable linkages with biomolecules [2,8–10]. Among the promising molecules, oxadiazolyl methyl sulfone-based compounds showed not only rapid and selective reaction with thiols in proteins but also the capacity of forming conjugates (via an oxadiazole–protein thiolate bond) which were more stable than those derived from maleimides [11,12]. The preparation of thiol-reactive bifunctional chelators and prosthetic groups containing the oxadiazolyl methylsulfone moiety (e.g., phenyloxadiazolyl methylsulfone or PODS) and their attachment to proteins and peptides have been examined [13–15]. Once radiolabeled, the conjugates were found to be more stable in vitro than the maleimide-derived counterparts. Moreover, when used in vivo, the oxadiazolyl methylsulfone linker-derived agents demonstrated reduced uptake in non-targeted tissues than the maleimide equivalents [13–15]. Based on these promising reports, we have prepared two novel phenyloxadiazolyl methylsulfone-containing (PODS) bifunctional chelators (NOTA-PODS and NODAGA-PODS) for the conjugation and aluminum fluoride ($[^{18}F]AlF$) radiolabeling of a cysteine-containing biomolecule. For our study, an EGFR-targeting affibody molecule ($Z_{EGFR:03115}$) was used as a thiol-bearing protein representative. The effect of the chelator structure on the synthesis and radiolabeling of the two conjugates ($[^{18}F]AlF$-NOTA-PODS-$Z_{EGFR:03115}$ and $[^{18}F]AlF$-NODAGA-PODS-$Z_{EGFR:03115}$) was studied together with their in vitro and in vivo profile. Additionally, a comparison was carried out with the maleimide-bearing $[^{18}F]AlF$-NOTA-$Z_{EGFR:03115}$ to benchmark in vitro and in vivo performance.

2. Results and Discussion

2.1. Preparation of Bifunctional Chelators NOTA-PODS and NODAGA-PODS

The aluminum fluoride-18 ($[^{18}F]AlF$) radiolabeling procedure has generated significant interest by combining the convenient and straightforward metal-based radiochemistry (e.g., one-pot reaction, minimal purification steps, aqueous solution compatibility) and the excellent decay characteristics of fluorine-18 (97% positron emission, low positron energy, half-life of 109.8 min). This method was used to directly radiolabel small molecules, peptides and proteins containing macrocyclic chelators such as NOTA or NODA [16,17]. Accordingly, in this work, the NOTA and NODAGA chelators were selected for the $[^{18}F]AlF$ radiolabeling of affibody molecule $Z_{EGFR:03115}$. However, instead of the conventional maleimide, thiol-reactive bifunctional chelators for the conjugation to the cysteine-containing small protein were developed by functionalizing the macrocycles with PODS. The reaction between PODS and the NHS ester of either NOTA or NODAGA was performed in DMF in the presence of base (Scheme 1). The products were obtained in a quantitative yield (Figure S1); NOTA-PODS and NODAGA-PODS were used for the subsequent conjugation reaction without further purification. Alternatively, pure bifunctional chelators could be isolated by semi-preparative RP–HPLC using formic acid instead of TFA in the mobile phase (Figures S2–S4). Of note, when NOTA-PODS solutions in DMF were stored at −20 °C, noticeable signs of degradation were detected already after 2 months. Conversely, NODAGA-PODS in DMF showed good stability for at least 10 months (Figure S5).

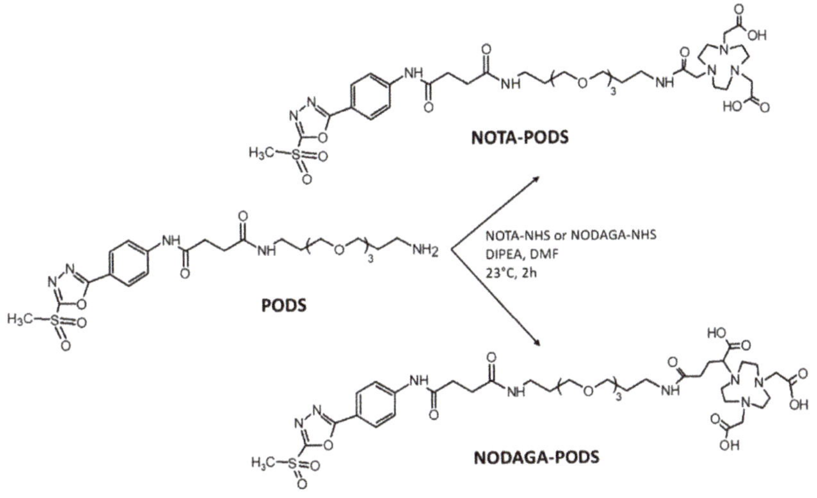

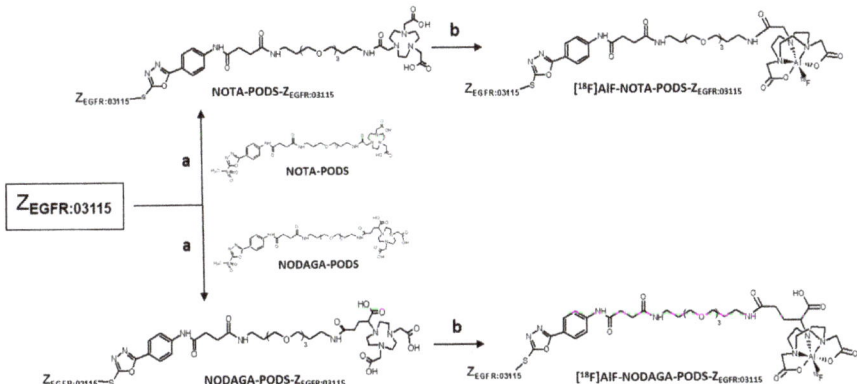

Scheme 1. Synthesis of NOTA-PODS and NODAGA-PODS.

2.2. Preparation of NOTA-PODS-$Z_{EGFR:03115}$ and NODAGA-PODS-$Z_{EGFR:03115}$

The bifunctional chelators were conjugated to $Z_{EGFR:03115}$ in a one-pot reaction using TCEP-HCl as the reducing agent (Scheme 2, Figure S6). A quantitative yield was achieved using a lower molar excess of PODS-bifunctional chelator to protein compared to a conventional maleimide-based chelator (15 vs. 35 equivalents). Being the great molar excess necessary as a consequence of the inhibition of the reactivity of maleimide-based compounds by TCEP, this result suggests that PODS is not as susceptible to TCEP as maleimides [18]. To obtain extremely pure products for the subsequent radiolabeling reaction, a single final purification of the conjugates by semi-preparative HPLC was performed (Figure 1). The pure compounds were isolated in a ca. 38% and ca. 24% yield for NOTA-PODS-$Z_{EGFR:03115}$ and NODAGA-PODS-$Z_{EGFR:03115}$, respectively. All products were characterized by ESI mass spectrometry (Figure S7).

Scheme 2. Preparation of [^{18}F]AlF-NOTA-PODS-$Z_{EGFR:03115}$ and [^{18}F]AlF-NODAGA-PODS-$Z_{EGFR:03115}$. Reaction conditions: (a) TCEP-HCl, 1 M phosphate buffer pH 7; (b) AlCl$_3$ in sodium acetate pH 4, [^{18}F]F$^-$ aq./ethanol 1:1 (v/v), for 15 min, and at 100 °C.

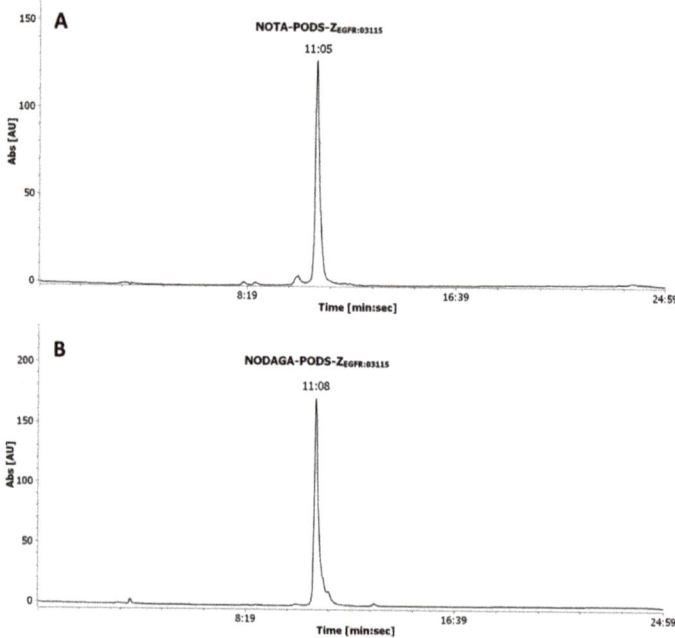

Figure 1. RP–HPLC analysis of NOTA-PODS-$Z_{EGFR:03115}$ (**A**), and NODAGA-PODS-$Z_{EGFR:03115}$ (**B**). The absorbance was recorded at the wavelength of 280 nm. The retention time (Rt) is indicated as min:sec.

2.3. [^{18}F]AlF Radiolabeling and In Vitro Stability

The affibody conjugates were radiolabeled with [^{18}F]AlF in a one-pot reaction at pH 4, 100 °C for 15 min using ethanol as organic co-solvent (50% v/v) (Scheme 2). Consistent with previously reported results for maleimide-derived affibody conjugates, some thermal degradation products were formed during the reaction (Figures S8,S9) [17]. Therefore, a RP–HPLC purification procedure, followed by HLB-SPE, was required to produce [^{18}F]AlF-NOTA-PODS-$Z_{EGFR:03115}$ and [^{18}F]AlF-NODAGA -PODS-$Z_{EGFR:03115}$ with a RCP > 98%, verified by RP–HPLC (Figure 2). As shown in Table 1, the radiochemical yields and the apparent specific/molar activities (SA/MA) achieved for [^{18}F]AlF-NODAGA -PODS-$Z_{EGFR:03115}$ were lower than for [^{18}F]AlF-NOTA-PODS-$Z_{EGFR:03115}$. Similar findings have been described by other research groups and attributed to the chelator structure and chelation capacity towards the [^{18}F]AlF complex [19,20].

Table 1. Summary of radiolabeling reactions. Radiochemical yields (RCY) and apparent specific (SA) and molar activities (MA) are decay corrected.

Radioconjugate	RCY	SA/MA	Protein Recovery
[^{18}F]AlF-NOTA-PODS-$Z_{EGFR:03115}$	11.0%–12.7%	0.40–0.59 MBq/µg 3.0–4.4 MBq/nmol	37.1%–38.0%
[^{18}F]AlF-NODAGA-PODS-$Z_{EGFR:03115}$	4.3%–8.1%	0.11–0.23 MBq/µg 0.8–1.7 MBq/nmol	23.9%–26.0%
[^{18}F]AlF-NOTA-$Z_{EGFR:03115}$	10.7%–38.0%	0.57–1.09 MBq/µg 4.1–7.8 MBq/nmol	10.6%–34.5%

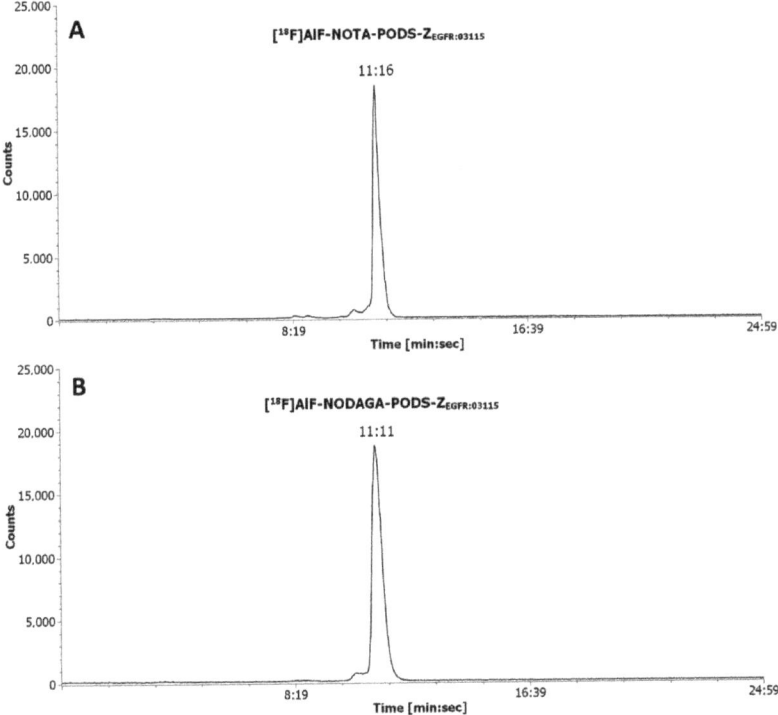

Figure 2. Radiochromatograms of pure [^{18}F]AlF-NOTA-PODS-Z$_{EGFR:03115}$ (**A**), and [^{18}F]AlF-NODAGA-PODS-Z$_{EGFR:03115}$ (**B**). The retention time (Rt) is indicated as min:sec.

To investigate the impact of the PODS linker on the radioconjugate properties, the distribution coefficient (logD$_{7.4}$) values of the two products were determined (Table S1). The logD$_{7.4}$ for [^{18}F]AlF-NOTA-PODS-Z$_{EGFR:03115}$ and [^{18}F]AlF-NODAGA-PODS-Z$_{EGFR:03115}$ was found to be hydrophilic, with values of −1.73 ± 0.07 and −3.62 ± 0.06, respectively. Including the maleimide-bearing product [^{18}F]AlF-NOTA-Z$_{EGFR:03115}$ logD$_{7.4}$ value of −1.13 ± 0.1, the NODAGA-radioconjugate proved to be the most hydrophilic of the tested compounds.

The stability of the PODS-bearing radioconjugates, with respect to change in radiochemical purity (RCP) and loss of radioactivity from the affibody molecule, was determined in mouse serum at 37 °C for 1 h (Table S2). As affibody molecules have short blood half-life, and the in vivo studies (imaging and biodistribution) are typically performed 1 h post-injection, a longer incubation time was not investigated [21]. According to RP–HPLC analysis, both conjugates exhibited good stability as 92.7 ± 2.6% of [^{18}F]AlF-NOTA-PODS-Z$_{EGFR:03115}$ and 97.2 ± 1.2% of [^{18}F]AlF-NODAGA-PODS-Z$_{EGFR:03115}$ remained intact after 1 h (Figure S10, Table S2). As previously observed for [^{18}F]AlF-NOTA-Z$_{EGFR:03115}$, a residual activity associated with the pelleted protein indicates some non-specific affinity of the radioconjugates towards the serum proteins (Table S2). However, amongst the tested compounds, the maleimide-bearing [^{18}F]AlF-NOTA-Z$_{EGFR:03115}$ demonstrated significantly higher protein-associated radioactivity than [^{18}F]AlF-NOTA-PODS-Z$_{EGFR:03115}$ and [^{18}F]AlF-NODAGA-PODS-Z$_{EGFR:03115}$ (30.5 ± 2.1% vs. 24.2 ± 3.4% and 20.4 ± 0.7%, respectively). Studies to determine whether this effect is connected to the maleimide linker are under investigation.

2.4. In Vivo Studies

To investigate the chelator effect on the pharmacokinetics and targeting properties of the conjugates, the two PODS-bearing radioactive agents were assessed in vivo in high EGFR-expressing U87MGvIII tumor-bearing mice. To determine the influence of PODS, a comparison to the maleimide derivative [^{18}F]AlF-NOTA-Z$_{EGFR:03115}$ was performed. Previous studies found that 12 µg of radiolabeled EGFR-targeting affibody conjugate with an apparent specific activity of 0.09–0.15 MBq/µg was able to partially saturate the endogenous EGFR expression (e.g., liver) and allowed for clear visualization of the tumors 1 h post-injection [22]; therefore, a dose of 12 µg (1.1–1.8 MBq) of each radioconjugate was injected. The biodistribution data indicate that the nature of the linker (i.e., PODS or maleimide) minimally influenced the radioconjugate pharmacokinetics, since similar distribution profiles were observed by both the NOTA-based products [^{18}F]AlF-NOTA-PODS-Z$_{EGFR:03115}$ and [^{18}F]AlF-NOTA-Z$_{EGFR:03115}$ (Figure 3, Table S3). However, compared to the NOTA-bearing products, [^{18}F]AlF-NODAGA-PODS-Z$_{EGFR:03115}$ showed an inconsistent radioactivity accumulation in the liver. This result suggests that the chelator can influence the radioconjugate pharmacokinetics and that chelator-customized injected protein doses and specific activities should be investigated further by dose-escalation studies. Among the non-targeted organs, the highest uptake of all three conjugates was found in the kidneys, which is due to renal excretion of the affibody molecule, and the reabsorption of the radioactive metabolites followed by their retention in the proximal tubular cells [23]. Based on this effect, the kidney uptake value for [^{18}F]AlF-NODAGA-PODS-Z$_{EGFR:03115}$ suggests a possible higher renal clearance of the NODAGA conjugate compared to [^{18}F]AlF-NOTA-PODS-Z$_{EGFR:03115}$. Moreover, both PODS-bearing radioconjugates show significantly lower kidney uptakes than the maleimide analogue, indicating a possible effect of the PODS linker on the renal retention (Figure 3 and Table S3). More experiments will be needed to confirm this observation. High accumulation of the conjugates was found in the tumors, with uptakes of 14.1 ± 5.3 and 16.7 ± 4.5 %ID/g for [^{18}F]AlF-NOTA-PODS-Z$_{EGFR:03115}$ and [^{18}F]AlF-NODAGA-PODS-Z$_{EGFR:03115}$, respectively (Figure 4, Table S3). The favorable tumor-to-muscle ratios for the two PODS-bearing radioconjugates resulted in high-contrast PET images already 1 h post-injection (Table 2). In contrast to Adumeau et al., there was no difference in both the biodistribution profile and the tumor-to-tissue ratios between the [^{18}F]AlF PODS-bearing conjugates and the maleimide-based product (Table 2, Table S3) [15].

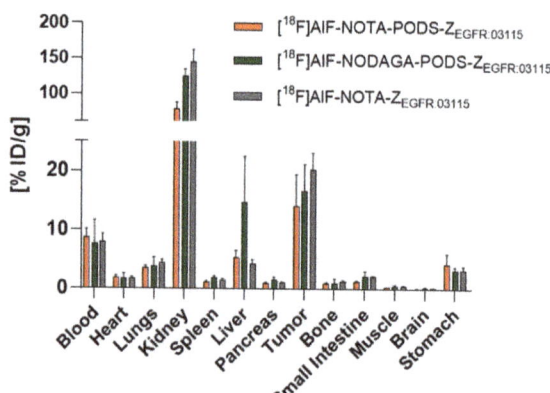

Figure 3. Biodistribution results for [^{18}F]AlF-NOTA-PODS-Z$_{EGFR:03115}$ and [^{18}F]AlF-NODAGA-PODS-Z$_{EGFR:03115}$ at 1 h p.i. [^{18}F]AlF-NOTA-Z$_{EGFR:03115}$ was used as a control. The data are reported as the mean percentage of the injected dose per gram of tissue (%ID/g) ± SD (for each group, n = 3).

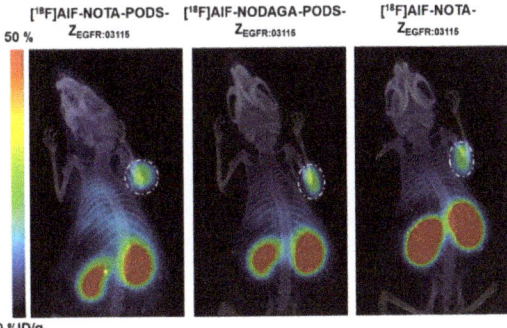

Figure 4. PET/CT images of U87MGvIII tumor-bearing mice using [^{18}F]AlF-NOTA-PODS-Z$_{EGFR:03115}$ and [^{18}F]AlF-NODAGA-PODS-Z$_{EGFR:03115}$. [^{18}F]AlF-NOTA-Z$_{EGFR:03115}$ was used as control. High-contrast images were acquired as early as 1 h p.i. The tumors are indicated by a white circle.

Table 2. Tumor-to-organ ratios at 1 h p.i. of the three examined [^{18}F]AlF conjugates. No differences were identified.

	Tumor-to-Organ Ratio			
	Blood	Kidney	Liver	Muscle
[^{18}F]AlF-NOTA-PODS-Z$_{EGFR:03115}$	1.6 ± 0.6	0.2 ± 0.1	2.7 ± 1.2	37.2 ± 12.4
[^{18}F]AlF-NODAGA-PODS-Z$_{EGFR:03115}$	2.6 ± 1.2	0.1 ± 0.0	1.3 ± 0.6	31.3 ± 8.2
[^{18}F]AlF-NOTA-Z$_{EGFR:03115}$	2.8 ± 0.9	0.1 ± 0.0	4.7 ± 0.8	34.9 ± 3.4

3. Materials and Methods

3.1. General Materials and Methods

Chemicals and solvents were purchased and used without further purification unless otherwise stated. The 2,2'-(7-{2-[(2,5-dioxopyrrolidin-1-yl)oxy]-2-oxoethyl}-1,4,7-triazonane-1,4-diyl)diacetic acid (NOTA-NHS) and 2,2'-(7-{1-carboxy-4-[(2,5-dioxopyrrolidin-1-yl)oxy]-4-oxobutyl}-1,4,7-triazonane-1,4-diyl}diacetic acid (NODAGA-NHS) were purchased from Chematech (Dijon, France). Tris(2-carboxyethyl)-phosphine hydrochloride (TCEP-HCl) was purchased from Thermo Fisher Scientific (Loughborough, UK). Ethanol (EtOH) and HPLC grade acetonitrile, trifluoroacetic acid (TFA) and formic acid were purchased from Thermo Fisher Scientific (Loughborough, UK). Phosphate-buffered saline (PBS) was purchased from Gibco (Life Technologies, Paisley, UK). Ethylenediaminetetraacetic acid (EDTA), N,N-diisopropylethylamine (DIPEA), mouse serum, dimethylformamide (DMF), n-octanol and Iso-Disc PVDF syringe filters (13 mm, 0.2 μm) were purchased from Sigma-Aldrich (Gillingham, UK). Aluminum chloride hexahydrate (AlCl$_3$, 99.9995%) was purchased from Alfa Aesar (Heysham, UK). Sodium acetate (AnalR Normapur) was purchased from VWR International (Lutterworth, UK). The Z$_{EGFR:03115}$-Cys affibody molecule was provided by Affibody AB (Solna, Sweden http://www.affibody.com) as a solution in 0.2 M sodium acetate. Phenyloxadiazolyl methylsulfone (PODS) was synthesized as described in the literature [15]. The maleimide-bearing affibody analogue NOTA-Z$_{EGFR:03115}$ was prepared and radiolabeled as previously reported [22]. Low-protein-binding microcentrifuge tubes (1.5 mL) were purchased from Eppendorf (Stevenage, UK). Dry bifunctional chelators and affibody conjugates were achieved using a Concentrator Plus (Eppendorf, Stevenage, UK). Incubation of the reaction mixtures was performed using a Grant Bio thermo shaker (Camlab, Stevenage, UK). Protein conjugate concentration was determined by measuring the UV absorbance at 280 nm on a NanoDrop 2000 spectrophotometer (Thermo Scientific, Loughborough, UK) using a molar extinction coefficient (ε_{280}) of 36345 cm^{-1} M^{-1}. Oasis HLB (1 mL, 30 mg sorbent) SPE cartridges were purchased from Waters

(Elstree, UK). [^{18}F]Fluoride was produced a GE PETrace cyclotron by 16 MeV irradiation of an enriched [^{18}O]H$_2$O target, supplied by Alliance Medical Radiopharmarcy Ltd. (Warwick, UK) and used without further purification.

Analytical and semi-preparative HPLC were carried out on an Agilent Infinity 1260 quaternary pump system equipped with a 1260 Diode array (Agilent Technologies, Didcot, UK). Elution profiles were recorded using Laura 4 software (Lablogic, Sheffield, UK, 2013). NOTA-PODS and NODAGA-PODS were analyzed on an Eclipse XDB C18 column, 4.6 × 150 mm, 5 µm (Agilent Technologies, Didcot, UK) using Gradient 1: 0–20 min 3%–60% B, at flow rate of 1 mL/min with 0.1% TFA in water as eluent A and 0.1% TFA in acetonitrile as eluent B. The bifunctional chelators were purified by semi-preparative RP–HPLC on a Ultracarb ODS C18 column, 10 × 250 mm, 7 µm (Phenomenex, Macclesfield, UK), using Gradient 1a: 0–20 min 3%–60% B, at a flow rate of 3mL/min with 0.1% formic acid in water as eluent A and 0.1% formic acid in acetonitrile as eluent B and. Affibody conjugates and radioconjugates were analyzed on a Zorbax 300SB C18 column, 4.6 × 250 mm, 5 µm (Agilent Technologies, Didcot, UK) using Gradient 2: 0–20 min. 30%–65% B, with 0.1% TFA in water as eluent A and 0.1% TFA in acetonitrile as eluent B at a flow rate of 1 mL/min. Affibody conjugates were purified by semi-preparative RP–HPLC on a Jupiter C18 column, 10 × 250 mm, 10 µm, 300 Å (Phenomenex, Macclesfield, UK), using Gradient 2 at a 3 mL/min flow rate. The radioactivity of the eluate was monitored using an IN/US Systems Gamma-ram Model 4 NaI radiodetector (Lablogic, Sheffield, UK). Retention times (Rt) are expressed as minutes:seconds (min:sec). Electro spray ionization high-resolution mass spectrometry (ESI–HRMS) was performed using an Agilent 1200 series LC pump with a 6210 time-of-flight (TOF) mass analyzer. Protein MS was performed on 6520 a Series qToF mass spectrometer fitted with a dual ESI ionization source (Agilent, Santa Clara, CA, USA).

3.2. Preparation of NOTA-PODS and NODAGA-PODS

The macrocyclic chelator NHS ester (10.32 µmol), was added solid to a solution of PODS (6.88 µmol) in DMF (400 µL) and DIPEA (7.2 µL, 41.3 µmol). The clear solution was incubated in a thermo shaker at 23 °C for 2 h (900 rpm). The product was analyzed by RP–HPLC and either purified by RP–HPLC (Gradient 1a) or used for the next reaction without further purification.

NOTA-PODS = Analytical RP–HPLC (Gradient 1): Rt: 12:15 min:sec; ESI–HRMS: [M + H]$^+$ (m/z) calcd: 827.9, found 827.3633.

NODAGA-PODS = Analytical RP–HPLC (Gradient 1): Rt: 12:13 min:sec; ESI–HRMS: [M + H]$^+$ (m/z) calcd: 899.9, found 899.3895.

3.3. Preparation of NOTA-PODS-Z$_{EGFR:03115}$ and NODA-PODS-Z$_{EGFR:03115}$

In a 1.5 mL low-protein-binding centrifuge tube, a freshly prepared solution of TCEP-HCl in 1 M phosphate buffer pH 7 (0.025 mg/µL, 9.8 µL, 1.71 µmol) was added to a solution of Z$_{EGFR:03115}$-Cys (200 µL, 2.3 mg/mL, 68.4 nmol). The mixture was incubated in a thermo shaker at 85 °C for 5 min (900 rpm) followed by 25 min at room temperature. Either NOTA-PODS or NODAGA-PODS solution in DMF was then added (1.03 µmol) and the solution was incubated in a thermo shaker at 37 °C for 1 h (900 rpm). The reaction mixture was analyzed by RP–HPLC. The product was purified by semi-preparative RP–HPLC. The collected fractions containing the product were dried and quantified by measuring the UV absorbance at 280 nm.

NOTA-PODS-Z$_{EGFR:03115}$ = yield 38.5%. Analytical RP–HPLC (Gradient 2): Rt: 11:06 min:sec; ESI–HRMS (m/z): [M + H]$^+$ expected: 7470, found: 7470.91.

NODAGA-PODS-Z$_{EGFR:03115}$ = yield 39.9%. Analytical RP–HPLC (Gradient 2): Rt: 11:08 min:sec; ESI–HRMS (m/z): [M + H]$^+$ expected: 7543, found: 7543.07.

3.4. Radiosynthesis of [^{18}F]AlF-NOTA-PODS-Z$_{EGFR:03115}$ and [^{18}F]AlF-NODAGA-PODS-Z$_{EGFR:03115}$

To a 1.5 mL low-protein-binding plastic tube containing either NOTA-PODS-Z$_{EGFR:03115}$ or NODAGA-PODS-Z$_{EGFR:03115}$ (lyophilized, 10–20 nmol), 2 mM AlCl$_3$ (4.0–6.5 µL, 7–13 nmol) in 0.5 M sodium acetate buffer pH 4, 50 mM ascorbic acid in 25 mM sodium acetate pH 4.5 (to a final concentration of 1 mM), and aqueous non-purified [^{18}F]fluoride (180–200 MBq) were added. A volume of ethanol was then added to achieve a final 1:1 v/v aqueous to organic solvent ratio. The mixture was heated at 100 °C for 15 min. After cooling to ambient temperature, the solution was purified by RP–HPLC using Gradient 2. The collected fraction containing the product was diluted with 0.1% aq TFA (3 mL) and loaded on an HLB-SPE cartridge (1 mL, 30 mg sorbent). The trapped radioactivity was washed with 0.1% aq TFA (4 mL) and then eluted with 60% ethanol/water (v/v, 120 µL). The product was quantified by measuring the UV absorbance at 280 nm. Synthesis time (from the beginning of the reaction) = ca. 45 min.

[^{18}F]AlF-NOTA-PODS-Z$_{EGFR:03115}$ = Analytical RP–HPLC (Gradient 2): Rt: 11:16 min:sec; RCY (decay corrected at the beginning of reaction): 11%–12.7%; apparent specific activity: 0.40–0.59 MBq/µg (apparent molar activity: 3.0–4.4 MBq/nmol).

[^{18}F]AlF-NODAGA-PODS-Z$_{EGFR:03115}$ = Analytical RP–HPLC (Gradient 2): Rt: 11:11 min:sec; RCY (decay corrected at the beginning of reaction): 4.3%–8.1%; apparent specific activity: 0.11–0.23 MBq/µg (apparent molar activity: 0.8–1.7 MBq/nmol).

3.5. Determination of Distribution Coefficient (LogD) at pH 7.4

To 0.5 mL of PBS (pH 7.4), either [^{18}F]AlF-NOTA-PODS-Z$_{EGFR:03115}$ or [^{18}F]AlF-NODAGA-PODS-Z$_{EGFR:03115}$ (0.08 MBq) was added followed by 0.5 mL of n-octanol. The mixture was vortexed for 10 min followed by centrifugation at 100× g for 10 min. The experiments were performed in triplicate. Three 100 µL samples were taken from each layer and the amount of activity was measured in a 2480 WIZARD^2 Automatic Gamma Counter (Perkin Elmer, Beaconsfield, UK) as counts per minutes (cpm). The distribution coefficient at pH 7.4 (logD$_{7.4}$) was expressed as the mean ± standard deviation (SD) and calculated using the following formula (Equation 1):

$$\text{LogD} = \log[(\text{counts}_{octanol})/(\text{counts}_{PBS})] \tag{1}$$

3.6. In Vitro Serum Stability Assay

The stability of [^{18}F]AlF-NOTA-PODS-Z$_{EGFR:03115}$ and [^{18}F]AlF-NODAGA-PODS-Z$_{EGFR:03115}$ was assessed as previously described by incubating the purified [^{18}F]AlF-radioconjugates (3.5–4 MBq) in mouse serum (500 µL) in a thermo shaker at 37 °C for 1 h (850 rpm) [17,24]. Each experiment was performed in triplicate. The data are expressed as the average of n = 3 measurements ± SD.

3.7. In Vivo Evaluation

All experiments were performed in compliance with license issued under the UK Animals (Scientific Procedures) Act 1986 and following local ethical review (project license PCC916B22, Animals in Science Regulation Unit, Home Office Science, London, UK). The studies followed the United Kingdom National Cancer Research Institute Guidelines for Animal Welfare in Cancer Research [25]. Female NCr athymic mice (6–8 weeks old) were subcutaneously injected on the right shoulder with U87MGvIII cells (0.5 × 10^6/mouse) suspended in 30% Matrigel. Tumors were allowed to grow to 100 mm^3. PET/CT studies were conducted on an Albira PET/SPECT/CT imaging system (Bruker, Coventry, UK). Mice were administered the radioconjugate (12 µg in 100 µL of 0.9% sterile saline, 1.1–1.8 MBq/mouse) by intravenous tail vein injection and were anesthetized using an isoflurane/O$_2$ mixture (1.5%–2.0% v/v) approximately 5 min prior to imaging. Whole-body static PET images were acquired 1 h post-radioconjugate injection for the duration of 10 min, with a 358 to 664 keV energy window, followed by CT acquisition as previously described [26]. The image data were processed and reconstructed as previously reported [26].

Immediately after image data acquisition, the mice were euthanized by cervical dislocation for the biodistribution studies. The major organs/tissues were dissected and weighed, and the radioactivity was measured in 2480 WIZARD^2 Automatic Gamma Counter (Perkin Elmer, Beaconsfield, UK).

The percentage of the injected dose per gram of tissue (%ID/g) was determined for each organ/tissue. The data are expressed as the average of $n = 3$ mice ± SD.

4. Conclusions

This study describes the preparation and attachment of two novel thiol-reactive PODS-bearing bifunctional chelators (NOTA-PODS and NODAGA-PODS) to the EGFR-targeting affibody molecule $Z_{EGFR:03115}$. When radiolabeled with [^{18}F]AlF, a RP–HPLC purification procedure followed by HLB-SPE was required to produce radioconjugates with a RCP > 98%. Overall, the radiolabeling efficiency for [^{18}F]AlF-NODAGA-PODS-$Z_{EGFR:03115}$ was found to be lower than for [^{18}F]AlF- NOTA-PODS-$Z_{EGFR:03115}$, a factor attributable to the chelator structure. Once purified, both radioconjugates showed a good serum stability. When injected in high EGFR-expressing tumor-bearing mice, [^{18}F]AlF-NOTA-PODS-$Z_{EGFR:03115}$ and [^{18}F]AlF-NODAGA-PODS-$Z_{EGFR:03115}$ showed similar pharmacokinetics and allowed for clear visualization of the tumors already 1 h post-injection. Additionally, the radiolabeling procedure, purification requirements, in vitro stability and in vivo behavior (at 1 h) of both PODS- and maleimide-bearing radioconjugates were found to be comparable. However, based on reports in the literature, it is possible that the benefits from the superior stability of PODS in vivo would be more noticeable at later time points. In conclusion, this investigation showed that PODS-based reagents are a viable alternative to maleimide for thiol-selective conjugation to cysteine-bearing proteins.

Supplementary Materials: The following are available online, Figure S1: RP–HPLC analysis (Gradient 1) of PODS (**A**), and the NODA-PODS (**B**) and NODAGA-PODS (**C**) reaction mixtures. NOTA-NHS and NODAGA-NHS elute with the mobile phase front, together with DMF (ca 3 min). The absorbance was recorded at the wavelength of 254 nm. The retention time (Rt) is indicated as min:sec. Figure S2: When the bifunctional chelators were purified by semi-preparative RPHPLC and subsequently dried using a speed-vacuum concentrator, the products showed clear signs of degradation RP–HPLC analysis (Gradient 1) of isolated NOTAPODS (**A**) and NODAGA-PODS (**B**). Each chromatogram shows the presence of one major degradation product (ca 10:40 min:sec). The absorbance was recorded at the wavelength of 254 nm. ESI–MS analysis of NOTA-PODS and the degradation product shows the expected mass of m/z 827 and a peak having a smaller mass (m/z 765) which could be associated to the hydrolysis derivative at the sulfone group (**C**). The prolonged presence of TFA in solution together with the type of drying process were possibly the cause. Figure S3: RP–HPLC analysis (Gradient 1) of pure NOTA-PODS (**A**) and NODAGAPODS (**B**) isolated by semi-preparative RP–HPLC using formic acid in the mobile phase instead of TFA. The absorbance was recorded at the wavelength of 254 nm. Figure S4: ESI–HRMS of NOTA-PODS (top) and NODAGA-PODS (bottom). Figure S5: RP–HPLC analysis of solutions of NOTA-PODS and NODAGA-PODS in DMF after being stored at −20 °C. Signs of degradation (peak at 10:36 min:sec) were detected already after 2 months for NOTA-PODS (**A**). Conversely, NODAGA-PODS showed good stability for at least 10 months (**B**). Figure S6: RP–HPLC analysis (Gradient 2) of NOTA-PODS-$Z_{EGFR:03115}$ (**A**), and NODAGA-PODS-$Z_{EGFR:03115}$ (**B**) reaction mixtures. NOTA-PODS and NODAGA-PODS elute with the mobile phase front, together with DMF (ca 3 min). The absorbance was recorded at the wavelength of 280 nm. The retention time (Rt) is indicated as min:sec. Figure S7: ESI–MS of purified NOTA-PODS-$Z_{EGFR:03115}$ (top) and NODAGA-PODS-$Z_{EGFR:03115}$ (bottom). Figure S8: Radiochromatograms (Gradient 2) of [^{18}F]AlF-NOTA-PODS-$Z_{EGFR:03115}$ (**A**), and [^{18}F]AlF-NODAGA-PODS-$Z_{EGFR:03115}$ (**B**) reaction mixtures. Free fluorine-18 elutes at ca 3 min. Labels on each peak on the chromatograms indicate the retention time (top) and the %ROI (bottom). Figure S9: Radiochromatograms (Gradient 2) of [^{18}F]AlF-NOTA-PODS-$Z_{EGFR:03115}$ (**A**), and [^{18}F]AlF-NODAGA-PODS-$Z_{EGFR:03115}$ (**B**) after purification by just HLB-SPE. As for [^{18}F]AlF-NOTA-$Z_{EGFR:03115}$ (**C**), the HLB-SPE-only purification step successfully removed the free fluorine-18 leaving the radioconjugate and the thermolysis products which elute at ca 3 min. The retention times (Rt) are expressed as min:sec. Figure S10: Representative radiochromatograms (Gradient 2) of [^{18}F]AlF-NOTA-PODS-$Z_{EGFR:03115}$ (**A**) and [^{18}F]AlF-NODAGA-PODS-$Z_{EGFR:03115}$ (**B**) after incubation in mouse serum for 1 h. The intact radioconjugates elute at ca 11 min. Activity non-associated with the conjugate elutes at ca 3 min. Labels on each peak on the chromatograms indicate the retention time (top) and the %ROI (bottom). Table S1: Summary of $LogD_{7.4}$ values measured for the three radioconjugates. Table S2: Summary of serum stability determined by RP–HPLC. The three radioconjugates were incubated in mouse serum at 37 °C for 1 h. The data are shown as the mean values of $n = 3$ experiments ± SD. Statistical analysis was performed using one-way ANOVA with Tukey correction using GraphPad Prism v8. Table S3: Summary of serum stability determined by RP–HPLC. The three radioconjugates were incubated in mouse serum at 37 °C for 1 h. The data are shown as the mean values of $n = 3$ experiments ± SD. Statistical analysis was performed using one-way ANOVA with Tukey correction using GraphPad Prism v8.

Author Contributions: Conceptualization, C.D.P., G.K.-M. and G.S.; methodology, C.D.P, A.M. J.M. and S.T.; formal analysis, C.D.P., S.T. and G.K.-M.; investigation, C.D.P., A.M., J.M., S.T., G.K.-M.; resources, G.S. and G.K.-M.; data curation, C.D.P., S.T. and G.K.-M.; writing—original draft preparation, C.D.P.; writing—review and editing, A.M., S.T., G.S. and G.K.-M.; visualization, C.D.P., S.T. and G.K.-M.; supervision, C.D.P. and G.K.-M.; project administration, G.S. and G.K.-M.; funding acquisition, A.M., G.S. and G.K.-M. All authors have read and agreed to the published version of the manuscript.

Funding: This work was supported by the Cancer Research UK-Cancer Imaging Centre (C1060/A16464) and German Cancer Aid (Deutsche Krebshilfe, project No: 70112043).

Acknowledgments: The authors gratefully thank Affibody AB for supplying the affibody molecule and the Structural Chemistry Facility (Cancer Therapeutics) for the provision of technical MS services.

Conflicts of Interest: The authors declare no conflict of interest.

References

1. Gunnoo, S.B.; Madder, A. Chemical protein modification through cysteine. *ChemBioChem* **2016**, *17*, 529–553. [CrossRef]
2. Koniev, O.; Wagner, A. Developments and recent advancements in the field of endogenous amino acid selective bond forming reactions for bioconjugation. *Chem. Soc. Rev.* **2015**, *44*, 5495–5551. [CrossRef] [PubMed]
3. Ravasco, J.M.J.M.; Faustino, H.; Trindade, A.; Gois, P.M.P. Bioconjugation with maleimides: A useful tool for chemical biology. *Chem. Eur. J.* **2019**, *25*, 43–59. [CrossRef] [PubMed]
4. Sabot, C.; Renard, P.-Y.; Renault, K.; Fredy, J.W. Covalent modification of biomolecules through maleimide-based labeling strategies. *Bioconjugate Chem.* **2018**, *29*, 2497–2513.
5. Alley, S.C.; Benjamin, D.R.; Jeffrey, S.C.; Okeley, N.M.; Meyer, D.L.; Sanderson, R.J.; Senter, P.D. Contribution of linker stability to the activities of anticancer immunoconjugates. *Bioconjugate Chem.* **2008**, *19*, 759–765. [CrossRef] [PubMed]
6. Baldwin, A.D.; Kiick, K.L. Tunable degradation of maleimide–thiol adducts in reducing environments. *Bioconjugate Chem.* **2011**, *22*, 1946–1953. [CrossRef] [PubMed]
7. Ponte, J.F.; Sun, X.; Yoder, N.C.; Fishkin, N.; Laleau, R.; Coccia, J.; Lanieri, L.; Bogalhas, M.; Wang, L.; Wilhelm, S.; et al. Understanding how the stability of the thiol-maleimide linkage impacts the pharmacokinetics of lysine-linked antibody-maytansinoid conjugates. *Bioconjugate Chem.* **2016**, *27*, 1588–1598. [CrossRef]
8. Szijj, P.A.; Bahou, C.; Chudasama, V. Minireview: Addressing the retro-Michael instability of maleimide bioconjugates. *Drug Discov. Today Technol.* **2018**, *30*, 27–34. [CrossRef]
9. Bernardim, B.; Cal, P.M.S.D.; Matos, M.J.; Oliveira, B.L.; Martínez-Sáez, N.; AlbuquerqueInês, S.; Perkins, E.; Corzana, F.; Burtoloso, A.C.B.; Jiménez-Osés, G.; et al. Stoichiometric and irreversible cysteine-selective protein modification using carbonylacrylic reagents. *Nat. Commun.* **2016**, *7*, 13128–13137. [CrossRef]
10. Kalia, D.; Malekar, P.V.; Parthasarathy, M. Exocyclic olefinic maleimides: Synthesis and application for stable and thiol-selective bioconjugation. *Angew. Chem. Int. Ed. Engl.* **2016**, *55*, 1432–1435. [CrossRef]
11. Toda, N.; Asano, S.; Barbas, C.F.I. Rapid, stable, chemoselective labeling of thiols with Julia-Kocieński-like reagents: A serum-stable alternative to maleimide-based protein conjugation. *Angew. Chem. Int. Ed. Engl.* **2013**, *52*, 12592–12596. [CrossRef] [PubMed]
12. Patterson, J.T.; Asano, S.; Li, X.; Rader, C.; Barbas, C.F.I. Improving the serum stability of site-specific antibody conjugates with sulfone linkers. *Bioconjugate Chem.* **2014**, *25*, 1402–1407. [CrossRef] [PubMed]
13. Zhang, Q.; Dall'Angelo, S.; Fleming, I.N.; Schweiger, L.F.; Zanda, M.; O'Hagan, D. Last-step enzymatic [^{18}F]-fluorination of cysteine-tethered RGD peptides using modified Barbas linkers. *Chem. Eur. J.* **2016**, *22*, 10998–11004. [CrossRef] [PubMed]
14. Chiotellis, A.; Sladojevich, F.; Mu, L.; Müller Herde, A.; Valverde, I.E.; Tolmachev, V.; Schibli, R.; Ametamey, S.M.; Mindt, T.L. Novel chemoselective ^{18}F-radiolabeling of thiol-containing biomolecules under mild aqueous conditions. *Chem. Commun.* **2016**, *52*, 6083–6086. [CrossRef] [PubMed]
15. Adumeau, P.; Davydova, M.; Zeglis, B.M. Thiol-reactive bifunctional chelators for the creation of site-selectively modified radioimmunoconjugates with improved stability. *Bioconjug. Chem.* **2018**, *29*, 1364–1372. [CrossRef]
16. McBride, W.J.; Sharkey, R.M.; Karacay, H.; D'Souza, C.A.; Rossi, E.A.; Laverman, P.; Chang, C.H.; Boerman, O.C.; Goldenberg, D.M. A novel method of 18F radiolabeling for PET. *J. Nucl. Med.* **2009**, *50*, 991–998. [CrossRef]

17. Da Pieve, C.; Allott, L.; Martins, C.D.; Vardon, A.; Ciobota, D.M.; Kramer-Marek, G.; Smith, G. Efficient [^{18}F]AlF radiolabeling of Z$_{HER3:8698}$ affibody molecule for imaging of HER3 positive tumors. *Bioconjug. Chem.* **2016**, *27*, 1839–1849. [CrossRef]
18. Kim, Y.; Ho, S.O.; Gassman, N.R.; Korlann, Y.; Landorf, E.V.; Collart, F.R.; Weiss, S. Efficient site-specific labeling of proteins via cysteines. *Bioconjug. Chem.* **2008**, *19*, 786–791. [CrossRef]
19. Liu, Y.; Hu, X.; Liu, H.; Bu, L.; Ma, X.; Cheng, K.; Li, J.; Tian, M.; Zhang, H.; Cheng, Z. A comparative study of radiolabeled bombesin analogs for the PET imaging of prostate cancer. *J. Nucl. Med.* **2013**, *54*, 2132–2138. [CrossRef]
20. D'Souza, C.A.; McBride, W.J.; Sharkey, R.M.; Todaro, L.J.; Goldenberg, D.M. High-yielding aqueous ^{18}F-labeling of peptides via Al^{18}F chelation. *Bioconjug. Chem.* **2011**, *22*, 1793–1803. [CrossRef]
21. Löfblom, J.; Feldwisch, J.; Tolmachev, V.; Carlsson, J.; Ståhl, S.; Frejd, F.Y. Affibody molecules: Engineered proteins for therapeutic, diagnostic and biotechnological applications. *Febs Lett.* **2010**, *584*, 2670–2680. [CrossRef] [PubMed]
22. Burley, T.A.; Da Pieve, C.; Martins, C.D.; Ciobota, D.M.; Allott, L.; Oyen, W.J.G.; Harrington, K.J.; Smith, G.; Kramer-Marek, G. Affibody-Based PET Imaging to Guide EGFR-Targeted Cancer Therapy in Head and Neck Squamous Cell Cancer Models. *J. Nucl. Med.* **2019**, *60*, 353–361. [CrossRef] [PubMed]
23. Behr, T.M.; Goldenberg, D.M.; Becker, W. Reducing the renal uptake of radiolabeled antibody fragments and peptides for diagnosis and therapy: Present status, future prospects and limitations. *Eur. J. Nucl. Med.* **1998**, *25*, 201–212. [CrossRef]
24. Su, X.; Cheng, K.; Jeon, J.; Shen, B.; Venturin, G.T.; Hu, X.; Rao, J.; Chin, F.T.; Wu, H.; Cheng, Z. Comparison of two site-specifically ^{18}F-labeled affibodies for PET imaging of EGFR positive tumors. *Mol. Pharm.* **2014**, *11*, 3947–3956. [CrossRef] [PubMed]
25. Workman, P.; Aboagye, E.O.; Balkwill, F.; Balmain, A.; Bruder, G.; Chaplin, D.J.; Double, J.A.; Everitt, J.; Farningham, D.A.H.; Glennie, M.J.; et al. Guidelines for the welfare and use of animals in cancer research. *Br. J. Cancer* **2010**, *102*, 1555–1577. [CrossRef]
26. Martins, C.D.; Da Pieve, C.; Burley, T.A.; Smith, R.; Ciobota, D.M.; Allott, L.; Harrington, K.J.; Oyen, W.J.G.; Smith, G.; Kramer-Marek, G. HER3-mediated resistance to Hsp90 inhibition detected in breast cancer xenografts by affibody-based PET imaging. *Clin. Cancer Res.* **2018**, *24*, 1853–1865. [CrossRef]

Sample Availability: Samples of the compounds are not available.

© 2020 by the authors. Licensee MDPI, Basel, Switzerland. This article is an open access article distributed under the terms and conditions of the Creative Commons Attribution (CC BY) license (http://creativecommons.org/licenses/by/4.0/).

Article

The Effects of Intramuscular Naloxone Dose on Mu Receptor Displacement of Carfentanil in Rhesus Monkeys

Peter J. H. Scott [1,*], Robert A. Koeppe [1], Xia Shao [1], Melissa E. Rodnick [1], Alexandra R. Sowa [1], Bradford D. Henderson [1], Jenelle Stauff [1], Phillip S. Sherman [1], Janna Arteaga [1], Dennis J. Carlo [2] and Ronald B. Moss [2,*]

1. Department of Radiology, University of Michigan, Ann Arbor, MI 48105, USA; koeppe@med.umich.edu (R.A.K.); xshao@umich.edu (X.S.); topperm@umich.edu (M.E.R.); sowa@umich.edu (A.R.S.); bkhend@umich.edu (B.D.H.); jrstauff@umich.edu (J.S.); psherman@umich.edu (P.S.S.); jannaa@umich.edu (J.A.)
2. Adamis Pharmaceuticals, 11682 El Camino Real, Suite # 300, San Diego, CA 92130, USA; Dcarlo@adamispharma.com
* Correspondence: pjhscott@umich.edu (P.J.H.S.); rmoss@adamispharma.com (R.B.M.)

Academic Editors: Anne Roivainen and Xiang-Guo Li
Received: 4 March 2020; Accepted: 15 March 2020; Published: 17 March 2020

Abstract: Naloxone (NLX) is a mu receptor antagonist used to treat acute opioid overdoses. Currently approved doses of naloxone to treat opioid overdoses are 4 mg intranasal (IN) and 2 mg intramuscular (IM). However, higher mu receptor occupancy (RO) may be required to treat overdoses due to more potent synthetic opioids such as fentanyl and carfentanil that have entered the illicit drug market recently. To address this need, a higher dose of NLX has been investigated in a 5 mg IM formulation called ZIMHI but, while the effects of intravenous (IV) and IN administration of NLX on the opioid mu receptor occupancy (RO) have been studied, comparatively little is known about RO for IM administration of NLX. The goal of this study was to examine the effect of IM dosing of NLX on mu RO in rhesus macaques using [^{11}C]carfentanil positron emission tomography (PET) imaging. The lowest dose of NLX (0.06 mg/kg) approximated 51% RO. Higher doses of NLX (0.14 mg/kg, 0.28 mg/kg) resulted in higher mu RO of 70% and 75%, respectively. Plasma levels were 4.6 ng/mL, 16.8 ng/mL, and 43.4 ng/mL for the three IM doses, and a significant correlation between percent RO and plasma NLX level was observed (r = 0.80). These results suggest that higher doses of IM NLX result in higher mu RO and could be useful in combating overdoses resulting from potent synthetic opioids.

Keywords: opioid; naloxone; overdose; fentanyl; carfentanil; [^{11}C]carfentanil; positron emission tomography; receptor occupancy; pharmacokinetics

1. Introduction

Mortality from drug overdoses has reached epidemic proportions in the U.S. (>70,000 in 2017), exceeding the number of yearly deaths during the peak of the AIDS epidemic [1]. One of the key drivers of the overdoses has been the abuse of synthetic opioids [2]. Naloxone (NLX) is the first line of treatment and an effective countermeasure in the event of an opioid overdose because, as a mu receptor antagonist, it is capable of rapid reversal of opioid toxicity [3]. Currently approved doses of NLX include 4 mg intranasal (IN) and 2 mg intramuscular (IM) [4] but, to date, there have been numerous reports suggesting that multiple doses of NLX may be required for successful reversal of opioid toxicity, especially when treating overdoses due to more potent synthetic opioids such as fentanyl and carfentanil that have entered the illicit drug market recently [5–9]. To address this issue, we previously reported that a higher dose of intramuscular naloxone (5 mg ZIMHI) has greater systemic exposure compared to the current community dose of naloxone (2 mg intramuscular and 4 mg intranasal) [10]. However, while the effects of intravenous (IV) and IN administration of NLX on mu opioid receptor occupancy (RO) have been studied, comparatively little is known about RO resulting from IM administration [11,12]. The goal of this study was, therefore, to examine the effects of IM NLX on mu RO in rhesus macaques. Using [^{11}C]carfentanil ([^{11}C]CFN, [13]), we employed preclinical positron emission tomography (PET) imaging to investigate RO of NLX administered by IM injection at three different doses in two mature rhesus monkeys, and we compared our imaging findings with the plasma levels of NLX in the monkeys determined by quantitative LC-MS/MS.

2. Results

Initially, two rhesus monkeys received baseline PET scans with [^{11}C]CFN (Figure 1 and Table 1). The scans showed the expected distribution of [^{11}C]CFN corresponding to the known mu opioid receptor density in the primate brain, with pronounced uptake in the basal ganglia (BG) and thalamus (THAL), as well as moderate uptake throughout the cortex (CTX). Following baselines studies, [^{11}C]CFN PET scans were repeated after dosing the monkeys IM with low (0.06 mg/kg), medium (0.14 mg/kg), and high (0.28 mg/kg) doses of NLX (Table 1). To enable comparison of results from this study to existing human data from the literature (see Discussion below), human equivalent IM doses were also estimated from the monkey doses using Ahmad's approach (Table 2) [14]. The radiotracer was injected 10 min after IM NLX and dose-dependent blockade of [^{11}C]CFN by NLX in the BG, THAL, and CTX regions of the monkey brain was apparent (see representative images from one animal in Figure 1).

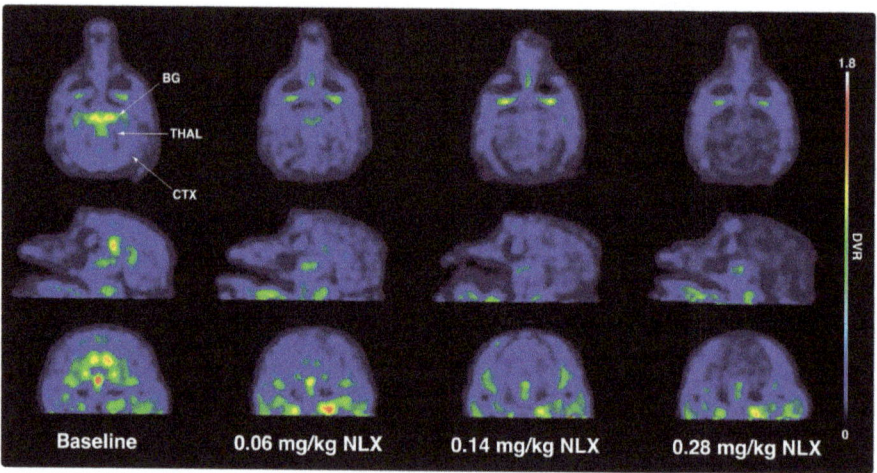

Figure 1. Representative transverse (top row), sagittal (middle row), and coronal (bottom row) monkey positron emission tomography (PET) images of a [^{11}C]CFN baseline scan and following blocking studies with low (0.06 mg/kg), medium (0.14 mg/kg), and high (0.28 mg/kg) doses of naloxone (NLX). Images are distribution volume ratio (DVR) images summed 0–60 min following intravenous (IV) injection of the radiotracer. BG = basal ganglia, THAL = thalamus, CTX = cortex.

Table 1. Study Metrics.

Monkey Naloxone Dose (mg/kg)	Monkey 1 [^{11}C]CFN PET	PK	Monkey 2 [^{11}C]CFN PET	PK
0.00 (baseline)	n = 1	n = 1	n = 2	n = 1
0.06	n = 2	n = 1	n = 2	n = 1
0.14	n = 2	n = 1	n = 1	n = 1
0.28	n = 1	n = 1	n = 1	n = 1

Table 2. Naloxone dosing [14].

Monkey Naloxone IM Dose (mg/kg)	Human Equivalent IM Dose (mg/kg) [14]	IM Dose to Average 60 kg Human (mg)
0.06	0.02	1.2
0.14	0.045	2.7
0.28	0.09	5.4

The mean % RO by naloxone in the basal ganglia and thalamus is shown in Table 3 and Figure 2. During blocking studies, blood samples were taken from the primates at 0, 30, and 60 min post-injection of naloxone. Plasma was separated and a quantitative LC-MS/MS method was utilized to determine plasma concentrations of NLX, and mean plasma levels 30 and 60 min post-IM injection of the different doses are shown in Table 4 and Figure 3. A significant correlation was observed between plasma NLX concentration and mu RO (r = 0.80) for both the basal ganglia and thalamus (Figure 4).

Table 3. Mean % receptor occupancy (RO) by Dose. [1]

Dose of Naloxone mg/kg	Basal Ganglia Mean %-RO (± SD)	Thalamus Mean %-RO ± SD
0.06	56 ± 17	47 ± 23
0.14	74 ± 7	65 ± 9
0.28	76 ± 3	74 ± 4

[1] Data is mean ± SD for all studies (at least n = 1 per monkey, as summarized in Table 1).

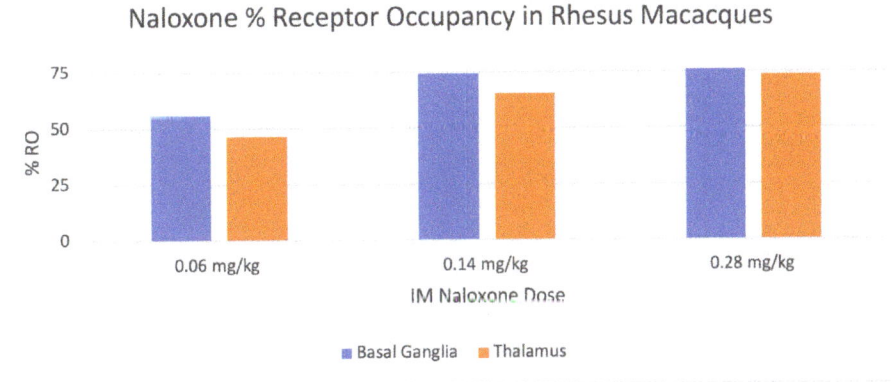

Figure 2. Mean % RO by intramuscular (IM) Dose.

Table 4. Mean plasma concentration of naloxone at 30 and 60 min post-injection. [1]

Dose of Naloxone mg/kg	Mean Plasma Conc NLX ± SD [Range] (ng/mL) 30 min Post-IM [1]	Mean Plasma Conc NLX ± SD [Range] (ng/mL) 60 min Post-IM
0.06	4.6 ± 1.9 [3.3–5.9]	2.3 ± 2.4 [0.7–4.0]
0.14	16.8 ± 2.3 [15.1–18.4]	8.1 ± 1.0 [7.4–8.8]
0.28	43.4 ± 19.0 [30.0–56.8]	23.2 ± 1.6 [22.0–24.3]

[1] Data is mean ± SD [range] for PK studies (n = 1 per monkey, as summarized in Table 1).

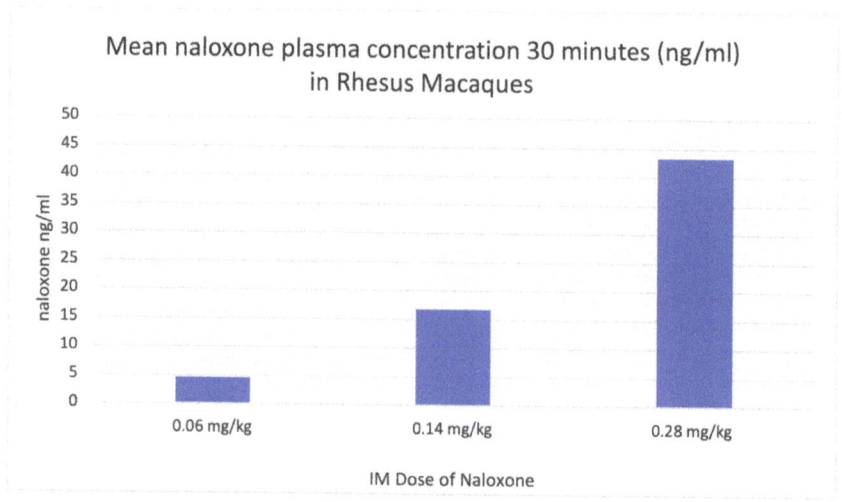

Figure 3. Plasma NLX levels 30 min after three different IM doses.

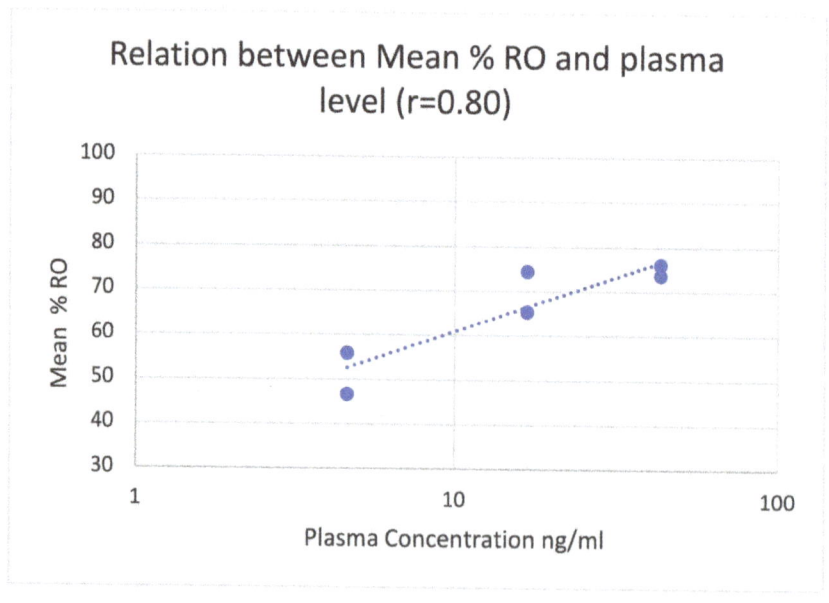

Figure 4. The relationship (r = 0.80) between NLX plasma level and RO in the basal ganglia and thalamus.

3. Discussion

Previous studies have examined the pharmacokinetics of NLX in different species, as well as various doses and dosage forms. These studies have revealed minimal species differences in, for example, metabolism [15], while establishing that ~50% of the drug is bound to plasma proteins (albumin) and the half-life in plasma is 1–2 h [16]. Prior studies in humans and primates, for example, have examined IV or IN delivery of NLX, including our study comparing the % RO for IN and IV administration in rhesus monkeys. Following a dose of NLX given by two different routes, IV administration resulted in greater RO (75%) compared to the same dose administered by the IN route (65%) [12]. Previous studies of IV administration of NLX in humans suggested that 1 mg IV resulted in approximately 50% occupancy [11]. Other studies in humans suggested that 2 mg IV of NLX resulted in approximately 80% RO [17]. Furthermore, the 2 mg IV dose was associated with a C_{max} plasma level of 38.7 ng/mL. A recent study in humans suggested that 1 mg and 2 mg IV NLX resulted in a mean RO of 54–82% and 71–96%, respectively [18]. The 1 mg and 2 mg IN doses of NLX in that study were associated with C_{max} pharmacokinetic levels of 1.83 ng/mL and 4.33 ng/mL, respectively. Human IM naloxone doses of 0.4, 0.8, and 2 mg have resulted in C_{max} levels of 1.2–1.3, 2.2, and 7.9 nm/mL, respectively [19–21]. However, to the best of our knowledge, mu RO following IM NLX has not been investigated in humans or primates. This paper reports the first study in rhesus macaques examining mu RO using [^{11}C]carfentanil PET imaging. RO was compared to NLX plasma concentrations quantified using an LC-MS/MS method.

Rhesus monkeys initially received baseline PET scans with [^{11}C]CFN (Figure 1 and Table 1). Logan analysis was conducted to determine distribution volume ratio (DVR) [22]. Subsequently, [^{11}C]CFN PET scans were repeated 10 min after dosing the monkeys IM with low (0.06 mg/kg), medium (0.14 mg/kg), and high (0.28 mg/kg) doses of NLX. Changes in RO were estimated from the distribution DVR calculated in the baseline scan compared with the DVRs obtained in the NLX blocking studies. Since Logan analysis needs to be conducted after [^{11}C] CFN reaches equilibrium (30–40 min post-injection of the radiotracer), RO values determined in this study are averaged for the PET scan but weighted toward 30–60 min post-injection of [^{11}C]CFN (40–70 min post-injection of naloxone). At the lowest dose used in this study (0.06 mg/kg), IM NLX averaged 52% RO with a plasma level of 4.6 ng/mL 30 min post-injection, while the 0.14 mg/kg dose resulted in a mean RO by NLX of 70% and a plasma concentration of 16.8 ng/mL. Lastly, the highest dose of IM NLX (0.28 mg/kg) resulted in a mean RO of 75% and a plasma level of 43.4 ng/mL. The plasma levels determined in this study (Table 3) are consistent with known values following IM administration to humans [19–21]. It is not always straightforward comparing data between dosage forms (IM/IV), across species (monkey/human), and analytical techniques (imaging/plasma concentrations). However, by extrapolation, the lowest IM dose in this study (0.06 mg/kg) appears to have similar RO and plasma levels to a 1 mg IV dose of NLX in humans [11], which is consistent with the estimated human equivalent dose of 1.2 mg IM (Table 2). Similarly, the highest IM dose of NLX examined in this study (0.28 mg/kg), which corresponded to an estimated dose of 5.4 mg IM to an average 60 kg human (Table 2), appears to result in a comparable RO and plasma level to a 2 mg IV NLX dose in humans [17]. As stated above, it has been previously reported that C_{max} values are lower for IM doses of naloxone than for the equivalent IV doses [23], and our data support these findings.

Most importantly, we observed a direct correlation between the plasma levels of NLX and the RO for IM injections of NLX. Therefore, the plasma level of NLX, which is dependent on the administered dose, appears to be a good predictor of RO. By comparison in humans, the plasma levels associated with the lowest dose of IM NLX in this study are comparable to levels observed with the 4 mg IN and 2 mg IM doses of NLX [10]. Interestingly, the middle-dose group in this study (0.14 mg/kg) was found to have plasma NLX levels comparable to those resulting from a higher 5 mg IM dose of NLX administered to human subjects [7].

A number of factors may affect the competitive binding of NLX to the mu receptor. For example, K_i values for mu receptors for both NLX and the opioid-involved are important factors to consider.

In addition, we have previously shown in a predictive model that the level of mu receptor-bound opioid impacts the level of NLX needed for opioid toxicity reversal [unpublished results]. In the current macaque model, a subclinical dose of radiolabeled [^{11}C]carfentanil was used to estimate RO resulting from an IM injection of naloxone. Thus, the results of this study may underestimate the dose and plasma level of NLX required to achieve a clinically significant level of RO, as very high levels of synthetic opioids have been found in overdose patients [24–28]. However, this study does suggest that for an invariable dose of carfentanil, increasing doses of IM NLX result in greater RO and, furthermore, increasing doses of IM NLX results in higher systemic levels. Lastly, we observed a significant correlation between percent RO and plasma NLX plasma levels and, taken together, these results support the notion that higher doses of IM NLX result in higher mu RO. Given that higher doses of naloxone could be needed to treat overdoses due to more potent synthetic opioids such as fentanyl and carfentanil [8,9], development of higher dose forms of naloxone is warranted in light of the findings in this study.

4. Materials and Methods

4.1. General Considerations

Primate imaging studies were performed at the University of Michigan in accordance with the standards set by the Institutional Animal Care And Use Committee (IACUC) (Protocol 000008103, Biodistribution and Pharmacokinetics of Radiolabeled Compounds, 1/16/2018–1/16/2021) and all applicable federal, state, local, and institutional laws or guidelines governing animal research.

4.2. Subjects

Two intact, mature female rhesus monkeys were used in this study, aged 19.5 ± 0.7 years and weighing 9.4 ± 0.4 kg, without controlling for the phase of their menstrual cycle. Both monkeys were individually housed in steel cages (83.3 cm high × 76.2 cm wide × 91.4 cm deep) on a 12-h light/12-h dark schedule. Monkeys were fed Laboratory Fiber Plus Monkey Diet (PMI Nutrition International LLC, St. Louis, MO, USA) that was supplemented with fresh fruit daily. Water and enrichment toys were available continuously in the home cage. Each monkey has served in previous PET imaging and blocking studies, corresponding to a fairly extensive drug administration history.

4.3. Nonhuman Primate PET Imaging Studies

Primate PET imaging studies were performed using a Concorde Microsystems P4 PET scanner (Siemens, Knoxville, TN, USA). The animals were anesthetized in the home cage with ketamine (Par Sterile Products, Chestnut Ridge, NY, USA) and transported to the PET facility. Subjects were intubated for mechanical ventilation, and anesthesia was continued with isoflurane (Patterson Veterinary Supply Inc., Devens, MA, USA). Anesthesia was maintained throughout the duration of the PET scan. A venous catheter was inserted into one hind limb and the monkey was placed on the PET gantry with its head secured to prevent motion artifacts. Following a transmission scan, IM NLX (or saline in the case of the baseline studies) was administered. Ten minutes later, 3.4 ± 1.4 mCi of [^{11}C]CFN was administered in a bolus dose over one minute. Mass of CFN administered was ≤0.03 µg/kg, consistent with our clinical dose limit [13]. Emission data were collected beginning with the injection and continued for 60.0 min (12 × 5-min frames). Data were corrected for attenuation and scatter and reconstructed using the three dimensional–maximum a priori method (3D MAP algorithm).

4.4. PET Data Analysis

The dynamic sequence of PET images was summed for the baseline scans and regions-of-interest (ROIs) were drawn manually on multiple planes to obtain volumes-of-interest (VOIs) for the thalamus and basal ganglia of the baseline scan for each monkey. The VOIs were then applied to the full dynamic datasets to obtain the regional tissue time-radioactivity data curve. Images from all subsequent

[^{11}C]CFN scans following NLX blocking studies were registered to that monkey's baseline scan, using the NeuroStat package freely available on the internet (https://neurostat.neuro.utah.edu/documents/NEUROSTAT2016.RTF), to allow image data to be extracted from the same set of VOIs. These data were used to construct brain tissue–radioactivity curves that were then analyzed with the method of Logan, with the occipital cortex being the reference region [22]. Changes in receptor occupancy were estimated from the distribution volume ratio (DVR) calculated from each VOI in the baseline scan compared with the DVRs obtained from the NLX blocking studies, using Equation (1).

$$\text{Occupancy}(\%) = 100 \times (1 - (\text{DVR}_{block} - 1)/(\text{DVR}_{base} - 1)) \tag{1}$$

4.5. Study Drugs

Naloxone Hydrochloride provided in pre-filled syringes (International Medication Systems Ltd., South El Monte, CA, USA) was utilized for IM injections of the two primates at doses of 0.06 mg/kg, 0.14 mg/kg, and 0.28 mg/kg. Synthesis and quality control testing of [^{11}C]carfentanil was conducted as previously described [13]. Briefly, [^{11}C]CFN was prepared in a TRACER lab FX$_{C-Pro}$ synthesis module (GE Healthcare, Uppsala, Sweden) fitted with an Agilent Bond Elut C2 cartridge (Agilent Technologies, Santa Clara, CA, USA). [^{11}C]MeOTf (~1 Ci) was bubbled into a solution of desmethyl carfentanil TBA salt (0.5 mg) (MilliporeSigma, Burlington, MA, USA, prepared as described in [13]) in ethanol, USP (100 µL) (Akorn, Lake Forest, IL, USA) at room temperature for 3 min. After this time, 1% NH$_4$OH (1 mL) (Fisher Scientific, Hampton, NH, USA) was added to the reaction vessel. This crude reaction mixture was further diluted with 1% NH$_4$OH (5 mL), and the resulting mixture was passed through the Agilent Bond Elut C2 cartridge to trap [^{11}C]CFN. The cartridge was washed with 20% EtOH (3 mL) (Decon Laboratories, King of Prussia, PA, USA), followed by Milli-Q water (10 mL), to remove unreacted precursor and impurities from the cartridge, and dried for 1.0 min with He gas. [^{11}C]CFN was eluted with EtOH, USP (0.5 mL) and diluted with Sterile Water for Injection, USP (9.5 mL) (Hospira, Lake Forest, IL). The formulated product was then passed through a Millipore-GV 0.22-µm filter (MilliporeSigma, Burlington, MA, USA) into a sterile dose vial (Jubilant Hollister-Stier, Spokane, WA) and analyzed for pH, radiochemical purity, mass of CFN, and molar activity as previously described [13].

4.6. Plasma Naloxone Assay

A quantitative LC-MS/MS method with an internal standard was developed for use in determination of the concentration of NLX in monkey plasma. Specificity, range, and linearity were assessed. For specificity, extracted ion chromatograms of monkey plasma, and monkey plasma spiked with an internal standard, demonstrated that monkey plasma does not interfere with NLX or internal standard quantification. For linearity and range, the analytical calibration curve was constructed with 8 non-zero standards by plotting the peak area ratio of NLX to the internal standard versus the concentration. The concentration range was evaluated from 0.5 to 100 ng/mL for quantification. A blank sample (matrix sample processed without internal standard) was used to exclude contamination or interference. The curve was assessed with weighted linear regression ($1/X^2$). The linearity of the relationship between peak area ratio and concentration was demonstrated by the correlation coefficient ($r = 0.9974$).

5. Conclusions

Mu receptor occupancy and plasma concentrations of naloxone following IM delivery of the drug have been investigated in rhesus monkeys. A significant correlation between percent RO and plasma naloxone level was found and confirms that higher doses of IM naloxone result in higher mu RO. These results support the notion that higher doses of IM naloxone may be useful for treatment of opioid overdoses due to potent synthetic opioids.

Author Contributions: Conceptualization, P.J.H.S. and R.B.M.; methodology, J.S., J.A., P.S.S., X.S., B.D.H., M.E.R., and A.R.S.; formal analysis, X.S., B.D.H., R.A.K., P.J.H.S., R.B.M., M.E.R., and A.R.S.; investigation, J.S., J.A., P.S.S., X.S., B.D.H., M.E.R., and A.R.S.; resources, P.J.H.S. and R.B.M.; data curation, X.S., B.D.H., M.E.R., A.R.S., R.A.K., and P.J.H.S.; writing—original draft preparation, P.J.H.S. and R.B.M.; writing—review and editing, all authors; supervision, P.J.H.S., R.B.M., and D.J.C.; project administration, P.J.H.S. and R.B.M.; funding acquisition, R.B.M. and P.J.H.S. All authors have read and agreed to the published version of the manuscript.

Funding: The funding for this work was provided by Adamis Pharmaceuticals.

Acknowledgments: The authors would like to thank Bo Wen and Lu Wang of the University of Michigan Pharmacokinetic and Mass Spectrometry Core for assistance with the plasma naloxone assay.

Conflicts of Interest: D.J.C. and R.B.M. are employees of Adamis Pharmaceuticals who funded this study and provided naloxone.

References

1. Drug Overdose Deaths. Available online: https://www.cdc.gov/drugoverdose/data/statedeaths.html (accessed on 13 December 2019).
2. Provisional Drug Overdose Death Counts. Available online: https://www.cdc.gov/nchs/nvss/vsrr/drug-overdose-data.htm (accessed on 13 December 2019).
3. U.S. Surgeon General's Advisory on Naloxone and Opioid Overdose. Available online: https://www.surgeongeneral.gov/priorities/opioid-overdose-prevention/naloxone-advisory.html (accessed on 13 December 2019).
4. Rzasa, L.R.; Galinkin, J.L. Naloxone dosage for opioid reversal: Current evidence and clinical implications. *Ther. Adv. Drug Saf.* **2018**, *9*, 63–88. [CrossRef] [PubMed]
5. Schumann, H.; Erickson, T.; Trevonne, M.; Thompson, J.; Zautcke, L.; Denton, J.S. Fentanyl epidemic in Chicago, Illinois and surrounding Cook County. *Clin. Toxicol.* **2008**, *46*, 501–506. [CrossRef] [PubMed]
6. Bell, A.; Bennett, A.; Jones, T.S.; Simkins, M.; Williams, L.D. Amount of naloxone used to reverse opioid overdoses outside of medical practice in a city with increasing illicitly manufactured fentanyl in illicit drug supply. *Subst. Abus.* **2019**, *40*, 52–55. [CrossRef] [PubMed]
7. Somerville, N.J.; O'Donnell, J.; Gladden, R.M.; Zibbell, J.E.; Green, T.C.; Younkin, M.; Ruiz, S.; Babakhanlou-Chase, H.; Chan, M.; Callis, B.P.; et al. Characteristics of Fentanyl Overdose - Massachusetts, 2014–2016. *Morb. Mortal. Wkly. Rep.* **2017**, *66*, 382–386. [CrossRef]
8. Bardsley, R. Higher naloxone dosing may be required for opioid overdose. *Am. J. Health-Syst. Pharm.* **2019**, *76*, 1835–1837. [CrossRef]
9. Moss, R.B.; Carlo, D.J. Higher doses of naloxone are needed in the synthetic opioid era Higher doses of naloxone are needed in the synthetic opioid era. *Subst. Abus. Treat. Prev. Policy* **2019**, *14*, 6. [CrossRef]
10. Moss, R.B.; Carleton, F.; Lollo, C.P.; Carlo, D.J. Comparative Pharmacokinetic Analysis of Community Use Naloxone formulations for Acute Treatment of Opioid Overdose. *J. Addict. Adolesc. Behav.* **2019**, *2*. [CrossRef]
11. Melihar, J.K.; Nutt, D.J.; Malizia, A.L. Naloxone displacement at opioid receptor sites measured in vivo in the human brain. *Eur. J. Pharmacol.* **2003**, *459*, 217–219. [CrossRef]
12. Saccone, P.A.; Lindsey, A.M.; Koeppe, R.A.; Zelenock, K.A.; Shao, X.; Sherman, P.; Quesada, C.A.; Woods, J.H.; Scott, P.J.H. Intranasal opioid administration in Rhesus Monkeys: PET imaging and antincicpetion. *J. Pharmacol. Exp. Ther.* **2016**, *359*, 366–373. [CrossRef]
13. Blecha, J.B.; Henderson, B.D.; Hockley, B.G.; VanBrocklin, H.F.; Zubieta, J.K.; DaSilva, A.F.; Kilbourn, M.R.; Koeppe, R.A.K.; Scott, P.J.H.; Shao, X. An updated synthesis of [^{11}C] carfentanil for positron emission tomography (PET) imaging of the μ-opioid receptor. *J. Label. Compd. Radiopharm.* **2017**, *60*, 375–380. [CrossRef]
14. Reagan-Shaw, S.; Nihal, M.; Ahmad, N. Dose translation from animal to human studies revisited. *FASEB* **2007**, *22*, 659–661. [CrossRef] [PubMed]
15. Weinstein, S.H.; Pfeffer, M.; Schor, J.M. Metabolism and Pharmacokinetics of Naloxone. *Adv. Biochem. Psychopharmacol.* **1973**, *8*, 525–535. [PubMed]
16. Koyyalagunta, K. Opioid Analgesics. *Pain Manag.* **2007**, *2*, 939–964.
17. Kim, S.; Wagner, H.N.; Villemagne, V.L.; Kao, P.F.; Dannals, R.F.; Ravert, H.T.; Joh, T.; Dixon, R.B.; Civelek, C. Longer Occupancy of Opioid Receptors by Nalmephene compared to Naloxone as measured by In Vivo by a Dual Detector System. *J. Nucl. Med.* **1997**, *38*, 1727–1731.

18. Johansson, J.; Hirvonen, J.; Lovró, Z.; Ekblad, L.; Kaasinen, V.; Rajasilta, O.; Helin, S.; Tuisku, J.; Sirén, S.; Pennanen, M.; et al. Intranasal naloxone rapidly occupies brain mu-opioid receptors in human subjects. *Neuropsychopharmacol* **2019**, *44*, 1667–1673. [CrossRef]
19. EVISO®Prescribing Information. Available online: https://dailymed.nlm.nih.gov/dailymed/fda/fdaDrugXsl.cfm?setid=5fbe8d17-a72f-406d-a736-48e61620f9d8&type=display (accessed on 13 March 2020).
20. Edwards, E.S.; Gunn, R.; Kelley, G.; Smith, A.; Goldwater, R. American Academy of Pain Medicine 2015 Abstract 216: Naloxone 0.4 mg bioavailability following a single injection with a novel naloxone auto-injector, EVZIO®, in healthy adults, with reference to a 1 mL standard syringe and intramuscular needle. *Pain Med.* **2015**, *16*, 608–609.
21. Ryan, S.A.; Dunne, R.B. Pharmacokinetic properties of intranasal and injectable formulations of naloxone for community use: A systematic review. *Pain Manag.* **2018**, *8*, 231–245. [CrossRef]
22. Logan, J.; Fowler, J.S.; Volkow, N.D.; Wang, G.J.; Ding, Y.S.; Alexoff, D.L. Distribution volume ratios without blood sampling from graphical analysis of PET data. *J. Cereb. Blood Flow Metab.* **1996**, *16*, 834–840. [CrossRef]
23. McDonald, R.; Lorch, U.; Woodward, J.; Bosse, B.; Dooner, H.; Mundin, G.; Smith, K.; Strong, J. Pharmacokinetics of concentrated naloxone nasal spray for opioid overdose reversal: Phase I healthy volunteer study. *Addiction* **2018**, *113*, 484–493. [CrossRef]
24. Tomassoni, A.J.; Hawk, K.F.; Jubanyik, K.; Nogee, D.P.; Durant, T.; Lynch, K.L.; Patel, R.; Ding, D.; Ulrich, A.; D'Onofrio, G. Multiple Fentanyl Overdoses—New Haven, Connecticut, June 23, 2016. *Morb. Mortal. Wkly. Rep.* **2017**, *66*, 107–111. [CrossRef]
25. Lee, D.; Crhonister, C.W.; Broussard, W.A.; Utley-Bobak, S.R.; Schultz, D.; Vega, R.S.; Golderberger, B.A. Illicit fentanyl-related fatalaties in florida: Toxicological findings. *J. Anal. Toxicol.* **2016**, *40*, 588–594. [CrossRef] [PubMed]
26. Dwyer, J.B.; Jannsen, J.; Luckasevic, T.M.; Williamns, K.E. Report of increasing overdose deaths that acetyl fentanyl in multiple counties in the southwestern region of the commonwealth of Pennsylvania in 2015–2016. *J. Forensi. Sci.* **2018**, *63*, 195–200. [CrossRef] [PubMed]
27. Fogarty, M.F.; Papsun, D.M.; Logan, B.K. Analysis of Fentanyl and 18 Novel Fentanyl Analogs and Metabolites by LC–MS-MS, and report of Fatalities Associated with Methoxyacetylfentanyl and Cyclopropylfentanyl. *J. Anal. Toxicol.* **2018**, *42*, 592–604. [CrossRef] [PubMed]
28. Sutter, M.E.; Gerona, R.R.; Davis, M.T.; Roche, B.M.; Colby, D.K.; Chenoweth, J.A.; Adams, A.J.; Owen, K.P.; Ford, J.B.; Black, H.B.; et al. One Pill can Kill. *Acad. Emerg. Med.* **2017**, *24*, 106–113. [CrossRef] [PubMed]

© 2020 by the authors. Licensee MDPI, Basel, Switzerland. This article is an open access article distributed under the terms and conditions of the Creative Commons Attribution (CC BY) license (http://creativecommons.org/licenses/by/4.0/).

Article

Evaluation of Organo [^{18}F]Fluorosilicon Tetrazine as a Prosthetic Group for the Synthesis of PET Radiotracers

Sofia Otaru [1], Surachet Imlimthan [1], Mirkka Sarparanta [1], Kerttuli Helariutta [1], Kristiina Wähälä [1,2] and Anu J. Airaksinen [1,3,*]

[1] Department of Chemistry, Radiochemistry, University of Helsinki, 00100 Helsinki, Finland; sofia.otaru@helsinki.fi (S.O.); surachet.imlimthan@helsinki.fi (S.I.); mirkka.sarparanta@helsinki.fi (M.S.); Kerttuli.Helariutta@helsinki.fi (K.H.); kristiina.wahala@helsinki.fi (K.W.)
[2] Department of Biochemistry and Developmental Biology, Faculty of Medicine, University of Helsinki, 00100 Helsinki, Finland
[3] Turku PET Centre, Department of Chemistry, University of Turku, 20500 Turku, Finland
* Correspondence: anu.airaksinen@helsinki.fi

Academic Editors: Anne Roivainen and Xiang-Guo Li
Received: 19 February 2020; Accepted: 5 March 2020; Published: 7 March 2020

Abstract: Fluorine-18 is the most widely used positron emission tomography (PET) radionuclide currently in clinical application, due to its optimal nuclear properties. The synthesis of ^{18}F-labeled radiotracers often requires harsh reaction conditions, limiting the use of sensitive bio- and macromolecules as precursors for direct radiolabeling with fluorine-18. We aimed to develop a milder and efficient in vitro and in vivo labeling method for trans-cyclooctene (TCO) functionalized proteins, through the bioorthogonal inverse-electron demand Diels-Alder (IEDDA) reaction with fluorine-18 radiolabeled tetrazine ([^{18}F]SiFA-Tz). Here, we used TCO-modified bovine serum albumin (BSA) as the model protein, and isotopic exchange (IE) (^{19}F/^{18}F) chemistry as the labeling strategy. The radiolabeling of albumin-TCO with [^{18}F]SiFA-Tz ([^{18}F]**6**), providing [^{18}F]fluoroalbumin ([^{18}F]**10**) in high radiochemical yield (99.1 ± 0.2%, $n = 3$) and a molar activity (MA) of 1.1 GBq/µmol, confirmed the applicability of [^{18}F]**6** as a quick in vitro fluorination reagent for the TCO functionalized proteins. While the biological evaluation of [^{18}F]**6** demonstrated defluorination in vivo, limiting the utility for pretargeted applications, the in vivo stability of the radiotracer was dramatically improved when [^{18}F]**6** was used for the radiolabeling of albumin-TCO ([^{18}F]**10**) in vitro, prior to administration. Due to the detected defluorination in vivo, structural optimization of the prosthetic group for improved stability is needed before further biological studies and application of pretargeted PET imaging.

Keywords: fluorine-18; positron emission tomography (PET); defluorination; isotopic exchange; silicon-based fluoride acceptor; bioorthogonal chemistry; tetrazine; inverse electron-demand Diels-Alder ligation

1. Introduction

Fluorine-18 is an ideal radionuclide for labeling of radiopharmaceuticals for positron emission tomography (PET), due to its nuclear and physical characteristics, including the relatively long half-life (109.7 min), the low energy levels of emitted positrons ($E_{max} = 0.635$ MeV), and high positron decay probability (97%) [1]. Fluorine-18 can be produced via the ^{18}O(p,n)^{18}F reaction, by irradiating ^{18}O-enriched water with protons, yielding high molar activity [^{18}F]fluoride, in aqueous solutions. Several advances in ^{18}F-fluorinations, such as metal-mediated (e.g., Pd, Cu, Ag) aromatic, aliphatic, and aryl boronic ester radiofluorinations, as well as TiO$_2$-catalyzed ^{18}F-fluorinations in aqueous media, have been recently presented [2–4]. The incorporation of nucleophilic [^{18}F]fluoride into molecules generally requires alkaline conditions and elevated temperatures. Faster, milder, and more selective radiolabeling methodologies are desired, especially for the radiolabeling of compounds sensitive to temperature and higher pH. Fast and efficient catalyst-free click-reactions, such as the inverse electron-demand Diels-Alder (IEDDA) reaction, have been applied as effective tools for the selective incorporation of radiolabels, such as ^{18}F, into bio- and macromolecules, via small radiolabeled prosthetic groups. Other widely known click-reactions, such as the copper(I)-catalyzed azide-alkyne cycloaddition (CuAAC) and the stainpromoted azide-alkyne cycloaddition (SPAAC) have been the starting point for the modification of chemoselective biomolecules [5]. The CuAAC reaction was first reported in 2002 by Sharpless et al. [6] and Meldal et al. [7]. The aim of these studies was to utilize the CuAAC reaction for the formation of an enormous variety of five-membered heterocycles, triazoles, and peptide derivatives. The use of CuAAC led to the investigation on the utility of these highly efficient reactions for the labeling of biomolecules in living systems. However, the toxicity of copper limited the feasibility of CuAAC in biological applications. SPAAC was developed in 2004 by Bertozzi et al. who demonstrated the chemical modification of live Jurkat cells with an azide-modified sugar for the subsequent conjugation of alkyne-biotin for fluorescent labeling with FITC-avidin, without any apparent decline in cell viability [8]. Since then, SPAAC has served as a catalyst-free alternative to overcome the cytotoxicity concerns of the CuAAC reaction. However, both CuAAC and SPAAC have relatively slow reaction kinetics, which renders the reactions unsuitable for labeling of biomolecules in a living system, for applications lile in vivo pretargeting. In 1959, Lindsey et al. reported the ability of tetrazines to react chemoselectively with unsaturated compounds through a 1,4-cycloaddition reaction [9]. These findings introduced a new and highly reactive bioorthogonal IEDDA click-reaction as a pivotal tool for synthetic modification of biomolecules.

The bioorthogonal IEDDA-reaction has been successfully used for various pretargeted in vivo radiolabeling applications [10–12]. Pretargeted imaging has found exceptional utility with imaging agents with slow pharmacokinetics, such as antibodies and nanomaterials, which when directly radiolabeled would require the use of long-lived radioisotopes (such as ^{89}Zr or ^{111}In, both with ~3-day physical half-lives), to track their biodistribution in vivo. In the pretargeted approach, the targeting vector (such as an IgG antibody) is first modified by one reactant of the IEDDA reaction, and allowed to distribute in the body after administration. Next, it is tracked using the other reactant that is radiolabeled with a short-lived radioisotope, with improved image contrast and lower radiation burden to the subject [13,14]. The fastest IEDDA-reaction reported so far is from the conjugation between tetrazine (Tz) and trans-cyclooctene (TCO) ($k \approx 10^6$ M^{-1} s^{-1}) [5], rendering this reactive pair of utmost interest in the fields of chemical biology, nuclear imaging, and radiotracer development. However, the sensitivity of tetrazines towards alkaline conditions and elevated temperature renders the direct radiolabeling of tetrazines with fluorine-18 challenging. Therefore, the use of prosthetic groups for radiolabeling, such as the glycoconjugate [^{18}F]-5-fluoro-5-deoxyribose (FDR) or Al[^{18}F]F [15–17], is necessary to ensure that radiolabeling conditions preserve the reactivity of the Tz.

The silicon-fluoride acceptor (SiFA) chemistry relies on ^{19}F/^{18}F-isotopic exchange for introducing fluorine-18 into radiotracers, and has emerged as a fast and mild radiolabeling tool, especially for sensitive molecules [18]. The small lipophilic SiFA compounds mainly utilized in the radiolabeling of larger constructs such as peptides, proteins, and nanoparticles, in most cases have demonstrated excellent stability against in vivo defluorination [19–23]. The lipophilic character of the SiFA-derivatives can be used to tailor the pharmacokinetics of biomolecular tracers. To our knowledge, there is only one compound containing SiFA bound to a tetrazine, SiFA-OTz, which has been reported to date, but its enzymatic stability in vitro and in vivo has not yet been studied [24]. Nevertheless, the tracer SiFA-OTz demonstrated good stability under the radiolabeling conditions. We have reported the development of highly stable and highly hydrophilic ^{18}F-tetrazines from sugar analogues, such as [^{18}F]fluorodeoxyribose ([^{18}F]FDR-Tz) and 2-deoxy-2-[^{18}F]fluoro-D-glucose ([^{18}F]FDG-Tz) [17,25]. We also previously studied the feasibility of utilizing the ^{18}F-FDR-Tz for the in vivo IEDDA pretargeting of antibodies and nanoparticles and were able to prove this approach to be highly successful [26,27].

Here, we investigated the use of a SiFA as a reaction strategy for the [^{18}F]fluorination of tetrazines, under mild reaction conditions, with the possibility to yield a more hydrophobic tetrazine variant for the regulation of the pharmacokinetics of biomolecules labeled with [^{18}F]SiFA-Tz. The aim of this study was to investigate [^{18}F]SiFA-Tz as a standalone tracer, to reveal its potential for pretargeted imaging and its applicability for the rapid in vitro radiolabeling of TCO-containing biomolecules, under physiological conditions.

2. Results

2.1. Chemistry

SiFA-Tz (6) was synthesized via a three-step route, providing good yields for each step (Scheme 1) [14]. The first step comprised of an amide coupling reaction under argon, between a carboxylic acid and an amine, forming the t-Boc protected aminooxy-tetrazine 3, in 65% yield after silica gel column chromatography purification. The next reaction step was the deprotection of 3 with hydrochloric acid in methanol, to give 4 as a pink solid in cold diethyl ether in 65% yield. This hydrochloric salt intermediate was used such for the next reaction, the oxime bond formation between compounds 4 and 5 generating an imine double bond such as E- and Z-isomers of SiFA-Tz (6) [28]. The reaction mixture was purified with semi-preparative HPLC, providing the desired product 6 in 90% yield. ^{1}H-NMR, ^{13}C-NMR, ^{19}F-NMR spectra and ESI-TOF-MS were acquired for characterization of the final product, SiFA-Tz (6).

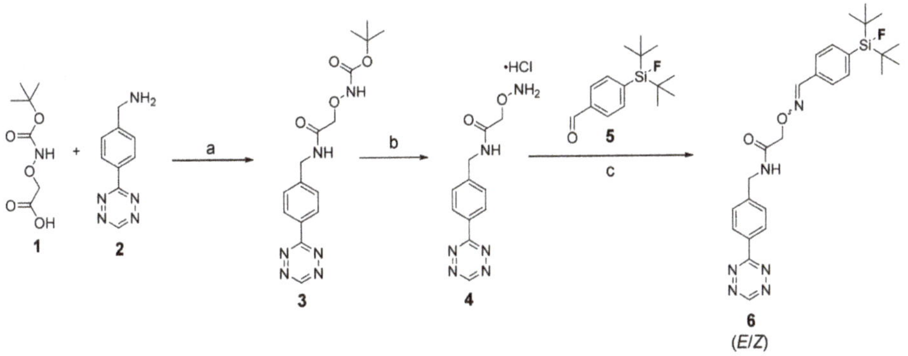

Scheme 1. Synthesis of the precursor SiFA-Tz (6). Reagents and conditions: (a) HATU, DIPEA, DMF, room temperature, 20 h, Argon (65%), (b) 1 M HCl Et$_2$O 25 °C, 24 h, MeOH (65%), and (c) Aniliniumacetate-buffer pH 4.6, 25 °C, 15 min (90%). (HATU; 1-[Bis(dimethylamino)methylene]-1H-1,2,3-triazolo[4,5-b]pyridinium 3-oxid hexafluorophosphate, DIPEA; N,N-Diisopropylethylamine, DMF; Dimethylformamide).

Bovine serum albumin (7) was functionalized with N-hydroxysuccinimide (NHS) ester of transcyclooctene (TCO) (200 eq), from the lysine residues available in the protein structure (Scheme 2). The analysis of the TCO:albumin ratio was carried out with MALDI-MS, which revealed a TCO:albumin ratio 40:1. The MALDI-MS analysis indicates that 65% of the total 62 lysine residues were functionalized with a TCO in one albumin molecule. The IEDDA cycloaddition product of SiFA-Tz (6) and albumin-TCO (9), fluoroalbumin (10), was synthesized as a reference for the radiolabeling studies by mixing the two reagents together at room temperature.

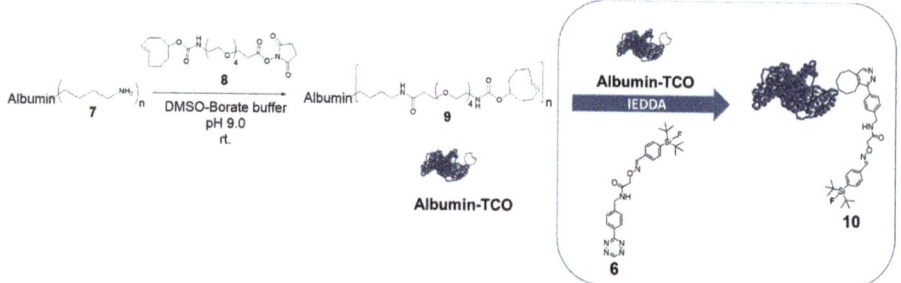

Scheme 2. Functionalization of bovine serum albumin (**7**) with trans-cyclooctene (**8**) to form albumin- trans-cyclooctene (TCO) (**9**), followed by the inverse-electron demand Diels-Alder (IEDDA) cycloaddition with SiFA-Tz (**6**) to form fluoroalbumin (**10**) (molar ratio 1:1.5 SiFA-Tz:fluoroalbumin).

2.2. Radiochemistry

Two different synthetic sequences were investigated for the radiosynthesis of [^{18}F]**6**—a one–step direct radiolabeling of compound **6** and a two–step method where we first radiolabeled the SiFA moiety **5**, before linking it to the tetrazine **4** (Scheme 3). The one-step radiolabeling of **6**, resulted in 21% radiochemical yield ($n = 1$), but the yield was detected to decrease rapidly as a function of time, indicating decomposition of the precursor **6** in the reaction mixture, under the alkaline conditions (pH 8.5–9). In the two-step method, the radiolabeling of the SiFA-moiety **5** resulted in incorporation yields ranging from 89 to 99.6 ± 0.5% ($n = 4$), at the optimal time point (2 min) analyzed by radio-TLC. The radiochemical impurities (maximum 11% of total radioactivity) formed in the first step were analyzed by radio-HPLC and are shown in the Supplementary Data (Figure S6). The SiFA radiolabeling was followed by an oxime bond formation between the [^{18}F]SiFA ([^{18}F]**5**) and tetrazine oxyamine **4** at 42.7 ± 14.2% ($n = 8$) yield in the reaction mixture (Figure S7). Radio-TLC analysis of product [^{18}F]**6** revealed the formation of two isomers, with the (E)-isomer of [^{18}F]**6** being predominant with the amount of the radiolabeled **6** (Z)-isomer only 1.45 ± 0.35% ($n = 16$). The final product [^{18}F]**6** was isolated at >98% radiochemical purity (Figure S9, radio-TLC) and was subsequently used as a prosthetic group to chemoselectively radiolabel the TCO-functionalized albumin in vitro. The [^{18}F]fluorinated albumin [^{18}F]**10** was radiolabeled at 99.1 ± 0.2% radiochemical yield (RCY) ($n = 3$) (Figure S12, radio-TLC) and good molar activity (1.1 ± 0.2 GBq/µmol, $n = 2$). To achieve 99% RCY and the total consumption of added [^{18}F]**6** (0.26 nmol), minimum of 0.27 nmol of albumin-TCO was to be used. This resulted in 2.5% of the TCOs in the albumin-TCO labeled with [^{18}F]**6**. A radiochemical yield of >99% of [^{18}F]**10** was achieved by incubating increasing amounts of albumin–TCO with [^{18}F]**6** (0.27 nmol) (Figure 1). The radio-HPLC chromatograms of the purified tracers [^{18}F]**6** and [^{18}F]**10** are presented in the Supplementary Data (Figures S8 and S12).

2.3. In Vitro Studies of [^{18}F]6 and [^{18}F]10.

The stability of [^{18}F]**6** in 0.01M PBS pH 7.4 was shown to be excellent at 90 min, with minimal detachment of fluorine-18 (<1%) during the incubation (Figure S10). Plasma protein binding and metabolic stability of [^{18}F]**6** were also evaluated by incubating the radiotracer in plasma and analyzing the deproteinized samples through radio-TLC and radio-HPLC methods, at selected time-points after incubation (Figure 2).

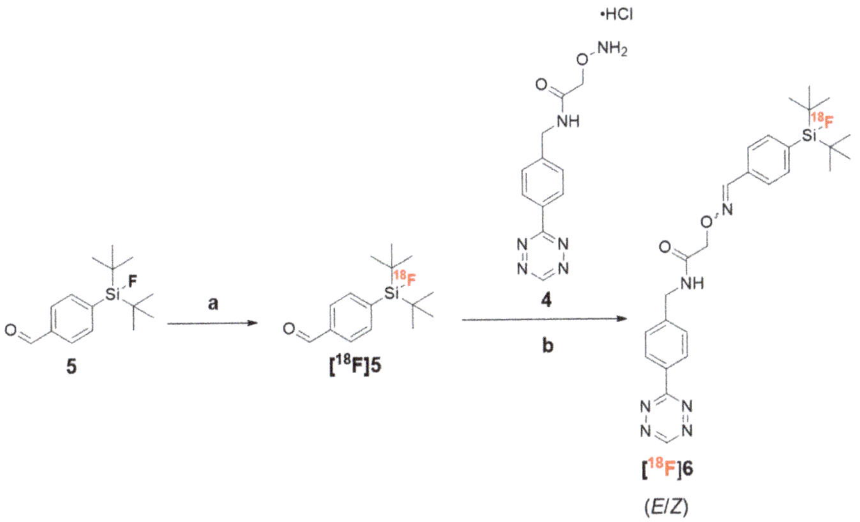

Scheme 3. Radiosynthesis of [^{18}F]SiFA-Tz ([^{18}F]6) by a two-step method. Reagents and conditions: (**a**) ACN, K[^{18}F]F-[K2.2.2], (RCY 99%) and (**b**) anilinium acetate buffer pH 4.6, 25 °C, 15 min (RCY 43%).

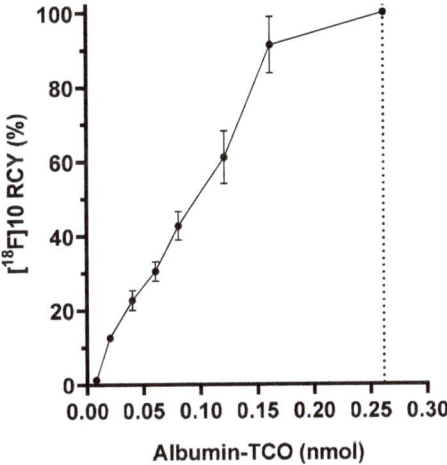

Figure 1. Radiochemical yield (RCY) of [^{18}F]fluoroalbumin ([^{18}F]10) during titration of [^{18}F]SiFA-Tz ([^{18}F]6), with albumin-TCO revealed a maximal radiochemical yield of >99% with 0.26 nmol of albumin-TCO.

Through radio-TLC analysis, [^{18}F]6 demonstrated good stability in plasma with up to 6% detachment of the radiolabel over 180 min (94.9 ± 1.6 % intact tracer, $n = 2$). The slight difference in the stability profiles between radio-TLC and radio-HPLC analysis can be explained due to their different ability to quantify free fluoride from the samples. The retention of free fluoride on the silica-based C18 HPLC-column material makes the quantification of free fluoride with HPLC less accurate. Representative radio-TLC and radio-HPLC chromatograms from the experiment are shown in the Supplementary Data (Figures S15 and S16). A radio-HPLC analysis revealed no other radiometabolites of [^{18}F]6 during the 180 min incubation (Figure 3), in addition to a highly polar component, most likely the free fluoride. Despite the observed minor defluorination, the enzymatic stability was found to be sufficient for proceeding into in vivo evaluation.

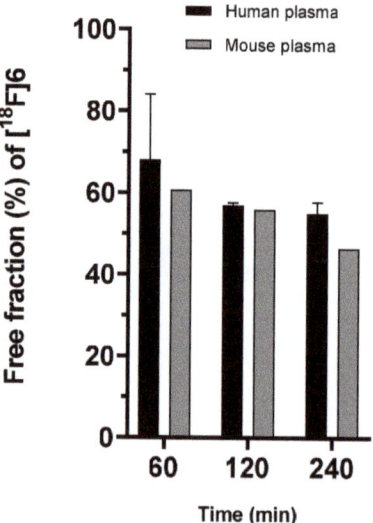

Figure 2. Free fraction of the radiotracer [^{18}F]SiFA-Tz ([^{18}F]6) during 240 min incubation in human and mouse plasma demonstrated comparable profiles with ~50% of the radioactivity in the free fraction at the end of incubation.

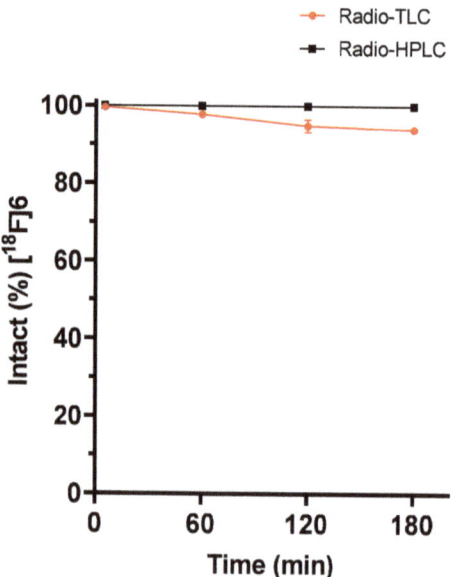

Figure 3. In vitro stability of [^{18}F]SiFA-Tz ([^{18}F]6) after incubation in human plasma at 37 °C, demonstrated excellent stability with ~95% intact tracer at 180 min after start of incubation. The amount of parent [^{18}F]6 was measured through radio-TLC and radio-HPLC analysis.

The LogD$_{pH7.4}$ of [^{18}F]6 (1.6 ± 0.2, n = 9) was determined by the shake-flask method, as previously reported [16]. It has been shown that lipophilic compounds tend to exhibit a higher plasma-protein binding than the more hydrophilic compounds [29], which support the findings of the measured LogD-value and the value obtained for the plasma protein-bound [^{18}F]6, in this study.

2.4. Biodistribution of [^{18}F]6 and [^{18}F]10

The biodistribution of [^{18}F]6 was investigated in healthy 11 to 12 week-old female CD-1 mice. After intravenous administration into the lateral tail vein, [^{18}F]6 (14.4 ± 0.5 MBq/animal in ~200 µL of 10% EtOH and 0.5% Solutol HS 15 in 0.01 M PBS pH 7.4) exhibited hepatobiliary excretion and a fast clearance from the blood (Figure 4, Figure S13). The highest percentage of the injected dose per gram of tissue (%ID/g) for urine (229.5 ± 204.5) and gallbladder (143.9 ± 103.1) was found to be at 60 min post-injection. In addition to the high radioactivity detected in the urine, gallbladder, liver, and the feces, a high bone uptake of ^{18}F$^-$ was observed 60 min post-injection (13.4 ± 1.6% ID/g). No major passage of [^{18}F]6 through the blood–brain barrier was observed (0.7 ± 0.2% ID/g) at 5 min post-injection. In order to study the metabolism of [^{18}F]6, the blood samples were collected at (t = 5, 30, and 60 min) post-injection. The deproteinized plasma samples were analyzed through radio-HPLC and radio-TLC methods, which revealed the formation of a highly polar metabolite (R$_f$ 0.00 on TLC) that was retained at the origin. On HPLC, a metabolite ([^{18}F]M1) eluting at 7 min was also detected. Furthermore, there was no indication of more lipophilic metabolites (Figure S17).

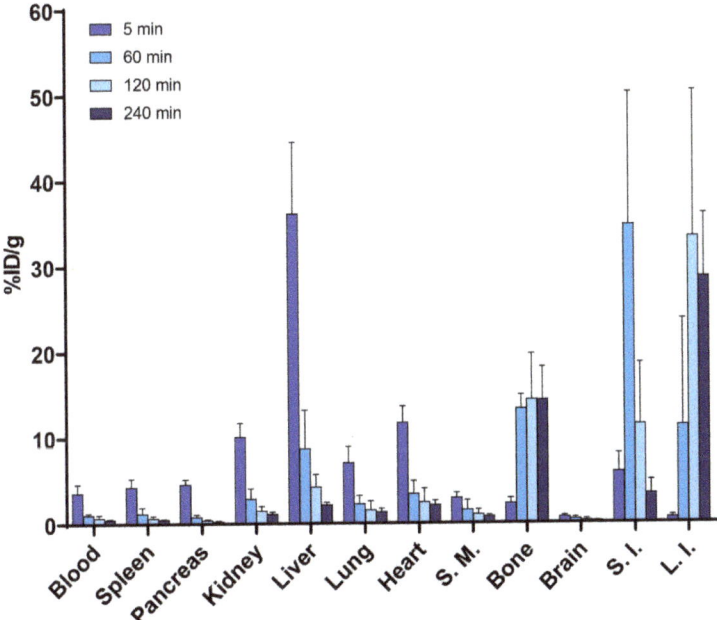

Figure 4. Biodistribution of radioactivity after the intravenous injection of [^{18}F]SiFA-Tz ([^{18}F]6, n = 3), demonstrating a high bone uptake of radioactivity, 60 min post-injection. (S.M.—skeletal muscle; bone—tibia, S.I.—small intestine; L.I.—large intestine).

In order to investigate how the conjugation of [^{18}F]6 onto a macromolecule influences its in vivo defluorination rate, we used [^{18}F]6 to synthesize [^{18}F]fluoroalbumin ([^{18}F]10) and evaluated its biodistribution in CD-1 mice. Intravenously injected [^{18}F]10 (0.6 ± 0.1 MBq/animal in ~100 µL of 0.01 M PBS, pH 7.4) had a prolonged residence time in blood with 7.1 ± 0.3% ID/g, at 60-min post-injection and a plasma half-life of 49 min (Figure 5). The highest % ID/g for urine (54.7 ± 25.8) and gallbladder (275.3 ± 185.6), after intravenous injection of [^{18}F]10 was shown to be at 60 min post-injection, for both tissues.

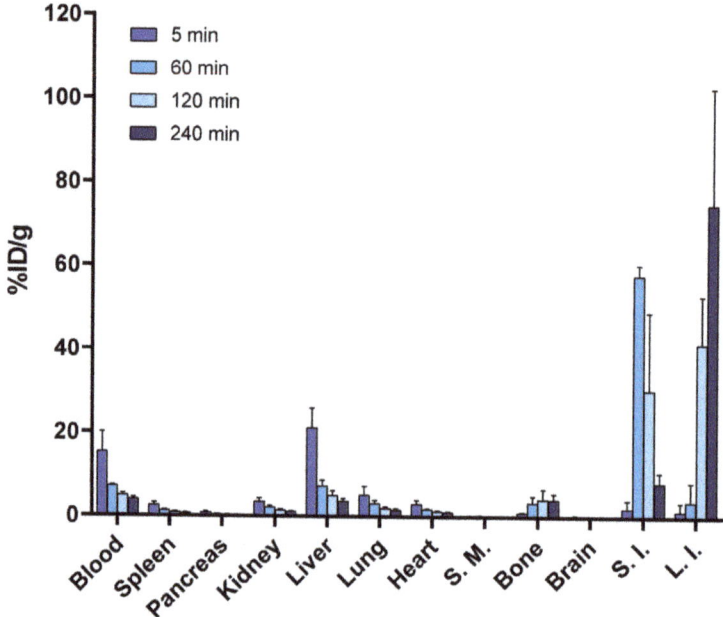

Figure 5. Biodistribution of radioactivity after the intravenous administration of [^{18}F]fluoroalbumin. ([^{18}F]10, n = 5) demonstrating a prolonged residence time of radioactivity in whole blood and reduced bone accumulation, indicating resistance to defluorination in vivo. (S.M.—skeletal muscle; bone—tibia; S.I.—small intestine; L.I.—large intestine).

Stability of [^{18}F]10 was further investigated by separating molecules with molecular weight of ≥30 kDa from plasma samples collected after intravenous administration of [^{18}F]10 through ultrafiltration. The proportion of radiolabeled molecules with a MW of ≥30 kDa (presenting intact [^{18}F]10) was over 90% until 180 min post-injection, after which it decreased to 40% over the next 60 min (240 min incubation in total), as shown in Figure 6.

By comparing the bone uptake of ^{18}F in these two biodistribution studies, it was seen that the ^{18}F-Si bond in [^{18}F]10 was noticeably more stable than in the [^{18}F]SiFA-Tz ([^{18}F]6) alone (Figure 7). The bone uptake for both tracers peaked around 120 min after administration and the radioactivity persisted in the bone until the last time point of the biodistribution study, 240 min post-injection. This was an indication of the uptake of free [^{18}F]fluoride, which was a result of defluorination of the SiFA moiety, in vivo [30].

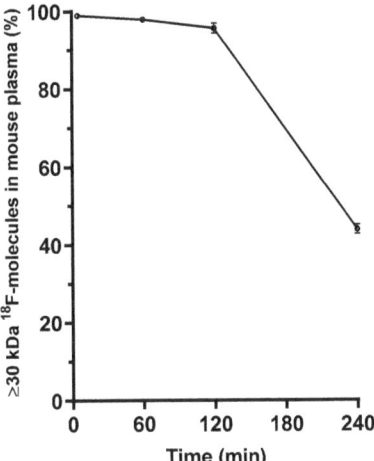

Figure 6. The ex vivo distribution (%) of ≥30 kDa ^{18}F-radiolabeled molecules in mouse plasma after intravenous injection of [^{18}F]fluoroalbumin ([^{18}F]10), separated with a molecular weight cut-off filter revealed a highly stable albumin tracer until 2 h post-injection. At 240 min post-injection around 40% of radioactivity was of ≥30 kDa size, with almost 60% of the radioactivity being comprised of species of lower molecular weight.

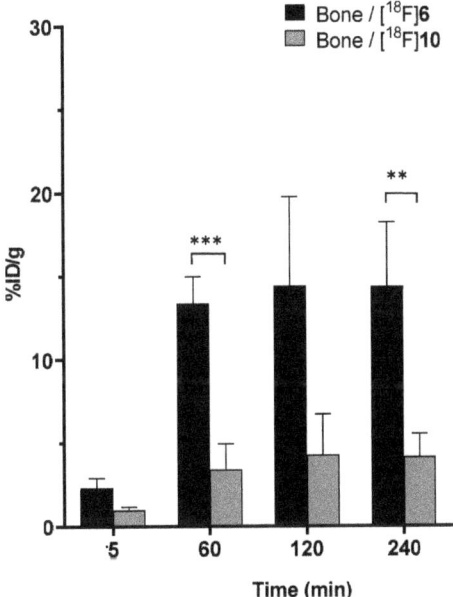

Figure 7. Comparison of the radioactivity accumulated in bone (tibia) in the biodistribution studies of [^{18}F]SiFA-Tz ([^{18}F]6) and [^{18}F]fluoroalbumin ([^{18}F]10). A significant enhancement in stability was seen for the albumin-bound [^{18}F]SiFA group in [^{18}F]10 (** $p \leq 0.01$, *** $p \leq 0.001$), compared to [^{18}F]6.

Furthermore, the rate of defluorination was substantially diminished when [^{18}F]6 was used to chemoselectively radiolabel albumin in vitro through the IEDDA ligation. The plasma protein radiotracer [^{18}F]10, which acted as a model protein in this study, demonstrated a significantly ($p \leq$

0.0001) longer blood circulation time in healthy CD-1 mice than the ^{18}F-labeled tetrazine [^{18}F]6 alone (Figure 8), with a biological half-life similar to what has been reported for other [^{18}F]SiFA-radiolabeled serum albumins.

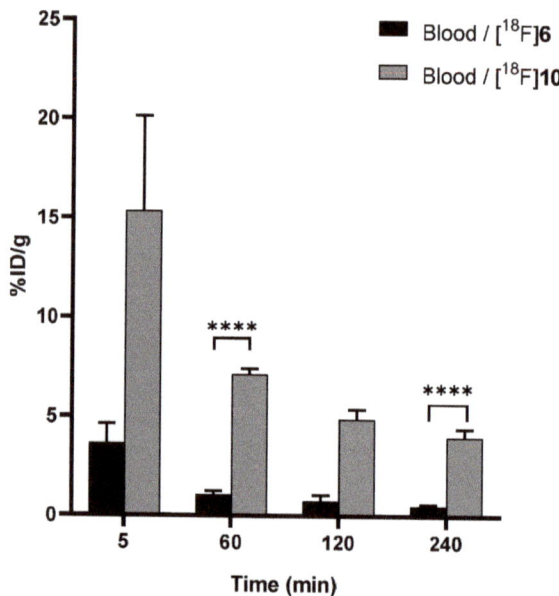

Figure 8. Radioactivity (%ID/g) in ex vivo blood samples at selected time–points, post-injection for [^{18}F]SiFA-Tz ([^{18}F]6) and [^{18}F]fluoroalbumin ([^{18}F]10), demonstrated a significantly prolonged circulation time for the plasma protein tracer ([^{18}F]10) (**** $p \leq 0.0001$).

3. Discussion

Two PET-tracer candidates, [^{18}F]SiFA-Tz ([^{18}F]6) and [^{18}F]fluoroalbumin ([^{18}F]10) were synthesized and evaluated in vivo. Compound [^{18}F]6 was synthesized with two different synthesis methods, from which the two-step approach was selected, due its higher RCY and good reproducibility. In the one-step radiolabeling approach, the RCY of the product was observed to decrease rapidly as a function of time, indicating decomposition of the precursor in the alkaline reaction mixture. In the two-step method, aminooxy tetrazine 4 was introduced into the reaction mixture at pH 4.6, which was found to be an advantage to avoid any unnecessary decomposition of the tetrazine group. In addition to the two-step method presented in this study, alterative elution protocols with milder reagents, such as copper salts or weak base solutions, as described by Scott et al., should be investigated for the radiolabeling of base sensitive precursors [31]. The in vivo metabolic profile of [^{18}F]6 displayed hepatobiliary elimination, which is characteristic for compounds with low hydrophilicity. Nevertheless, no observable passage of [^{18}F]6 through the blood–brain barrier was detected (0.7 ± 0.2% ID/g) at 5 min post-injection, despite the favorable lipophilicity of the tracer (LogD = 1.56 ± 0.20). Based on the radio-HPLC metabolite analysis (Supplement. Figure S17) from ex vivo blood samples and the detection of radioactivity in the bone, we concluded that [^{18}F]6 underwent rapid biotransformation, generating highly polar metabolites, one of which was most likely free [^{18}F]fluoride detached from the radiotracer. Furthermore, the accumulation of radioactivity in bone is a characteristic indication of fast defluorination in vivo. After defluorination, the free fluoride is sequestered rapidly from circulation and either binds into the surface of the bone or accumulates irreversibly into the hydroxyapatite $Ca_{10}(PO_4)_6(OH)_2$, forming fluorapatite ($Ca_{10}(PO_4)_6F_2$) [30]. Thus, it was evident that unexpected and relatively fast defluorination was observed in vivo. Free fluoride was also excreted into the urine, in vivo. Defluorination of ^{18}F-radiolabeled tracer [^{18}F]6 could be detected in the bone as early as 10–20 min after injection [30]. The observed rapid defluorination in vivo limited the utility of [^{18}F]6 for pretargeted PET imaging, and further structural optimization was warranted to stabilize the structure towards the defluorination. However, since the stability of biomacromolecular SiFA conjugates has been reported to be good, the possibility of using [^{18}F]6 as a prosthetic group for the in vitro bioorthogonal radiolabeling of proteins was investigated through administration of [^{18}F]10, to healthy mice. Stability of the [^{18}F]SiFA-Tz group against in vivo defluorination was dramatically improved when the group was bound to albumin (13.4 ± 1.6% ID/g in bone for [^{18}F]6 vs. 3.4 ± 1.5% ID/g in bone for [^{18}F]10, at 60 min post-injection). Furthermore, the blood circulation half-life was 48 min, which is in the order of the reported plasma half-life of 60 min, for the bovine serum albumin in mice [32,33].

There are some examples of small molecular SiFA derivatives and a SiFA-conjugated peptide that have exhibited detectable in vivo defluorination, but not at the level observed in our study [34,35]. Rat serum albumin (RSA) radiolabeled with [^{18}F]SiFA, through isothiocyanate modification of lysine residues has been shown to be relatively stable with only a low rate of defluorination, until 90 min after administration [36]. It has also been shown that the conjugation position of the [^{18}F]SiFA-moiety on the albumin could have an influence on the rate of defluorination in vivo. A more stable maleimido-[^{18}F]SiFA conjugated to RSA via thiol groups is an example of the enhanced stability of the radiolabel in a [^{18}F]SiFA-radiolabeled serum albumin [37]. Thus, this radiolabeling system could be further improved by using a more selective conjugation chemistry (maleimide over N-hydroxysuccinimide) for the addition of the TCO to albumin, while simultaneously optimizing the TCO:albumin ratio and availability of the TCO moiety to the IEDDA reaction, with [^{18}F]6. Nevertheless, our results demonstrated the feasibility of using the highly selective and rapid bioorthogonal reaction strategy for the radiolabeling of biomacromolecules with fluorine-18, under mild reaction conditions.

4. Materials and Methods

All reagents and solvents were purchased from commercial providers and used as received without further purification. Hyox-18 ^{18}O–enriched water (98%) was purchased from Rotem Industries Limited (Arava, Israel). Ultrapure water (18.0 MΩ) was produced with a Milli-Q Integral Water Purification System (Merck Millipore, Burlington, MA, USA). HATU, DMF, DIPEA, DMSO, LiCl, methanol, aniline, Kryptofix 2.2.2, 1 M HCl in diethylether, formic acid, ethylacetate and boc-aminooxy acetic acid were purchased from Sigma-Aldrich (St. Louis, MO, USA). MgSO4 was purchased from Merck Millipore (Darmstadt, Germany). TCO-PEG4-NHS ester was purchased from either Jena Biosciences (Jena, Germany) or Conju-Probe (San Diego, CA, USA). Tetrazine amine was purchased from either BroadPharm (San Diego, CA, USA) or Conju-Probe (San Diego, CA, USA). DNA synthesis quality anhydrous acetonitrile (max. 10 ppm water) was purchased from Merck (Kenilworth, NJ, USA). SiFA-aldehyde was purchased from Enamine (Monmouth, NJ, USA). Bovine serum albumin was purchased from Merck (Kenilworth, NJ, USA). Moisture or air sensitive reactions were carried out under an argon atmosphere in oven-dried glassware. Reactions were monitored by TLC Silica gel 60 F254 Merck Millipore (Darmstadt, Germany). Silica gel TLC-plates were run in EtOAc:heptane (7:3) as eluent. [^{18}F]SiFA-Tz ([^{18}F]6) R_f = 0.59, [^{18}F]fluoroalbumin ([^{18}F]10) R_f = 0.00.

^{1}H-, ^{13}C-, and ^{19}F-NMR spectra were acquired with a Varian Mercury spectrometer (300 MHz, 500 MHz, 600 MHz) (Palo Alto, CA, USA). Chemical shifts (δ) are reported in ppm units, using the solvent residual signal as a reference. Coupling constants (J) are expressed in hertz (Hz). The purities of radiolabeled compounds were determined through RP-HPLC with photodiode array (PDA)-, and radiodetector and through silica TLCs analyzed with a Fujifilm FLA 5100 scanner (Fujifilm Life sciences, Cambridge, MA, USA). The excised tissue samples were measured with 1480 Wallac Wizard® 3" (PerkinElmer™ Life Sciences, Waltham, MA, USA) gamma counter for 60 s per sample.

High performance liquid chromatography was carried out with a Shimadzu HPLC system consisting of a DGU-20A degasser, an LC-20AD UPLC LC unit, a SIL-20A HT autosampler, a CTO20 AC column oven, a CBM-20A communications bus module, a Scionix Holland scintillation detector with a 51 BP 51/2 NaI(Tl) crystal and an SPD-M20A diode array detector. For the [^{18}F]SiFATz, a Waters Symmetry semi-preparative C18 column (300 × 7.8 mm, 7 µm) was used, with 0.01 M H_3PO_4:ACN (20:80, 3 mL/min) as the eluent. Phenomenex BioSep SEC s3000 size exclusion column was used, with 0.1 M phosphate buffer pH 7 (0.8 mL/min) as the eluent, to analyze the conjugated albumin-TCO and radiolabeled protein tracer [^{18}F]fluoroalbumin ([^{18}F]10).

Preparative high performance liquid chromatography was carried out using a system consisting of a Phenomenex Degassex™ DG-4400 degasser, Merck LaChrom L-7100 pump, in-house prepared remote-controlled injection system, Amersham pharmacia biotech REC 112 dual channel chart recorder, Carroll & Ramsey Associates 101-H-DC3 multi-channel radiation detector, and a Knauer Azura UVD 2.1S detector. A waters Symmetry semi-preparative C18 column (300 × 7.8 mm, 7 µm), with 0.01 M H3PO4:ACN (20:80, 3 mL/min flowrate) as the eluent was used for the preparative purification of the [^{18}F]SiFA-Tz radiotracer ([^{18}F]6).

4.1. Synthesis

Tert-butyl(2-{[4-(1,2,4,5-tetrazin-3-yl)ben-zyl]amino}-2-oxoethoxy)carbamate (**3**). HATU (77 mg, 201 µmol) in DMF (0.5 mL) was added to a solution of **1** (32 mg, 168 µmol) in dry DMF (0.7 mL), under argon. The mixture was stirred at room temperature for 10 min, then **2** (25 mg, 112 µmol) in DMF (2.5 mL) was added, followed by DIPEA (49 µL, 281 µmol). After 20 h mixing at room temperature, ethyl acetate (15 mL) was added and the organic phase was extracted with ultrapure water (10 mL). The organic phase was extracted with 5% LiCl solution (3 × 10 mL, 1 × 20 mL), dried over MgSO4, and concentrated in vacuo. The crude product was purified through silica gel column chromatography (EtOAc:heptane) to afford **3** as a pink solid (26 mg, 72 µmol, 65% yield).

^{1}H-NMR (300 MHz, CDCl$_3$, Supplement Figure S1) δ 10.20 (d, J = 0.6 Hz, 1H), 8.79 (s, 1H), 8.61–8.49 (m, 2H), 7.72 (s, 1H), 7.60–7.51 (m, 2H), 4.63 (d, J = 6.1 Hz, 2H), 4.40 (s, 2H), 1.42 (d, J = 0.6 Hz, 9H). ^{13}C-NMR (75 MHz, CDCl3, Supplement Figure S2) 169.29, 166.53, 158.19, 157.98, 144.09, 130.73, 128.00, 83.63, 68.00, 42.83, 28.27 ppm.

N-[4-(1,2,4,5-tetrazin-3-yl)benzyl]-2-(aminooxy)acetamide hydrochloride (**4**). 1 M HCl in diethyl ether (50 mL) was added to a solution of **3** (22 mg, 61 µmol) in MeOH (4 mL). After 24 h, the solution was concentrated in vacuo, to afford **4** as a pink solid. The solid was dissolved in 2 mL of methanol and 20 mL of cold diethyl ether was added to the solution. The closed flask was kept in +4 °C overnight, to facilitate the crystallization of the purified product. The recrystallized solid (10 mg, 39 µmol, 65% yield) was suction filtered and used as such in the next reaction step.

(E,Z)-N-[4-(1,2,4,5-tetrazin-3-yl)benzyl]-2-({[4-(di-tert-butylfluorosilyl)benzylidene]-amino}oxy)acetamide (**6**). Compound **5** (8.5 mg, 32 µmol) in ACN (0.5 mL) was added into a solution of **4** (4 mg, 15 µmol) in 0.3 M anilinium acetate buffer pH 4.6 (2 mL). After 15 min, ultrapure water (10 mL) was added to the reaction mixture, concentrated with two Sep-pak C18 Light cartridges (Milford, MA, USA) and eluted with ACN (5 mL). The crude product was purified through RP-HPLC (0.01 M H3PO4:CAN, 20:80, 3 mL/min) to afford **6** as a pink solid (7 mg, 14 µmol, 90% yield).

^{1}H-NMR (300 MHz, CDCl3, Supplement Figure S3) δ 10.21 (s, 1H), 8.57 (d, J = 8.5 Hz, 2H), 8.21 (s, 1H), 7.69–7.49 (m, 6H), 4.76 (s, 2H), 4.68 (d, J = 6.1 Hz, 2H), 1.05 (d, J = 1.1 Hz, 18H). ^{19}F-NMR (282 MHz, CDCl3 Supplement Figure S4) δ-189.13. ^{13}C-NMR (126 MHz, CDCl$_3$, Supplement Figure S5) δ 170.10, 166.58, 158.15, 151.49, 143.92, 134.80, 134.77, 132.32, 131.18, 129.02, 128.65, 126.67, 77.61, 77.56, 77.36, 77.11, 73.68, 42.93, 27.61, 20.64, 20.54. ESI-TOF MS: Calculated for $C_{26}H_{34}FN_6O_2Si$ [M + H]$^+$ 509.24911 m/z, found 509.2147 m/z. Calculated [M + Na]$^+$ 531.22378 m/z, found 531.1967 m/z. Calculated [M + K]$^+$ 547.19771 m/z, found 547.1706 m/z.

Albumin trans-cyclooctene (albumin-TCO (**9**)). TCO-PEG4-NHS ester (5 mg, 9.7 µmol) in DMSO : 0.5 M borate buffer (1:1, 1 mL) pH 9 was added to a solution of bovine serum albumin **7** (3.5 mg, 53 nmol) in borate buffer pH 9.0 (1 mL). After 1 h, the reaction mixture was purified with a PD-10 size-exclusion column (GE Healthcare, Chicago, IL, USA), using ultrapure water as the eluent. The collected fractions were analyzed by a SEC-column HPLC, using 0.1 M PBS as an eluent, with a flow rate of 0.8 mL/min, for identifying the fractions containing albumin-TCO. The fractions containing albumin-TCO were lyophilized to afford **9** as white solid ($n = 3$). Albumin-TCO (1 mg) was dissolved in 1 mL of ultrapure water and analyzed with MALDI-TOF-MS (calculated for bovine serum albumin 66,338 Da, measured for albumin-TCO 82039-82265 Da).

4.2. Radiochemistry

No-carrier-added ^{18}F-Fluoride was produced in-house with Cyclone 10/5 cyclotron (IBA, Louvain-la-Neuve, Belgium) through a ^{18}O$(p,n)^{18}$F nuclear reaction, by bombarding H$_2$^{18}O with 10 MeV protons. The radiosynthesis was carried out in a semiautomatic synthesis unit (DM Automation), with an integrated preparative HPLC system for the purification of the radiotracer. The nucleophilic ^{18}F$^-$ was trapped on a Waters QMA Light ion-exchange cartridge, followed by elution with a basic K[^{18}F]FK2.2.2-complex solution. Water residue was evaporated azeotropically by adding anhydrous ACN, followed by heating, under a 40-mL/min argon flow.

4.2.1. Radiosynthesis of [^{18}F]6

In the one-step method, precursor **6**, dissolved in 500 µL of anhydrous acetonitrile, was added into the dried K[^{18}F]F/K2.2.2 and incubated for 2 min (25 °C). The reaction mixture was diluted with an additional 500 µL of anhydrous acetonitrile for the radio-TLC and radio-HPLC analysis.

In the two-step method, after evaporation of the solvent, SiFA (**5**) in anhydrous ACN (0.5 mL) was added into the reaction vial containing K[^{18}F]F/K2.2.2 and incubated at room temperature for 2 min. Tetrazine oxyamine (**3**) in 0.3 M anilinium acetate buffer pH 4.6 (200 µL) and ACN (50 µL) was added into the reaction mixture, and the reaction was further incubated at room temperature, for 15 min (RCY 74%, radio-TLC), and purified by preparative HPLC, providing 385 MBq of [^{18}F]6 at 115 min from end-of-bombardment (EOB, 14.8 ± 1.6% DCY) and 5 GBq/µmol at end-of-bombardment (EOS). The purified (RCP 98.8 ± 1.3%, $n = 16$) product [^{18}F]6 was formulated (10% ethanol, 0.5% Solutol HS 15 in 0.01 M PBS), sterile filtered (0.22 µm), and used as such for injections.

4.2.2. Radiosynthesis of [^{18}F]10

The formulated [^{18}F]6 was added onto the lyophilized albumin-TCO (350 µg) and incubated at room temperature for 15 min. The produced [^{18}F]fluoroalbumin ([^{18}F]10) was purified by centrifugation (10,000 g, Eppendorf Centrifuge 5430, Ag, Hamburg, Germany) with molecular weight cut-off (MWCO) filters (30K, VWR®, Radnor, PA, USA), with 0.01 M PBS as the eluent, sterile-filtered and used as such for injections. The apparent molar activity was 1.1 GBq/µmol of protein at EOS. The radio-HPLC chromatogram of the purified [^{18}F]10 is presented in the Supplementary Data (Figure S11).

4.3. In Vitro Experiments

4.3.1. LogD$_{pH7.4}$ Determination for [^{18}F]6

A total of 25 µL of [^{18}F]6 was added to a mixture of 1-octanol and 0.02 M PBS (pH 7.4) in a 1.5 mL microtube. The mixture was shaken mechanically (500 rpm) for 10 min and centrifuged (1000 g, 5 min), and the layers were separated. A sample of 500 µL of each layer were pipetted into pre-weighed polypropylene tubes and the activity in the samples was measured with a Wizard gamma counter. The distribution of [^{18}F]6 between octanol and PBS was calculated, according to

$$LogD_{7.4} = Log \frac{Ac_{OCT}}{Ac_{PBS}}$$

where Ac_{OCT} = activity concentration of octanol and Ac_{PBS} = activity concentration of PBS. $LogD_{pH7.4}$ = 1.56 ± 0.20 ($n = 5$).

4.3.2. In Vitro Stability and Plasma Protein Binding for [^{18}F]6

A total of 40 µL of [^{18}F]6 was incubated in 0.01 M PBS at room temperature, in a microtube for 90 min, with mixing (400 rpm). At selected time-points (t = 5, 30, 60 and 90 min, n = 1), the samples were injected into an HPLC with PDA- and radiodetector, for analysis. Radiotracer [^{18}F]6 (40 µL) was incubated in 50% human plasma (anonymous donor FFP-24 plasma provided by the Finnish Red Cross Blood Service, Helsinki, Finland) in PBS at 37 °C, and in mouse plasma (separated from CD-1 mouse blood). At selected time-points (5, 60, 120, and 240 min, n = 2 for each), 100 µL of the samples were taken, diluted with 200 µL of cold acetonitrile, and centrifuged at 10,000 g for 5 min. The radioactivity in the precipitated pellet (protein-bound fraction) and supernatant (free fraction) were measured with a gamma counter and 100 µL of the samples were injected into HPLC, for radio-HPLC analysis.

4.4. Biological Studies

All animal experiments were conducted under a project license approved by the National Board of Animal Experimentation in Finland (license number ESAVI/12132/04.10.07/2017). The animals were group-housed in standard polycarbonate cages, on aspen bedding, in a HEPA-filtered housing unit (UniProtect, Ehret, Emmendingen, Germany), with food (Envigo Teklad Global Diet 2016) and tap water, available ad libitum. Conditions were maintained at 21 ± 1 °C and 55 ± 15% relative humidity, with a 12:12 lighting cycle. The biodistribution studies were conducted in healthy, female CD-1 mice (weight 25–33 g, 11 to 12 weeks, Charles River). The radiotracers [^{18}F]6 and [^{18}F]10 were injected via the tail vein to CD-1 mice, in the following formulations—10% EtOH and 0.5% Solutol HS 15 in 0.01 M PBS pH 7.4 for [^{18}F]6 and 0.01 M PBS pH 7.4 for [^{18}F]10. At selected time-points (5, 60, 120, and 240 min), the mice were euthanized with CO2 asphyxiation, followed by cervical dislocation, and selected tissues were collected, weighed, and the radioactivity was measured on an automated gamma counter.

4.4.1. Biodistribution of [^{18}F]6

14.3 ± 0.5 MBq (n = 12) (25.4 ± 1.4 µg, 44.2 ± 2.4 nmol) of 96.2% pure [^{18}F]6 was injected into the tail vein of healthy female CD-1 mice (n = 3 per time-point), to evaluate the biodistribution and stability of the tracer in vivo. The mice were euthanized at selected time-points (t = 5, 60, 120, and 240 min) and the tissues were collected and measured with a gamma counter, as described above. For metabolite studies, blood from a cardiac puncture was collected into an Eppendorf tube containing 2 µL of 1% heparin solution in 0.9% NaCl (aq.) and centrifuged at 1000 g, for 10 min, to separate the plasma from the blood cells. Cold acetonitrile (twice the volume of separated plasma) was added into the plasma and centrifuged (at 10 000 g for 5 min) to precipitate the proteins. A small sample (4 µL) was applied onto a silica TLC and analyzed with digital autoradiography.

4.4.2. Biodistribution of [^{18}F]10

A total of 0.6 ± 0.1 MBq (n = 15) (43.2 ± 1.4 µg of protein) of [^{18}F]fluoroalbumin was injected into the tail vein of female CD-1 mice (n = 3–4 per time point). The mice were euthanized at selected time-points and the tissues were collected and measured with a gamma counter, as described above. Blood from a cardiac puncture was collected into an Eppendorf tube containing 2 µL of 1% heparin solution in 0.9%

NaCl (aq) and was pretreated before analysis, as described above. After centrifugation, a small sample (4 µL) was applied onto a silica TLC plate and 100 µL of the supernatant was injected into HPLC for radiometabolite analysis. For the radiometabolite studies of [^{18}F]10, the blood was collected into an 1.5-mL microtube containing 1% heparin solution in 0.9% NaCl (aq), and was centrifuged (at 1000 g for 10 min). Plasma was separated from the cell pellet and added onto a 30-kDa molecular weight cut-off (MWCO) centrifugal filter (VWR® Radnor, PA, USA) and centrifuged (at 6500 rpm for 10 min). The filter with over 30 kDa ^{18}F-labeled molecules and the microtube containing the below-30-kDa ^{18}F-labeled molecules in the eluate, were measured with a gamma counter, to determine the percentage of small molecular weight metabolites from over 30 kDa molecules representing the intact BSA in the plasma.

4.4.3. Statistical Analysis

Statistical significance of ^{18}F bone accumulation after intravenous administration and the tracer blood circulation time were analyzed using an unpaired *t*-test (GraphPad Prism 8.0.1). The values presented in the synthesis, biodistribution studies, and Log*D*-measurements are mean ± standard deviation.

5. Conclusions

Despite of promising hydrolytic stability in vitro, [^{18}F]SiFA-Tz ([^{18}F]6) demonstrated fast defluorination in vivo, after intravenous administration in CD-1 mice limiting its utility as a standalone radiotracer, for pretargeted PET imaging. However, the fluorine-18 label in the biomacromolecular radiotracer [^{18}F]fluoroalbumin ([^{18}F]10), which was radiolabeled as a proof-of-concept model compound with [^{18}F]6, was found to be metabolically more stable, suggesting the utility of [^{18}F]SiFA-Tz as a prosthetic group for in vitro radiolabeling of biomolecules of higher molecular weight. Based on these findings, the structure of [^{18}F]SiFA-Tz warrants further optimization, before it can be considered for use as a radiolabeling tool for low molecular weight biomolecules or as a tracer for pretargeted PET imaging.

Supplementary Materials: The following are available online, NMR spectra, HPLC chromatograms, autoradiography profiles, and biodistributions of synthesized and radiolabeled compounds. Supplementary data are provided as a separate .pdf document.

Author Contributions: Conceptualization, S.O., M.S., K.H., K.W., and A.J.A.; methodology, S.O., S.I., M.S., K.H., K.W., and A.J.A.; software, S.O.; validation, S.O., S.I.; formal analysis, S.O.; investigation, S.O., S.I.; resources, M.S. and A.J.A.; data curation, S.O.; writing—original draft preparation, S.O.; writing—review and editing, S.O., S.I., M.S., K.H., K.W., and A.J.A.; visualization, S.O.; supervision, M.S., K.H., K.W., and A.J.A.; project administration, M.S. and A.J.A.; funding acquisition M.S. and A.J.A. All authors have read and agreed to the published version of the manuscript.

Funding: This work was supported by the Academy of Finland (decision numbers 306239, 298481, 278056) and a University of Helsinki three-year research grant. Open access funding provided by University of Helsinki.

Acknowledgments: The authors would like to thank Osku Alanen and Outi Keinänen for assistance in the experiments. The authors thank Eliza Lambidis and Petri Heinonen for the use of ESI–TOF MS and Dr. Sami Heikkinen for the use of NMR (600 MHz).

Conflicts of Interest: The authors declare no conflict of interest.

References

1. Jacobson, O.; Kiesewetter, D.O.; Chen, X. Fluorine-18 radiochemistry, labeling strategies and synthetic routes. *Bioconjug. Chem.* **2015**, *26*, 1–18. [CrossRef] [PubMed]
2. Deng, X.; Rong, J.; Wang, L.; Vasdev, N.; Zhang, L.; Josephson, L.; Liang, S.H. Chemistry for Positron Emission Tomography: Recent Advances in ^{11}C-, ^{18}F-, ^{13}N-, and ^{15}O-Labeling Reactions. *Angew. Chem. Int. Ed. Engl.* **2019**, *58*, 2580–2605. [CrossRef] [PubMed]
3. Preshlock, S.; Tredwell, M.; Gouverneur, V. ^{18}F-Labeling of Arenes and Heteroarenes for Applications in Positron Emission Tomography. *Chem. Rev.* **2016**, *116*, 719–766. [CrossRef]

4. Brooks, A.F.; Topczewski, J.J.; Ichiishi, N.; Sanford, M.S.; Scott, P.J. Late-stage [^{18}F]Fluorination: New Solutions to Old Problems. *Chem. Sci.* **2014**, *5*, 4545–4553. [CrossRef] [PubMed]
5. Oliveira, B.L.; Guo, Z.; Bernardes, G.J.L. Inverse electron demand Diels-Alder reactions in chemical biology. *Chem. Soc. Rev.* **2017**, *46*, 4895–4950. [CrossRef] [PubMed]
6. Rostovtsev, V.V.; Green, L.G.; Fokin, V.V.; Sharpless, K.B. A stepwise huisgen cycloaddition process: Copper(I)-catalyzed regioselective "ligation" of azides and terminal alkynes. *Angew. Chem. Int. Ed. Engl.* **2002**, *41*, 2596–2599. [CrossRef]
7. Tornoe, C.W.; Christensen, C.; Meldal, M. Peptidotriazoles on solid phase: [1,2,3]-triazoles by regiospecific copper(I)-catalyzed 1,3-dipolar cycloadditions of terminal alkynes to azides. *J. Org. Chem.* **2002**, *67*, 3057–3064. [CrossRef]
8. Agard, N.J.; Prescher, J.A.; Bertozzi, C.R. A strain-promoted [3 + 2] azide-alkyne cycloaddition for covalent modification of biomolecules in living systems. *J. Am. Chem. Soc.* **2004**, *126*, 15046–15047. [CrossRef]
9. Carboni, R.A.; Lindsey, R.V. Reactions of Tetrazines with Unsaturated Compounds - a New Synthesis of Pyridazines. *J. Am. Chem. Soc.* **1959**, *81*, 4342–4346. [CrossRef]
10. Edem, P.E.; Sinnes, J.P.; Pektor, S.; Bausbacher, N.; Rossin, R.; Yazdani, A.; Miederer, M.; Kjaer, A.; Valliant, J.F.; Robillard, M.S.; et al. Evaluation of the inverse electron demand Diels-Alder reaction in rats using a scandium-44-labelled tetrazine for pretargeted PET imaging. *Ejnmmi. Res.* **2019**, *9*, 49. [CrossRef]
11. Houghton, J.L.; Membreno, R.; Abdel-Atti, D.; Cunanan, K.M.; Carlin, S.; Scholz, W.W.; Zanzonico, P.B.; Lewis, J.S.; Zeglis, B.M. Establishment of the In Vivo Efficacy of Pretargeted Radioimmunotherapy Utilizing Inverse Electron Demand Diels-Alder Click Chemistry. *Mol. Cancer Ther.* **2017**, *16*, 124–133. [CrossRef] [PubMed]
12. Lappchen, T.; Rossin, R.; van Mourik, T.R.; Gruntz, G.; Hoeben, F.J.M.; Versteegen, R.M.; Janssen, H.M.; Lub, J.; Robillard, M.S. DOTA-tetrazine probes with modified linkers for tumor pretargeting. *Nucl. Med. Biol.* **2017**, *55*, 19–26. [CrossRef] [PubMed]
13. Membreno, R.; Cook, B.E.; Fung, K.; Lewis, J.S.; Zeglis, B.M. Click-Mediated Pretargeted Radioimmunotherapy of Colorectal Carcinoma. *Mol. Pharm.* **2018**, *15*, 1729–1734. [CrossRef] [PubMed]
14. Meyer, J.P.; Houghton, J.L.; Kozlowski, P.; Abdel-Atti, D.; Reiner, T.; Pillarsetty, N.V.; Scholz, W.W.; Zeglis, B.M.; Lewis, J.S. ^{18}F-Based Pretargeted PET Imaging Based on Bioorthogonal Diels-Alder Click Chemistry. *Bioconjug. Chem.* **2016**, *27*, 298–301. [CrossRef] [PubMed]
15. Li, Z.; Cai, H.; Hassink, M.; Blackman, M.L.; Brown, R.C.; Conti, P.S.; Fox, J.M. Tetrazine-trans-cyclooctene ligation for the rapid construction of ^{18}F-labeled probes. *Chem. Commun. (Camb)* **2010**, *46*, 8043–8045. [CrossRef] [PubMed]
16. Fersing, C.; Bouhlel, A.; Cantelli, C.; Garrigue, P.; Lisowski, V.; Guillet, B. A Comprehensive Review of Non-Covalent Radiofluorination Approaches Using Aluminum [^{18}F]fluoride: Will [^{18}F]AlF Replace ^{68}Ga for Metal Chelate Labeling? *Molecules* **2019**, *24*, 2866. [CrossRef] [PubMed]
17. Keinänen, X.G.; Li, N.K.; Chenna, D.; Lumen, J.; Ott, C.F.M.; Molthoff, M.; Sarparanta, K.; Helariutta, T.; Vuorinen, A.D.; Windhorst, A.J.; et al. New Highly Reactive and Low Lipophilicity Fluorine18 Labeled Tetrazine Derivative for Pretargeted PET Imaging. *ACS Med. Chem. Lett.* **2016**, *7*, 62–66. [CrossRef]
18. Schirrmacher, E.; Wangler, B.; Cypryk, M.; Bradtmoller, G.; Schafer, M.; Eisenhut, M.; Jurkschat, K.; Schirrmacher, R. Synthesis of p-(di-tert-butyl[^{18}F]fluorosilyl)benzaldehyde ([^{18}F]SiFA-A) with high specific activity by isotopic exchange: A convenient labeling synthon for the ^{18}F-labeling of N-aminooxy derivatized peptides. *Bioconjug. Chem.* **2007**, *18*, 2085–2089. [CrossRef]
19. Berke, S.; Kampmann, A.L.; Wuest, M.; Bailey, J.J.; Glowacki, B.; Wuest, F.; Jurkschat, K.; Weberskirch, R.; Schirrmacher, R. ^{18}F-Radiolabeling and In Vivo Analysis of SiFA-Derivatized Polymeric Core-Shell Nanoparticles. *Bioconjug. Chem.* **2018**, *29*, 89–95. [CrossRef]
20. Niedermoser, S.; Chin, J.; Wangler, C.; Kostikov, A.; Bernard-Gauthier, V.; Vogler, N.; Soucy, J.P.; McEwan, A.J.; Schirrmacher, R.; Wangler, B. In Vivo Evaluation of ^{18}F-SiFAlin-Modified TATE: A Potential Challenge for ^{68}Ga-DOTATATE, the Clinical Gold Standard for Somatostatin Receptor Imaging with PET. *J. Nucl. Med.* **2015**, *56*, 1100–1105. [CrossRef]
21. Wangler, B.; Quandt, G.; Iovkova, L.; Schirrmacher, E.; Wangler, C.; Boening, G.; Hacker, M.; Schmoeckel, M.; Jurkschat, K.; Bartenstein, P.; et al. Kit-like ^{18}F-labeling of proteins: Synthesis of 4-(di-tert-butyl[^{18}F]fluorosilyl)benzenethiol (Si[^{18}F]FA-SH) labeled rat serum albumin for blood pool imaging with PET. *Bioconjug. Chem.* **2009**, *20*, 317–321. [CrossRef] [PubMed]

22. Wangler, C.; Waser, B.; Alke, A.; Iovkova, L.; Buchholz, H.G.; Niedermoser, S.; Jurkschat, K.; Fottner, C.; Bartenstein, P.; Schirrmacher, R.; et al. One-step ^{18}F-labeling of carbohydrate-conjugated octreotate-derivatives containing a silicon-fluoride-acceptor (SiFA): In vitro and in vivo evaluation as tumor imaging agents for positron emission tomography (PET). *Bioconjug. Chem.* **2010**, *21*, 2289–2296. [CrossRef] [PubMed]
23. Zhu, J.; Chin, J.; Wangler, C.; Wangler, B.; Lennox, R.B.; Schirrmacher, R. Rapid ^{18}F-labeling and loading of PEGylated gold nanoparticles for in vivo applications. *Bioconjug. Chem.* **2014**, *25*, 1143–1150. [CrossRef] [PubMed]
24. Zhu, J.; Li, S.; Wangler, C.; Wangler, B.; Lennox, R.B.; Schirrmacher, R. Synthesis of 3-chloro-6-((4-(di-tertbutyl[18]F]fluorosilyl)-benzyl)oxy)-1,2,4,5-tetrazine (^{18}F]SiFA-OTz) for rapid tetrazine-based ^{18}F-radiolabeling. *Chem. Commun.* **2015**, *51*, 12415–12418. [CrossRef] [PubMed]
25. Keinänen, O.; Partelova, D.; Alanen, O.; Antopolsky, M.; Sarparanta, M.; Airaksinen, A.J. Efficient cartridge purification for producing high molar activity ^{18}F]fluoro-glycoconjugates via oxime formation. *Nucl. Med. Biol.* **2018**, *67*, 27–35. [CrossRef]
26. Keinänen, O.; Fung, K.; Pourat, J.; Jallinoja, V.; Vivier, D.; Pillarsetty, N.K.; Airaksinen, A.J.; Lewis, J.S.; Zeglis, B.M. Sarparanta, M. Pretargeting of internalizing trastuzumab and cetuximab with a ^{18}F-tetrazine tracer in xenograft models. *EJNMMI Res.* **2017**, *7*, 95. [CrossRef]
27. Keinänen, O.; Mäkilä, E.M.; Lindgren, R.; Virtanen, H.; Liljenback, H.; Oikonen, V.; Sarparanta, M.; Molthoff, C.; Windhorst, A.D.; Roivainen, A.; et al. Pretargeted PET Imaging of trans-Cyclooctene-Modified Porous Silicon Nanoparticles. *ACS Omega* **2017**, *2*, 62–69. [CrossRef]
28. Kolmel, D.K.; Kool, E.T. Oximes and Hydrazones in Bioconjugation: Mechanism and Catalysis. *Chem. Rev.* **2017**, *117*, 10358–10376. [CrossRef]
29. Fauber, B.P.; Rene, O.; Boenig, G.d.L.; Burton, B.; Deng, Y.; Eidenschenk, C.; Everett, C.; Gobbi, A.; Hymowitz, S.G. Reduction in lipophilicity improved the solubility, plasma-protein binding, and permeability of tertiary sulfonamide RORc inverse agonists. *Bioorg. Med. Chem. Lett.* **2014**, *24*, 3891–3897. [CrossRef]
30. Piert, M.; Zittel, T.T.; Becker, G.A.; Jahn, M.; Stahlschmidt, A.; Maier, G.; Machulla, H.J.; Bares, R. Assessment of porcine bone metabolism by dynamic. *J. Nucl. Med.* **2001**, *42*, 1091–1100.
31. Mossine, A.V.; Brooks, A.F.; Ichiishi, N.; Makaravage, K.J.; Sanford, M.S.; Scott, P.J. Development of Customized ^{18}F]Fluoride Elution Techniques for the Enhancement of Copper-Mediated Late-Stage Radiofluorination. *Sci. Rep.* **2017**, *7*, 233. [CrossRef] [PubMed]
32. Kallinen, A.M.; Sarparanta, M.P.; Liu, D.F.; Mäkilä, E.M.; Salonen, J.J.; Hirvonen, J.T.; Santos, H.A.; Airaksinen, A.J. In Vivo Evaluation of Porous Silicon and Porous Silicon Solid Lipid Nanocomposites for Passive Targeting and Imaging. *Mol. Pharmaceut.* **2014**, *11*, 2876–2886. [CrossRef] [PubMed]
33. Matsumura, Y.; Maeda, H. A new concept for macromolecular therapeutics in cancer chemotherapy mechanism of tumoritropic accumulation of proteins and the antitumor agent smancs. *Cancer Res.* **1986**, *6*, 6387–6392.
34. Dialer, L.O.; Selivanova, S.V.; Muller, C.J.; Muller, A.; Stellfeld, T.; Graham, K.; Dinkelborg, L.M.; Kramer, S.D.; Schibli, R.; Reiher, M.; et al. Studies toward the development of new silicon containing building blocks for the direct ^{18}F labeling of peptides. *J. Med. Chem.* **2013**, *56*, 7552–7563. [CrossRef] [PubMed]
35. Lindner, S.; Michler, C.; Leidner, S.; Rensch, C.; Wangler, C.; Schirrmacher, R.; Bartenstein, P.; Wangler, B. Synthesis and in vitro and in vivo evaluation of SiFA-tagged bombesin and RGD peptides as tumor imaging probes for positron emission tomography. *Bioconjug. Chem.* **2014**, *25*, 738–749. [CrossRef] [PubMed]
36. Rosa-Neto, P.; Wangler, B.; Iovkova, L.; Boening, G.; Reader, A.; Jurkschat, K.; Schirrmacher, E. [18]F]SiFA isothiocyanate: A new highly effective radioactive labeling agent for lysine-containing proteins. *Chembiochem* **2009**, *10*, 1321–1324. [CrossRef]
37. Iovkova, L.; Wangler, B.; Schirrmacher, E.; Schirrmacher, R.; Quandt, G.; Boening, G.; Schurmann, M.; Jurkschat, K. para-Functionalized aryl-di-tert-butylfluorosilanes as potential labeling synthons for ^{18}F-radiopharmaceuticals. *Chemistry* **2009**, *15*, 2140–2147. [CrossRef]

Sample Availability: Not available.

© 2020 by the authors. Licensee MDPI, Basel, Switzerland. This article is an open access article distributed under the terms and conditions of the Creative Commons Attribution (CC BY) license (http://creativecommons.org/licenses/by/4.0/).

Article

Evaluation of a ^{68}Ga-Labeled DOTA-Tetrazine as a PET Alternative to ^{111}In-SPECT Pretargeted Imaging

Patricia E. Edem [1,2,3,†], Jesper T. Jørgensen [1,2,†], Kamilla Nørregaard [1,2], Rafaella Rossin [4], Abdolreza Yazdani [5,6], John F. Valliant [5], Marc Robillard [4], Matthias M. Herth [1,3,*] and Andreas Kjaer [1,2,*]

1. Department of Clinical Physiology, Nuclear Medicine & PET, Rigshospitalet, Blegdamsvej 9, 2100 Copenhagen, Denmark; patredem@gmail.com (P.E.E.); jespertj@sund.ku.dk (J.T.J.); kamilla.noerregaard@gmail.com (K.N.)
2. Cluster for Molecular Imaging, Department of Biomedical Sciences, University of Copenhagen, Blegdamsvej 3, 2200 Copenhagen, Denmark
3. Department of Drug Design and Pharmacology, University of Copenhagen, Jagtvej 162, 2100 Copenhagen, Denmark
4. Tagworks Pharmaceuticals, Geert Grooteplein Zuid 10, 6525 GA Nijmegen, The Netherlands; raffaella.rossin@tagworkspharma.com (R.R.); marc.robillard@tagworkspharma.com (M.R.)
5. Department of Chemistry and Chemical Biology, McMaster University, 1280 Main St West, Hamilton, ON L8S 4M1, Canada; ayazdani.mcmaster@gmail.com (A.Y.); valliant@mcmaster.ca (J.F.V.)
6. Pharmaceutical Chemistry and Radiopharmacy Department, School of Pharmacy, Shahid Beheshti University of Medical Sciences, PO Box 14155–6153, Tehran, Iran
* Correspondence: matthias.herth@sund.ku.dk (M.M.H.); akjaer@sund.ku.dk (A.K.)
† Patricia E. Edem and Jesper T. Jørgensen contributed equally to this work.

Academic Editors: Anne Roivainen and Xiang-Guo Li
Received: 23 December 2019; Accepted: 15 January 2020; Published: 22 January 2020

Abstract: The bioorthogonal reaction between a tetrazine and strained trans-cyclooctene (TCO) has garnered success in pretargeted imaging. This reaction was first validated in nuclear imaging using an ^{111}In-labeled 1,4,7,10-tetraazacyclododecane-1,4,7,10-tetraacetic acid (DOTA)-linked bispyridyl tetrazine (Tz) ([^{111}In]In-DOTA-PEG$_{11}$-Tz) and a TCO functionalized CC49 antibody. Given the initial success of this Tz, it has been paired with TCO functionalized small molecules, diabodies, and affibodies for in vivo pretargeted studies. Furthermore, the single photon emission tomography (SPECT) radionuclide, ^{111}In, has been replaced with the β-emitter, ^{177}Lu and α-emitter, ^{212}Pb, both yielding the opportunity for targeted radiotherapy. Despite use of the 'universal chelator', DOTA, there is yet to be an analogue suitable for positron emission tomography (PET) using a widely available radionuclide. Here, a ^{68}Ga-labeled variant ([^{68}Ga]Ga-DOTA-PEG$_{11}$-Tz) was developed and evaluated using two different in vivo pretargeting systems (Aln-TCO and TCO-CC49). Small animal imaging and ex vivo biodistribution studies were performed and revealed target specific uptake of [^{68}Ga]Ga-DOTA-PEG$_{11}$-Tz in the bone (3.7 %ID/g, knee) in mice pretreated with Aln-TCO and tumor specific uptake (5.8 %ID/g) with TCO-CC49 in mice bearing LS174 xenografts. Given the results of this study, [^{68}Ga]Ga-DOTA-PEG$_{11}$-Tz can serve as an alternative to [^{111}In]In-DOTA-PEG$_{11}$-Tz.

Keywords: tetrazine ligation; PET; SPECT; gallium-68; indium-11

1. Introduction

Radioimmunoconjugates have gained interest as important tools in the treatment and management of cancer [1–3]. Their use has been applied in both radiotherapy and nuclear molecular imaging through single photon emission computed tomography (SPECT) and positron emission tomography (PET); where gamma emitting or positron emitting radionuclides are linked to a monoclonal antibody (mAb). Antibodies have the advantage of high target affinity, yet their use is hampered by poor target to non-target (T:NT) ratios due to slow target accumulation and blood clearance when compared to other targeting vectors at early time points [4]. This requires a long interval between the administration of the radioimmunoconjugate and image acquisition, thus necessitating the use of long-lived radionuclides such as indium-111 ($t_{1/2}$ = 67.3 h) or zirconium-89 ($t_{1/2}$ = 78.4 h) [5]. Pretargeted imaging can overcome these limitations by offering the temporal separation between administration of the mAb and the signaling agent, through a stepwise procedure.

In pretargeted imaging, a primary agent is administered and allowed to accumulate at the target site and clear from non-targeted tissues. Afterwards, a low-molecular weight radiolabeled secondary agent possessing high affinity (or reactivity) towards the primary agent is administered (Figure 1) [6]. In this way, the non-radioactive primary agent may accumulate at the target site and clear from the circulation without emitting any unnecessary radiation, while the radioactive secondary agent has the ability to clear from the circulation much faster after administration). To date, pretargeted imaging has only been validated in the clinic using the hapten-bispecific antibody, and the biotin–streptavidin interactions; however, these methods often give rise to an immunogenic response in patients [5]. The bioorthogonal chemical reaction between a tetrazine (Tz) and a trans-cyclooctene (TCO) is a chemical alternative that can mitigate these challenges. For this reason, radiolabeled tetrazines have been used extensively in pretargeted imaging with TCO functionalized mAbs (Figure 1) [6–16].

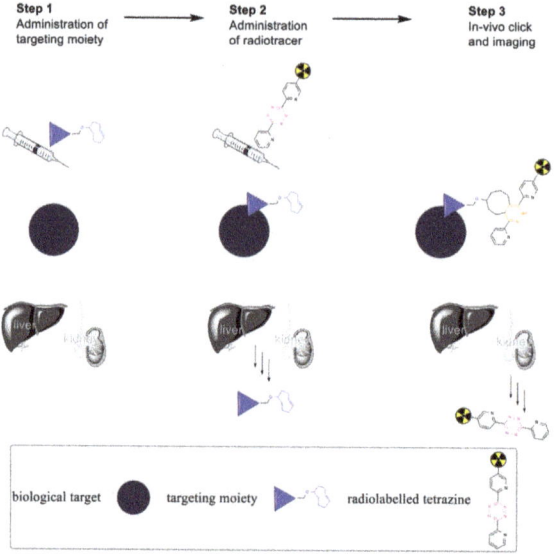

Figure 1. General strategy for in vivo pre-targeted imaging. In Step 1, the TCO functionalized targeted moiety is administered to the living system. In Step 2, the radiolabeled tetrazine is administered. The targeting moiety has been allowed to accumulate at the target site while any excess has cleared via the renal and/or hepatobiliary systems. In the third step in vivo imagining occurs. This is after the radioactivity has accumulated at the target site via in vivo click chemistry and any excess has cleared through the renal system.

The first radiolabeled tetrazine applied in pretargeted imaging used a 1,4,7,10-tetraazacyclododecane-1,4,7,10-tetraacetic acid (DOTA) functionalized bispyridyl tetrazine (**1**) (Figure 2) [6]. The indium labeled variant, [^{111}In]**2**, was used alongside a TCO modified CC49 mAb that targets the tumor associated glycoprotein 72 (TAG-72) antigen in human colorectal cancer cells to image tumors in mice [6]. Following this seminal work, radiolabeled variants of **1** have been used in combination with different

TCO functionalized primary agents such as antibodies [6,12–14,17,18], affibodies [19], diabodies [15], antibody drug conjugates [20] and small molecules [21,22]. These studies often incorporated long-lived radionuclides such as indium-111 [12–15,19,20], lutetium-177 ($t_{1/2}$ = 6.6 d) [12–15,18–20], and lead-212 ($t_{1/2}$ = 10.6 h) [17]. Despite DOTA forming stable complexes with a variety of metals, there are no reports of pretargeted imaging using TCO functionalized CC49 mAbs and **1** with a readily available PET based radionuclide. Given the advantage that the pretargeting strategy provides for slow clearing targeting agents, there is also an unmet need to produce an analogue of **1** radiolabeled with a short-lived radionuclide for pretargeted tumor imaging.

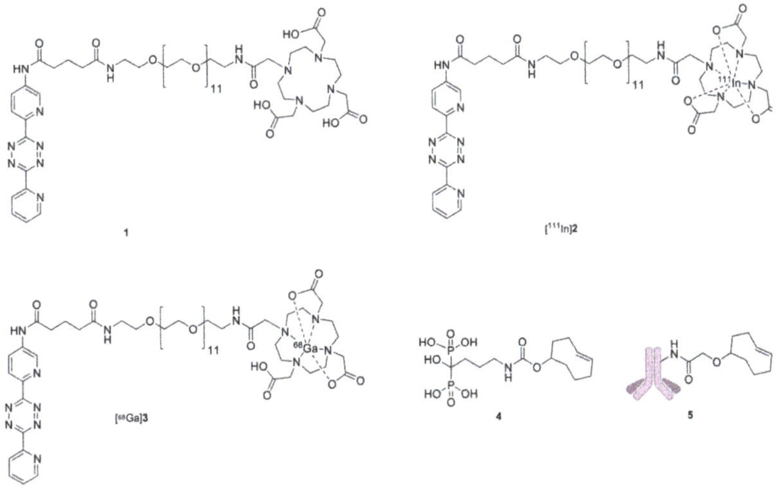

Figure 2. Pretargeting components used in this study: DOTA-PEG$_{11}$-Tz (**1**); [^{111}In]In-DOTA-PEG$_{11}$-Tz (**2**); [^{68}Ga]Ga-DOTA-PEG$_{11}$-Tz (**3**); TCO functionalized alendronate (**4**); CC49-TCO (**5**).

Here, we introduce a gallium-68 labeled variant of **1** ([^{68}Ga]**3**) as an alternative to [^{111}In]**2** (Figure 2). In order to evaluate [^{68}Ga]**3** in mice two different primary agents were selected. We used two primary agents to demonstrate that the ability of [^{68}Ga]**3** or [^{111}In]**2** to bind to the respective targeting vector is independent of the targeting vector itself. TCO functionalized alendronate (**4**) was selected as the first targeting vector (Figure 2). It binds to exposed hydroxyapatite surfaces in bone [23] and has been used in pretargeting experiments to evaluate tetrazines bearing the short-lived radioisotopes, technetium-99m, ($t_{1/2}$ = 6 h) and scandium-44 ($t_{1/2}$ = 4 h) [21–26]. The second agent was based on the mAb CC49, which targets the TAG 72 antigen in human colorectal cancer cells. The TCO modified CC49 (**5**) has been used with [^{111}In]**2** in pretargeted imaging (Figure 2) [13,14]. Initially, SPECT/CT images were obtained using [^{111}In]**2** to compare the image quality at early and late time points prior to evaluating [^{68}Ga]**3** at early time-points with PET/CT.

2. Results

2.1. Radiolabeling and In Vitro Stability

Radiosynthesis of [^{111}In]**2** [6] was performed in high radiochemical yield (RCY) (93–100%) with high radiochemical purity (RCP > 95% by radio-HPLC) (Supplementary Figure S1). The apparent molar activity (A_m) ranged from 6–13 GBq/μmol at the end of synthesis (EOS). The octanol-water partition coefficient (logD) of [^{111}In]**2** was −3.659 ± 0.005.

Radiosynthesis of [^{68}Ga]**3** was accomplished in a decay corrected (d.c.) RCY of 48–78% and RCP (> 95% by radio-HPLC) (Supplementary Figure S2). The apparent A_m ranged from 1–2 GBq/μmol EOS. The logD of [^{68}Ga]**3** was −2.19 ± 0.03. [^{68}Ga]**3**, was able to react quantitatively with **4** after 1.5 h in solution (Supplementary Figure S3A) and remained stable for at least 1.5 h at room temperature in 10% ethanol in phosphate buffered saline (PBS) (Supplementary Figure S3B).

2.2. Pretargeted SPECT/CT Bone Imaging

Prior to evaluating [^{68}Ga]3 in vivo, pretargeted bone imaging was performed using [^{111}In]2 and 4 (Figure 3). Groups of mice were pre-treated with 4 1 h prior to administration of [^{111}In]2. This gave a total in vivo TCO:Tz ratio of 76:1. As a control, mice that did not receive 4 were also administered [^{111}In]2. Small animal SPECT/CT images were obtained 2 and 22 h post injection (p.i.). Radioactivity accumulation was observed in the shoulders and the knees in the mice pre-treated with 4 at both time points (see representative images in Figure 3A). Regions of interest were manually drawn on target tissues and the percent injected dose per gram of tissue (%ID/g) was extracted after dose and decay correction. The uptake in muscle was included as a measure for background accumulation. Uptake in regions of interest placed on heart tissue and used as a surrogate for blood radioactivity content. The image derived analysis showed that the mean uptake in the shoulders (7.0 ± 1.9 %ID/g 2 h p.i.; 5.0 ± 1.2 %ID/g 22 h p.i.) and knees (7.7 ± 2.1 %ID/g 2 h; 5.6 ± 1.4 %ID/g 22 h p.i.), was substantially higher at both time points than the mean uptake in the muscle (0.16 ± 0.05 %ID/g) and the heart (0.13 ± 0.02 %ID/g) (Table 1). Overall, favorable target-to-background ratios (> 40) were achieved as early as 2 h p.i. In contrast, for the non-treated animals the mean uptake values for the shoulders (0.06 ± 0.03 %ID/g) and knees (0.06 ± 0.03 %ID/g) were lower than the muscle (0.2 ± 0.1 %ID/g) and the heart (0.12 ± 0.01 %ID/g) 2 h p.i. (Supplementary Table S1). Given that the image contrast was quite good for the pre-treated mice 2 h p.i., these results indicate that 4 would be a suitable primary agent for in vivo pretargeting with the short-lived radionuclide, gallium-68.

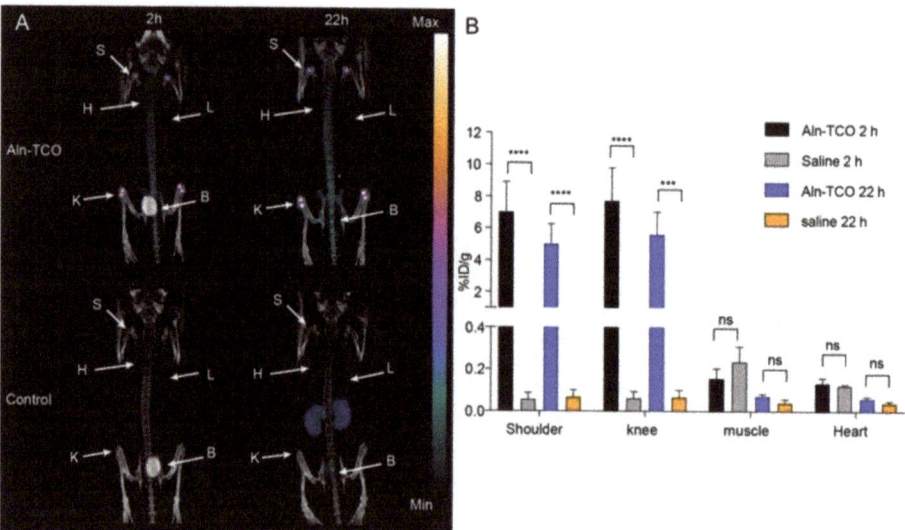

Figure 3. Pretargeted SPECT/CT bone imaging of [^{111}In]2 targeting TCO functionalized alendronate (4), which was administered 1 h prior to the administration of [^{111}In]2. (**A**) Respective SPECT/CT images after 2 h and 22 h. (**B**) Image-derived quantitative analysis of tissue uptake. ***, $p < 0.001$; and ****, $p < 0.0001$. Each image is scaled between its minimum and maximum pixel intensity.

Table 1. Uptake values in selected tissues from SPECT/CT scans 2 h and 22 h after injection of [^{111}In]2 in healthy BALB/c mice ($n = 3$) pretreated with **4** and from PET/CT scans and ex vivo (gammacounter) 2 h after injection of [^{68}Ga]3 in healthy BALB/c mice ($n = 4$) pretreated with **4**. Data is given as mean ± standard error of mean (SEM). * Image-derived uptake in heart from SPECT and PET images used as a surrogate for the blood radioactivity content. ** Not measured ex vivo.

	Pretargeted [^{111}In]2 (%ID/g)		Pretargeted [^{68}Ga]3 (%ID/g)	
	SPECT 2 h	SPECT 22 h	PET 2 h	Ex Vivo 2 h
Shoulder	7.0 ± 1.9	5.0 ± 1.2	1.20 ± 0.01	2.3 ± 0.5
Knee	7.7 ± 2.1	5.6 ± 1.4	2.0 ± 0.2	3.7 ± 0.6
Muscle	0.16 ± 0.05	0.07 ± 0.01	0.23 ± 0.01	0.10 ± 0.04
Heart *	0.13 ± 0.02	0.06 ± 0.01	0.36 ± 0.05	-**

2.3. Pretargeted PET/CT Bone Imaging

In a similar fashion [^{68}Ga]3 was evaluated in pretargeted PET imaging using **4** as the primary agent. Mice were injected with **4** 1 h prior to administration of [^{68}Ga]3, giving a total in vivo TCO:Tz ratio of 25:1. Mice that that did not receive **4** were also given [^{68}Ga]3 as a control. PET/CT images were obtained 2 h p.i. In the treatment group, the shoulders and knees were clearly visualized (Figure 4). The uptake was quantified and the mean uptake values in the shoulders and knees of the pre-treated mice were found to be 1.20 ± 0.01 %ID/g and 2.0 ± 0.2 %ID/g, respectively (Figure 4B; Table S1). High radioactivity uptake was also observed in the bladder, similar to what was observed in the SPECT/CT images 2 h after administering [^{111}In]2. There was no substantial uptake in the bone when there was no pre-treatment with **4** (≥ 0.4 %ID/g) (Supplementary Table S1). Although the target areas were clearly visualized, the PET derived target-to-background ratios were lower (< 8.6) than those obtained from the analogous SPECT experiment.

Following the last scan shoulder, knee, and muscle tissue was collected and counted for radioactivity. The mean uptake values were 2.3 ± 0.5 %ID/g for the shoulders and 3.7 ± 0.6 %ID/g for the knees. The uptake in the muscle was quite low (0.10 ± 0.04 %ID/g) (Table 1). Overall, pretargeting using **4** provided a fairly good image contrast with low background uptake using both [^{111}In]2 and [^{68}Ga]3.

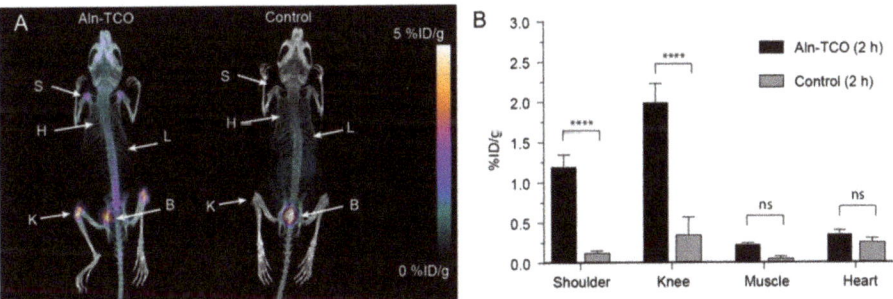

Figure 4. Pretargeted PET/CT bone imaging of [^{68}Ga]3 targeting TCO functionalized alendronate (**4**), which was administered 1 h prior to the administration of [^{68}Ga]3. (**A**) Respective PET/CT images after 2 h. (**B**) Image-derived quantitative analysis of tissue uptake. **** $p < 0.0001$.

2.4. Pretargeted SPECT/CT Tumor Imaging

Given that we were able to obtain high contrast images with [^{68}Ga]3 using **4** as the primary agent, we were encouraged to evaluate this tetrazine with a slow clearing primary agent. Prior to evaluating [^{68}Ga]3, [^{111}In]2 was tested as a positive control alongside **5**. Mice bearing LS174T tumor xenografts in the left flank were injected intravenously with **5** one day prior to administration of [^{111}In]**2**. The total in vivo TCO:Tz ratio was 1:2. As a control, mice bearing the LS174T tumor xenografts, without any pre-treatment, were also injected with [^{111}In]**2**. SPECT/CT images were obtained 2 and 22 h p.i. Representative SPECT/CT images and the quantified mean uptake values are shown in Figure 5. The tumor was clearly visualized in mice pre-treated with **5** (Figure 5A) at both time points, although the background was quite high at 2 h p.i. The mean heart uptake (8.7 ± 0.4 %ID/g) was used as a surrogate for activity in the blood pool and was found to be almost twice as high as the mean uptake determined for the tumor (4.8 ± 0.4 %ID/g) (Figure 5B; Table 2). After 22 h, the mean uptake was reduced in all tissues except for the tumors, where the uptake value increased to 12.3 ± 0.6 %ID/g. The tumor-to-background ratios did increase overtime with a maximal value of 9.2 for the tumor-to-muscle ratio 22 h p.i. (Table 2). The high background activity at 2 h p.i. combined with the increased tumor uptake at 22 h p.i. indicates a portion of **5** remained in the circulation and then reacted with [^{111}In]**2** in the blood following administration. In contrast, there was no substantial uptake in the tumor, muscle, or heart in the control animals (< 0.6 %ID/g) (Supplementary Table S2).

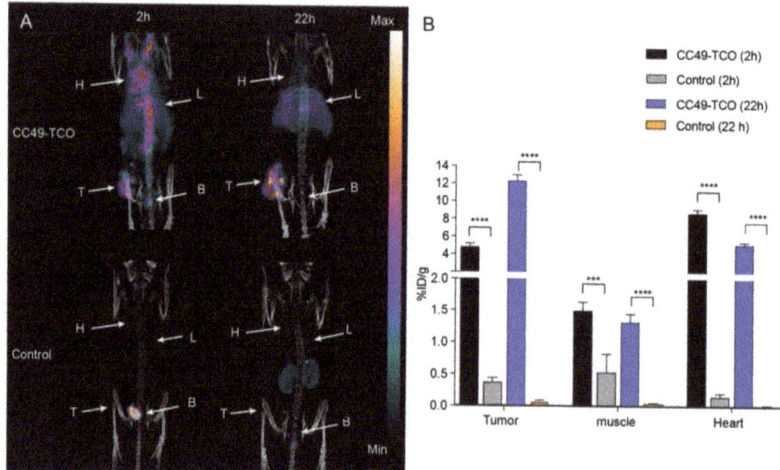

Figure 5. Pretargeted SPECT/CT imaging of [^{111}In]**2** targeting CC49-TCO (**5**), which was administered 24 h prior to the administration of [^{111}In]**2**. (**A**) Respective SPECT/CT images after 2 h and 22 h. (**B**) Image-derived quantitative analysis of tissue uptake. *** $p < 0.001$ and **** $p < 0.0001$. Each image is scaled between its minimum and maximum pixel intensity.

2.5. Pretargeted PET/CT Tumor Imaging

Following SPECT/CT imaging, pretargeted tumor imaging was performed using **5** as the primary agent and [^{68}Ga]**3** as the secondary agent. Mice bearing LS174T tumor xenografts in the left flank were administered **5** 24 h prior to administration of [^{68}Ga]**3**, giving a total in vivo TCO:Tz ratio of 1:2. Control mice bearing tumor xenografts without pre-treatment were also given [^{68}Ga]**3**. PET/CT images were obtained 2 h p.i. Similar to the indium-111 studies, the tumor was clearly visualized; however, high background activity was also observed (Figure 6A). Nevertheless, the mean tumor uptake was higher in the pre-treated mice (2.1 ± 0.1 %ID/g) when compared to the tumor uptake in the control mice (0.19 ± 0.03 %ID/g) (Figure 6B; Table S2). Again, a considerable amount of activity was observed in the heart (2.9 ± 0.1 %ID/g) of the pre-treated animals compared to the controls (0.18 ± 0.03 %ID/g) indicating that the tetrazine ligation occurs with residual **5** circulating in the blood (Table 2: Supplementary Table S2).

Table 2. Uptake values, tumor-to-muscle (T/M) ratios and tumor-to-blood (T/B) ratios from [^{111}In]2 SPECT/CT (n = 3) and [^{68}Ga]3 PET/CT and ex vivo (gammacounter) (n = 4) in selected tissues in nude BALB/c mice bearing subcutaneous LS174T tumor xenografts pretreated with 5. Data is given as mean ± standard error of mean (SEM). * Image-derived uptake in heart from SPECT and PET images used as a surrogate for the blood radioactivity content. ** Radioactivity content in blood measured using gamma counter. *** Tumor-to-blood ratio based on image-derived radioactivity content found in regions of interest created on tumor and heart tissue.

	Pretargeted [^{111}In]2 (%ID/g)		Pretargeted [^{68}Ga]3	
	SPECT 2 h	SPECT 22 h	PET 2 h	Ex Vivo 2 h
Tumor	4.8 ± 0.4	12.3 ± 0.6	2.1 ± 0.1	5.8 ± 0.3
Heart (blood)	8.7 ± 0.4 *	5.2 ± 0.2 *	2.9 ± 0.1 *	5.5 ± 0.1 **
Muscle	1.5 ± 0.1	1.3 ± 0.1	0.48 ± 0.02	0.6 ± 0.1
T/M ratio	3.2	9.2	4.4	9.8
T/B ratio	0.6 ***	2.4 ***	0.8 ***	1.1

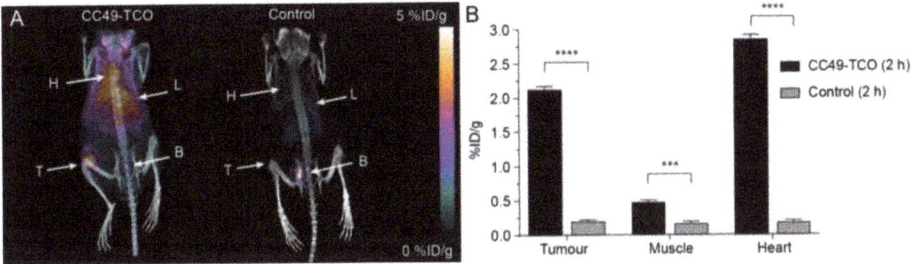

Figure 6. Pretargeted PET/CT imaging of [^{68}Ga]3 targeting CC49-TCO (5), which was administered 24 h prior to the administration of [^{68}Ga]3. (**A**) Respective PET/CT images after 2 h. (**B**) Image-derived quantitative analysis of tissue uptake. *** $p < 0.001$ and **** $p < 0.0001$.

Following the last scan, tumor, muscle, and blood samples were collected and radioactivity was measured. The mean uptake value for the tumor was 5.8 ± 0.3 %ID/g, the tumor-to-muscle ratio was 9.8 and the tumor-to-blood ratio was 1.1 (Table 2).

3. Discussion

We report the radiosynthesis of a ^{68}Ga-labeled DOTA tetrazine ([^{68}Ga]3) as a novel PET analogue to [^{111}In]2. The radiosynthesis of [^{68}Ga]3 was performed using methods similar to previous reports [27], however, modifications were put in place so that the reaction conditions were similar to that of the indium-111 labeling. The reaction was fast (10 min) and efficient, resulting in total incorporation of the gallium-68 available in solution. The less than quantitative RCY can be attributed to nonspecific binding of radioactive components to the reaction vessel and SPE cartridge during reformulation.

Pretargeting using a radiolabeled tetrazine and TCO functionalized mAb was established as an efficient method for imaging tumors in 2010 [6]. Since then, various pretargeting experiments involving the TCO modified CC49 antibodies have been optimized resulting in improved image quality for [^{111}In]2 [11–13]. One strategy that showed to significantly improve the image contrast involved the use of a tetrazine functionalized clearing agent to remove any residual mAb from the circulation. It has also been reported that increasing the amount of non-labeled tetrazine can reduce the blood capture of the radiolabeled tetrazine by any residual circulating TCO-functionalized antibodies [28]. Apart from this, great effort has been put into the development of additional radiolabeled tetrazines as potential pretargeting imaging agents. However, many reports of these new radiolabeled tetrazines lack in vivo evaluation of their use as pretargeted imaging agents, likely due to the lack of alternative pretargeting screening models to mAbs [29–34]. Although TCO-CC49 mAb serves as a positive control for pretargeting experiments because of its extensive optimization and evaluation, antibodies as a screening model is challenged by the need of meticulous pretargeting procedures, disease models to test them, and the fact that they and their TCO modification can be expensive to produce. One alternative to TCO-modified antibodies is TCO-linked bisphosphonates, alendronate (Aln), that can be used for pretargeted imaging of shoulder and knee joints in healthy mice and offers a much simpler experimental procedure for screening new tetrazines [21,35,36].

Hence, in this study, TCO-modified CC49 mAb and Aln were selected to evaluate the utility of [^{68}Ga]3 as a PET isotope labeled variant of [^{111}In]2. We found that healthy mice pre-treated with bisphosphonates gave reproducible ^{111}In-SPECT images with good contrast between the joints and background as early as 2 h p.i., which is a suitable time-point for ^{68}Ga-PET. Analysis showed that the uptake in the shoulder and the knees was nearly identical at 2 and 22 h p.i., respectively. This is expected as the IEDDA reaction is fast and irreversible, yielding a stable ligation product and because the unreacted TCO and Tz moieties are rapidly cleared from the system [5,21]. Using [^{68}Ga]3 and 2 h p.i. PET imaging, we again found that mice pre-treated with 4 had a significantly higher uptake in shoulder and knees compared to both control tissues (muscle and heart) and mice with no pre-treatment. Interestingly, the target uptake was higher using the [^{111}In]2 compared to [^{68}Ga]3. However, a higher TCO:Tz ratio was used for the [^{111}In]2 study potentially resulting in increased accumulation as a result of higher number of TCO in target tissue. Also, indium can form 7 or 8 coordinate DOTA complexes, whereas gallium forms 6 coordinate species [37–39]. This chemical difference can result in different biodistribution patterns for ^{68}Ga and ^{111}In-labeled radiotracers [40,41]. Furthermore, since Ga-DOTA complexes are less stable than In-DOTA complexes, there is a greater risk for in vivo demetallation [40,41]. All these factors may contribute to the fact that lower accumulation was observed with [^{68}Ga]3 compared to [^{111}In]2. Regardless, the image contrast of [^{68}Ga]3 was quite good 2 h and 22 h p.i.

Although previous optimization studies with 5 showed improved contrast with the use of a clearing agent, we opted to forgo this to simplify our approach. Instead, we employed a TCO:Tz ratio of 1:2 compared to a 1:1 ratio that was used in presence of a clearing agent. Mice bearing LS174T tumor xenografts were pre-treated with 5 24 h prior to receiving [^{111}In]2, and SPECT/CT scanned 2 and 22 h p.i. Tumors were clearly visualized at both time points, however, the image contrast significantly improved at 22 h p.i. Image derived uptake analysis also showed that the tumor uptake increased more than two-fold from 2 h to 22 h p.i., whereas the activity was reduced in the muscle and heart. In a similar way, PET/CT imaging with [^{68}Ga]3 2 h p.i. provided a fairly good tumor uptake, however, the image contrast was impaired by the very high background. In contrast, tumor-bearing mice that did not receive any pre-treatment, showed negligible uptake in all tissues with both [^{68}Ga]3 and [^{111}In]2. Given the higher blood activity levels in the pretargeted experiments in comparison to the controls (Figures 5 and 6), it appears that the radiolabeled tetrazines are captured by residual antibodies in the blood pool. Following IEDDA reaction the ligation product can slowly accumulate in the tumors, increasing the uptake over time.

For [^{68}Ga]3, there was some deviation between the uptake values from the ex vivo and the image derived analysis, e.g., most notably we found a mean tumor uptake of 2.1 %ID/g using PET and 5.8 %ID/g using a gamma counter. The tumors used in these studies were somewhat small ($\approx$125 mm^3) and therefor these deviations probably relate to the positron range of ^{68}Ga (R_{mean}: 2.9 mm; R_{max}: 8.2 mm (in water) [42]. For ^{68}Ga this is a particular a problem in small animal-imaging with high spatial resolution, and where the target tissue often is smaller or on the same scale as the positron range. This has also previously been reported by others [42]. However, in clinical settings this is less of a problem and ^{68}Ga, as a PET isotope, has the advantage of being produced from a relatively inexpensive and easily accessible generator system.

4. Material and Methods

4.1. General

All reagents and solvents were purchased from Sigma-Aldrich, unless otherwise, stated and used without further purification. [^{111}In]InCl$_3$ (non-carrier added) was purchased from Lægemiddel Styrelsen. [^{68}Ga]GaCl$_3$ was produced from ^{68}Ga/^{68}Ge generator (Eckert & Ziegler) and was eluted with 0.1 M HCl in fractions (1 mL). As these are both radioactive substance proper facilities, licenses, and procedures must be put in place prior to use. Water was distilled and deionized using a Milli-Q water filtration system (Millipore). Buffers used in the ^{111}In-labelings were treated with Chelex-100 resin overnight, filtered (0.22 µm) and stored at 4 °C. Analytical radioactive thin-layer chromatography (radio-TLC) was performed using silica gel plates (Sigma-Aldrich) or C18 silica gel (Merk) with fluorescent indicator UV254 and visualized using a TLC scanner. Analytical high-performance liquid chromatography (HPLC) was performed using a Thermo Scientific system equipped with a photodiode array (PDA) detector and a Gabi radioactivity detector (Raytest). Compounds were eluted with a Chromalith Performance RP-18 endcapped 100-4.6 HPLC column using the following elution conditions: Solvent A = water with 0.1% trifluoroactic acid (TFA), solvent B = MeCN with 0.1% trifluoracetic acid (TFA). Gradient: 5% B to 95% B, 0–10 min. The flow rate was set at 2.5 mL/min. UV monitoring occurred at 254 nm.

4.2. Statistics

Statistical analysis was performed in GraphPad Prism 8. Unless stated otherwise all values are mean ± s.e.m. The tissue uptake was compared with either unpaired t-test or one-way ANOVA with multiple comparion and Sidak's post hoc test. All statistical results were considered significant when p-value < 0.05.

4.3. Synthesis of Pretargeting Components

Aln-TCO (4) [21], CC49–TCO (5) [13,14], and the DOTA–tetrazine (1) [6] were prepared as previously described.

4.4. Radiochemistry—Labeling of 1 with Indium-111

[^{111}In]Indium-2,2′,2″-(10-(2,40,44-trioxo-44-((6-(6-(pyridin-2-yl)-1,2,4,5-tetrazx-yl)pyridx-yl)amino)-6,9,12,15,18,21,24,27,30,33,36-undecaoxa-3,39-diazatetratetracontyl)-1,4,7,10-tetraazacyclododecane-1,4,7-triyl)triacetate ([^{111}In]1). [^{111}In]Indium chloride (480–615 MBq) in 0.1 M HCl (1 mL) was added to a vial containing 1 (50–100 µg, 0.04–0.08 µmol) in 0.2 M ammonium acetate pH 7 (25–50 µL) and heated at 60 °C. After 10 min diethylenetriamine-pentaacetic acid (DTPA) (19 µg, 0.05 µmol) in PBS (5 µL) and gentisic acid (588 µg, 3.8 µmol) in saline (29.4 µL) was added and the solution was heated for an additional 5 min. RCY ndc = 93%-quantitative (n = 10); Radiochemical purity (RCP) > 95%; A$_m$: 6–16 GBq/µmol; HPLC (method A) R$_t$ = 3.26 min. Radio-TLC (200 mM EDTA in saline) R$_f$ = 0.

4.5. Labeling of 3 with Galium-68

[^{68}Ga]Gallium-2,2′,2″-(10-(2,40,44-trioxo-44-((6-(6-(pyridin-2-yl) 1,2,4,5-tetrazx-yl)pyridx-yl)amino)-6,9,12,15,18,21,24,27,30,33,36-undecaoxa-3,39-diazatetratetracontyl)-1,4,7,10-tetraazacyclododecane-1,4,7-triyl)triacetate ([^{68}Ga]3). [^{68}Ga]Gallium chloride (164–278 MBq) in 0.1 M HCl (1–2 mL) was eluted directly to a reaction vial containing 1 (50–100 µg, 0.04–0.08 µmol) in 0.2 M ammonium acetate pH 7 (25–50 µL), 70% ethanol (400 µL) and 0.2 M ammonium acetate pH 6 (1 mL). The reaction was heated at 90 °C for 10 min. After cooling to room temperature, the reaction was loaded on a solid phase extraction (SPE) cartridge. The SPE cartridge was pre-rinsed 70% ethanol (5 mL) and water (10 mL). Unreacted [^{68}Ga]GaCl$_3$ was eluted with water (10 mL) and the product was eluted with absolute ethanol (1 mL). RCY dc = 48–78% (n = 4); RCP > 90%; molar activity 2–6 GBq/µmol; HPLC R_t = 3.34 min; Radio-TLC (0.1 M citrate buffer, pH 4) R_f = 0.

4.6. Determination of Log D (pH 7.4)

The octanol–water partition coefficient (Log D) was measured in a similar manner previously described [43].

4.7. In Vitro Stability

Solutions of [^{111}In]2 were prepared in 0.2 M ammonium acetate at pH 5.5 and incubated at room temperature. Aliquots (1 µL) were taken at 5, 30, 45, 60, and 90 min. After 90 min an additional aliquot (1 µL) was added to 4 (10 µg, 0.02 µmol) in saline (10 µL). After 5 min samples were analyzed using RP radio-TLC. RP radioTLC (50% MeCN/H$_2$O) R_f([^{111}In]2) = 0.9; R_f([^{111}In]2 + 4) = 0.0.

Solutions of [^{68}Ga]3 were formulated in 10% EtOH/PBS and incubated at room temperature. Studies were performed as described for [^{111}In]2. RP radioTLC (50% MeCN/H$_2$O) R_f([^{68}Ga]3) = 0.2; R_f([^{68}Ga]3 + 4) = 0.0.

4.8. In Vivo Studies

All animal experiments were approved by the Danish Animal Welfare Council, Ministry of Justice and performed in accordance with the approved guidelines. Five-week-old female BALB/c nu/nu mice and female BALB/c mice were obtained from Charles River Laboratories and allowed to acclimatize for one week in the animal facility. At all times, they had access to water and feed ad libitum. LS174T human colon cancer cell line (ATCC) was cultured in minimum essential medium (MEM) supplemented with 10% fetal bovine serum (FBS), 1% L-Glutamine, 1% Sodium pyruvate, 1% non-essential amino acids, and 1% penicillin-streptomycin (Pen Strep) at 37 °C and 5% CO$_2$. At a confluence of 70–90%, cells were harvested by trypsination and subcutaneous tumors were established in the left flank of the 6 weeks old BALB/c nu/nu mice by inoculation of ~5 × 10^6 LS174T cells resuspended in sterile PBS (100 µL). Tumors were allowed to grow for 7–10 days before use. All animals were euthanized after their last scan. To determine tumor size, dimensions were measured using a caliper and the volume calculated as: volume = 1/2(length × width^2).

SPECT/CT imaging was conducted using a small animal dedicated nanoSPECT/CT system (Mediso). CT images were acquired using the semicircular multi field of view (FOV) method in Nucline Software and using the settings: 720 projections, 35 kVp, 980 µA, 450 ms, 1:4 binning. SPECT images were acquired at 245.4 keV (20% full width) as primary photopeak and 171.3.4 keV (20% full width) as secondary photopeak. Each projection was obtained with 50,000–100,000 counts, except for the control experiments with only 15,000 counts. In the Nucline Software, CT images were reconstructed using small voxel size, thin slice thickness, and cosine filter and SPECT images were reconstructed using Tera-Tomo^{TM} 3D SPECT reconstruction software. PET/CT imaging was conducted on a small animal dedicated Inveon PET/CT system (Siemens). Static PET images were acquired with an energy window of 350–650 KeV and a time resolution of 6 ns. CT scans were acquired using 360 projections, 65 kV, 500 µA, and 400 ms. PET images were reconstructed using a three-dimensional maximum a posteriori algorithm with CT-based attenuation correction. SPECT/PET and CT images were fused and co-registered and analyzed using either VivoQuant software (SPECT; Invicro) or Inveon software (PET; Siemens). The mean percentage of injected dose per grams of tissue (%ID/g) in the tissue volume was extracted by manually drawing regions of interest on the entire tissue and correct for decay and injected amount of radioactivity. During all scans, animals were kept anesthetized by breathing 3–5% sevoflurane mixed with 35% O$_2$ in N$_2$ and their body temperature maintained either by a temperature-controlled air flow or by being placed on a heating pad.

Regions of interest were manually drawn on target tissues and the percent injected dose per gram of tissue (%ID/g) was extracted after dose and decay correction.

4.8.1. Evaluation of [^{111}In]2 in Mice Pretreated with 5

Four female nude BALB/c mice bearing subcutaneous LS174T tumor xenografts (mean size of 111.8 ± 29.7 mm^3) were administered i.v. with **5** (100 µg) in PBS (50 µL). The control group for **5** was nude BALB/c mice bearing LS174T subcutaneous tumor xenografts (mean size of 136.7 ± 13.1 mm^3). After 24 h, [^{111}In]**2** (mean 59 MBq, 12 µg, 9.7 nmol) in 0.2 M ammonium acetate pH 5 (50 µL) was administered i.v. SPECT/CT images were obtained 2 and 22 h p.i.

4.8.2. Evaluation of [^{111}In]2 in Mice Pretreated with 4

Three female BALB/c mice were administered i.v. with **4** (100 µg, 0.25 µmol) in saline (100 µL). As a control group, we used three mice without pre-treatment. After 1 h [^{111}In]**2** (mean 43 MBq, 4.2 µg, 3.3 nmol) in 0.2 M ammonium acetate pH 5 (100 µL) was administered i.v. SPECT/CT images were obtained 2 and 22 h p.i.

4.8.3. Evaluation of [^{68}Ga]3 in Mice Bearing LS174T Tumor Xenografts

Four female nude BALB/c mice bearing subcutaneous LS174T tumor xenografts were administered i.v. with **5** (100 µg) in PBS (50 µL); four mice without pre-treatment were included as controls (mean size of 126.2 ± 26.0 mm^3). After 24 h [^{68}Ga]**3** (mean 6 MBq, 10–13 µg, 8–10 nmol) in < 10% EtOH/PBS (100µL) was administered i.v. PET/CT images were obtained 2 h p.i. Following last scan, the mice were euthanized and the tumor, blood, and muscle tissue were collected and weighed. Sample radioactivity was measured using a gamma counter and the %ID/g values were calculated after correcting for decay and injected amount of radioactivity.

4.8.4. Evaluation of [^{68}Ga]3 in Mice Pre-Treated with 4

Female BALB/c mice were administered i.v. with **4** (100 µg, 0.25 µmol) in saline (100 µL) (n = 4). Four mice without pre-treatment were included as controls. After 1 h [^{68}Ga]**3** (mean 11 MBq, 8–13 µg, 6.4–10 nmol) in < 10% EtOH/PBS (100 µL) was administered i.v. PET/CT images were obtained 2 p.i. Following the scan, the mice were euthanized, and the knee, shoulder, and muscle tissue were collected and weighed. Sample radioactivity was measured using a gamma counter and the %ID/g values were calculated after correcting for decay and injected amount of radioactivity.

5. Conclusions

Within this study, we demonstrated that [^{68}Ga]**3** can serve as a PET alternative for pretargeted imaging, even though it displayed lower accumulation values than the 'Gold-standard' for pretargeted imaging, [^{111}In]**2**.

Supplementary Materials: The following are available online. Analytical HPLC data of [111In]**2** and [68Ga]**3** (Figures S1 and S2), RadioTLC data of [68Ga]**3** (Figure S3), additional PET biodistribution data (Table S1) and SPECT biodistribution data (Table S2) are reported.

Author Contributions: Conceptualization, P.E.E., M.M.H., and A.K.; Synthesis, P.E.E., A.Y., J.F.V., R.R., and M.R.; Animal experiments, J.T.J. and K.N.; Writing—Original draft preparation, P.E.E., J.T.J. and K.N.; Writing—Review and editing, M.M.H. and A.K. All authors have read and agreed to the published version of the manuscript.

Funding: This project received funding from the European Union's Horizon 2020 research and innovation programme under grant agreements no. 668532 (Click-It) and 670261 (ERC Advanced Grant), the Lundbeck Foundation, the Novo Nordisk Foundation, the Innovation Fund Denmark, the Danish Cancer Society, Arvid Nilsson Foundation, Svend Andersen Foundation, the Neye Foundation, the Research Foundation of Rigshospitalet, the Danish National Research Foundation (grant 126), the Research Council of the Capital Region of Denmark, the Danish Health Authority, the Carlsberg Foundation, the John and Birthe Meyer Foundation and the Research Council for Independent Research.

Acknowledgments: This project has received funding from the European Union's EU Framework Programme for Research and Innovation Horizon 2020, under grant agreement no. 668532., from the European Union's EU Framework Programme for Research and Innovation Horizon 2020 (grant agreement no. 670261), the

Lundbeck Foundation, the Novo Nordisk Foundation, the Innovation Fund Denmark, and the Research Council for Independent Research.

Conflicts of Interest: No conflict of interest has to be reported.

References

1. Kraeber-Bodéré, F.; Bodet-Milin, C.; Rousseau, C.; Eugène, T.; Pallardy, A.; Frampas, E.; Carlier, T.; Ferrer, L.; Gaschet, J.; Davodeau, F.; et al. Radioimmunoconjugates for the treatment of cancer. *Semin. Oncol.* **2014**, *41*, 613–622. [CrossRef]
2. Bourgeois MBailly, C.; Frindel, M.; Guerard, F.; Chérel, M.; Faivre-Chauvet, A.; Kraeber-Bodéré, F.; Bodet-Milin, C. Radioimmunoconjugates for treating cancer: Recent advances and current opportunities. *Expert Opin. Biol. Ther.* **2017**, *17*, 813–819. [CrossRef] [PubMed]
3. Stéen, E.J.L.; Jørgensen, J.T.; Johann, K.; Norregaard, K.; Sohr, B.; Svatunek, D.; Birke, A.; Shalgunov, V.; Edem, P.E.; Rossin, R.; et al. Trans-Cyclooctene-Functionalized PeptoBrushes with Improved Reaction Kinetics of the Tetrazine Ligation for Pretargeted Nuclear Imaging. *ACS Nano* **2019**. [CrossRef] [PubMed]
4. Nayak, T.K.; Brechbiel, M.W. Radioimmunoimaging with longer-lived positron-emitting radionuclides: Potentials and challenges. *Bioconjug. Chem.* **2009**, *20*, 825–841. [CrossRef] [PubMed]
5. Steen, E.J.L.; Edem, P.E.; Nørregaard, K.; Jørgensen, J.T.; Shalgunov, V.; Kjaer, A.; Herth, M.M. Pretargeting in nuclear imaging and radionuclide therapy: improving efficacy of theranostics and nanomedicines. *Biomaterials* **2018**, *179*, 209–245. [CrossRef] [PubMed]
6. Rossin, R.; Verkerk, P.R.; van den Bosch, S.M.; Vulders, R.C.; Verel, I.; Lub, J.; Robillard, M.S. In vivo chemistry for pretargeted tumor imaging in live mice. *Angew. Chem. Int. Ed.* **2010**, *49*, 3375–3378. [CrossRef]
7. Zeglis, B.M.; Sevak, K.K.; Reiner, T.; Mohindra, P.; Carlin, S.D.; Zanzonico, P.; Weissleder, R.; Lewis, J.S. A Pretargeted PET Imaging Strategy Based on Bioorthogonal Diels-Alder Click Chemistry. *J. Nucl. Med.* **2013**, *54*, 1389–1396. [CrossRef]
8. Adumeau, P.; Carnazza, K.E.; Brand, C.; Carlin, S.D.; Reiner, T.; Agnew, B.J.; Lewis, J.S.; Zeglis, B.M. A pretargeted approach for the multimodal PET/NIRF imaging of colorectal cancer. *Theranostics* **2016**, *6*, 2267–2277. [CrossRef]
9. García, M.F.; Zhang, X.; Shah, M.; Newton-Northup, J.; Cabral, P.; Cerecetto, H.; Quinn, T. m99mTc-bio-orthogonal click chemistry reagent for in vivo pretargeted imaging. *Bioorgan. Med. Chem.* **2016**, *24*, 1209–1215.
10. Meyer, J.P.; Houghton, J.L.; Kozlowski, P.; Abdel-Atti, D.; Reiner, T.; Pillarsetty, N.V.; Scholz, W.W.; Zeglis, B.M.; Lewis, J.S. 18F-Based Pretargeted PET Imaging Based on Bioorthogonal Diels-Alder Click Chemistry. *Bioconjug. Chem.* **2016**, *27*, 298–301. [CrossRef]
11. Meyer, J.P.; Kozlowski, P.; Jackson, J.; Cunanan, K.M.; Adumeau, P.; Dilling, T.R.; Zeglis, B.M.; Lewis, J.S. Exploring Structural Parameters for Pretargeting Radioligand Optimization. *J. Med. Chem.* **2017**, *60*, 8201–8217. [CrossRef]
12. Rossin, R.; Lappchen, T.; van den Bosch, S.M.; Laforest, R.; Robillard, M.S. Diels-Alder Reaction for Tumor Pretargeting: In Vivo Chemistry Can Boost Tumor Radiation Dose Compared with Directly Labeled Antibody. *J. Nucl. Med.* **2013**, *54*, 1989–1995. [CrossRef] [PubMed]
13. Rossin, R.; van den Bosch, S.M.; Ten Hoeve, W.; Carvelli, M.; Versteegen, R.M.; Lub, J.; Robillard, M.S. Highly reactive trans-cyclooctene tags with improved stability for diels-alder chemistry in living systems. *Bioconjug. Chem.* **2013**, *24*, 1210–1217. [CrossRef] [PubMed]
14. Rossin, R.; Van Duijnhoven, S.M.J.; Läppchen, T.; Van Den Bosch, S.M.; Robillard, M.S. Trans-cyclooctene tag with improved properties for tumor pretargeting with the Diels-Alder reaction. *Mol. Pharm.* **2014**, *11*, 3090–3096. [CrossRef] [PubMed]
15. Van Duijnhoven, S.M.J.; Rossin, R.; van den Bosch, S.M.; Wheatcroft, M.P.; Hudson, P.J.; Robillard, M.S. Diabody Pretargeting with Click Chemistry In Vivo. *J. Nucl. Med.* **2015**, *56*, 1422–1428. [CrossRef]
16. Zeglis, B.M.; Brand, C.; Abdel-Atti, D.; Carnazza, K.E.; Cook, B.E.; Carlin, S.; Reiner, T.; Lewis, J.S. Optimization of a pretargeted strategy for the PET imaging of colorectal carcinoma via the modulation of radioligand pharmacokinetics. *Mol. Pharm.* **2015**, *12*, 3575–3587. [CrossRef]

17. Shah, M.A.; Zhang, X.; Rossin, R.; Robillard, M.S.; Fisher, D.R.; Bueltmann, T.; Hoeben, F.J.M.; Quinn, T.P. Metal-Free Cycloaddition Chemistry Driven Pretargeted Radioimmunotherapy Using α-Particle Radiation. *Bioconjug. Chem.* **2017**, *28*, 3007–3015. [CrossRef]
18. Läppchen, T.; Rossin, R.; van Mourik, T.R.; Gruntz, G.; Hoeben, F.J.M.; Versteegen, R.M.; Janssen, H.M.; Lub, J.; Robillard, M.S. DOTA-tetrazine probes with modified linkers for tumor pretargeting. *Nucl. Med. Biol.* **2017**, *55*, 19–26. [CrossRef]
19. Altai, M.; Perols, A.; Tsourma, M.; Mitran, B.; Honarvar, H.; Robillard, M.; Rossin, R.; ten Hoeve, W.; Lubberink, M.; Orlova, A.; et al. Feasibility of Affibody-Based Bio-orthogonal Chemistry-Mediated Radionuclide Pretargeting. *J. Nucl. Med.* **2016**, *57*, 431–436. [CrossRef]
20. Rossin, R.; Versteegen, R.M.; Wu, J.; Khasanov, A.; Wessels, H.J.; Steenbergen, E.J.; Ten Hoeve, W.; Janssen, H.M.; van Onzen, A.H.A.M.; Hudson, P.J.; et al. Chemically triggered drug release from an antibody-drug conjugate leads to potent antitumor activity in mice. *Nat. Commun.* **2018**, *9*, 1484. [CrossRef]
21. Yazdani, A.; Bilton, H.; Vito, A.; Genady, A.R.; Rathmann, S.M.; Ahmad, Z.; Janzen, N.; Czorny, S.; Zeglis, B.M.; Francesconi, L.C.; et al. A Bone-Seeking trans-Cyclooctene for Pretargeting and Bio-orthogonal Chemistry: A Proof of Concept Study Using 99mTc- and177Lu-Labeled Tetrazines. *J. Med. Chem.* **2016**, *59*, 9381–9389. [CrossRef] [PubMed]
22. Edem, P.E.; Sinnes, J.P.; Pektor, S.; Bausbacher, N.; Rossin, R.; Yazdani, A.; Miederer, M.; Kjær, A.; Valliant, J.F.; Robillard, M.S.; et al. Evaluation of the inverse electron demand Diels-Alder reaction in rats using a scandium-44-labeled tetrazine for pretargeted PET imaging. *EJNMMI Res.* **2019**, *9*, 4–11. [CrossRef] [PubMed]
23. Sato, M.; Grassser, W.; Endi, W.; Atkins, R.; Simmonis HThompsona, D.D.; Golub, E.; Rodan, G.A. Bisphosphonate action: Alendronate localization in rat bone and effects on osteoclast ultrastructure. *J. Clin. Investig.* **1991**, *88*, 2095–2105. [CrossRef] [PubMed]
24. Isabel, M.; Prata, M. Gallium-68: A New Trend in PET Radiopharmacy. *Curr. Radiopharm.* **2012**, *5*, 142–149. [CrossRef]
25. Vito, A.; Alarabi, H.; Czorny, S.; Beiraghi, O.; Kent, J.; Janzen, N.; Genady, A.R.; Al-Karmi, S.A.; Rathmann, S.; Naperstkow, Z.; et al. A 99mTc-labeled tetrazine for bio-orthogonal chemistry. Synthesis and biodistribution studies with small molecule trans-cyclooctene derivatives. *PLoS ONE* **2016**, *11*, 1–15. [CrossRef]
26. Yazdani, A.; Janzen, N.; Czorny, S.; Ungard, R.G.; Miladinovic, T.; Singh, G.; Valliant, J.F. Preparation of tetrazine-containing [2 + 1] complexes of 99mTc and: In vivo targeting using bioorthogonal inverse electron demand Diels-Alder chemistry. *Dalt. Trans.* **2017**, *46*, 14691–14699. [CrossRef]
27. Evans, H.L.; Carroll, L.; Aboagye, E.O.; Spivey, A.C. Bioorthogonal chemistry for 68Ga radiolabeling of DOTA-containing compounds. *J. Label. Compd. Radiopharm.* **2014**, *57*, 291–297. [CrossRef]
28. Albu, S.A.; Vito, A.; Dzandzi, J.P.; Zlitni, A.; Beckford-Vera, D.; Blacker, M.; Janzen, N.; Patel, R.M.; Capretta, A.; Valliant, J.F. 125I-Tetrazines and Inverse-Electron-Demand Diels-Alder Chemistry: A Convenient Radioiodination Strategy for Biomolecule Labeling, Screening, and Biodistribution Studies. *Bioconjug. Chem.* **2016**, *27*, 207–216. [CrossRef]
29. Fujiki, K.; Yano, S.; Ito, T.; Kumagai, Y.; Murakami, Y.; Kamigaito, O.; Haba, H.; Tanaka, K. A One-Pot Three-Component Double-Click Method for Synthesis of [67Cu]-Labeled Biomolecular Radiotherapeutics. *Sci. Rep.* **2017**, *7*, 1–9. [CrossRef]
30. Denk, C.; Svatunek, D.; Filip, T.; Wanek, T.; Lumpi, D.; Fröhlich, J.; Kuntner, C.; Mikula, H. Development of a 18F-labeled tetrazine with favorable pharmacokinetics for bioorthogonal PET imaging. *Angew. Chem. Int. Ed.* **2014**, *53*, 9655–9659. [CrossRef]
31. Herth, M.M.; Andersen, V.L.; Lehel, S.; Madsen, J.; Knudsen, G.M.; Kristensen, J.L. Development of a 11C-labeled tetrazine for rapid tetrazine-trans-cyclooctene ligation. *Chem. Commun.* **2013**, *49*, 3805–3807. [CrossRef]
32. Rashidian, M.; Wang, L.; Edens, J.G.; Jacobsen, J.T.; Hossain, I.; Wang, Q.; Victora, G.D.; Vasdev, N.; Ploegh, H.; Liang, S.H. Enzyme-Mediated Modification of Single-Domain Antibodies for Imaging Modalities with Different Characteristics. *Angew. Chem. Int. Ed.* **2016**, *55*, 528–533. [CrossRef]
33. Denk, C.; Svatunek, D.; Mairinger, S.; Stanek, J.; Filip TMatscheko, D.; Kuntner, C.; Wanek, T.; Mikula, H. Design, Synthesis, and Evaluation of a Low-Molecular-Weight 11C-Labeled Tetrazine for Pretargeted PET Imaging Applying Bio-orthogonal in Vivo Click Chemistry. *Bioconjug. Chem.* **2016**, *27*, 1707–1712. [CrossRef]

34. Keinänen, O.; Mäkilä, E.M.; Lindgren, R.; Virtanen, H.; Liljenbäck, H.; Oikonen, V.; Sarparanta, M.; Molthoff, C.; Windhorst, A.D.; Roivainen, A.; et al. Pretargeted PET Imaging of trans-Cyclooctene-Modified Porous Silicon Nanoparticles. *ACS Omega* **2017**, *2*, 62–69. [CrossRef]
35. Keinänen, O.; Fung, K.; Pourat, J.; Jallinoja, V.; Vivier, D.; Pillarsetty, N.K.; Airaksinen, A.J.; Lewis, J.S.; Zeglis, B.M.; Sarparanta, M. Pretargeting of internalizing trastuzumab and cetuximab with a 18F-tetrazine tracer in xenograft models. *EJNMMI Res.* **2017**, *7*, 95. [CrossRef]
36. Liu, S.; Pietryka, J.; Ellars, C.E.; Edwards, D.S. Comparison of Yttrium and indium complexes of DOTA-BA and DOTA-MBA: Models for 90Y- and 111in-labeled DOTA-biomolecule conjugates. *Bioconjug. Chem.* **2002**, *13*, 902–913. [CrossRef]
37. Broan, C.J.; Cox, J.P.L.; Craig, A.S.; Kataky, R.; Parker, D.; Harrison, A.; Randall, A.M.; Ferguson, G. Structure and solution stability of indium and gallium complexes of 1,4,7-triazacyclononanetriacetate and of yttrium complexes of 1,4,7,10-tetraazacyclododecanetetraacetate and related ligands: kinetically stable complexes for use in imaging and radioimmunotherapy. *J. Chem. Soc. Perkin Trans.* **1991**, *2*, 87–99. [CrossRef]
38. Heppeler, A.; Froidevaux, S.; Mäcke, H.R.; Jermann, E.; Béhé, M.; Powell, P.; Hennig, M. Radiometal-Labeled Macrocyclic Chelator-Derivatised Somatostatin Analogue with Superb Tumor-Targeting Properties and Potential for Receptor-Mediated Internal Radiotherapy. *Chem. A Eur. J.* **1999**, *5*, 1974–1981. [CrossRef]
39. Antunes, P.; Ginj, M.; Zhang, H.; Waser, B.; Baum, R.P.; Reubi, J.C.; Maecke, H. Are radiogallium-labeled DOTA-conjugated somatostatin analogues superior to those labeled with other radiometals? *Eur. J. Nucl. Med. Mol. Imaging* **2007**, *34*, 982–993. [CrossRef]
40. Wadas, T.J.; Wong, E.H.; Weisman, G.R.; Anderson, C.J. Coordinating radiometals of copper, gallium, indium, yttrium, and zirconium for PET and SPECT imaging of disease. *Chem. Rev.* **2010**, *110*, 2858–2902. [CrossRef]
41. Cutler, C.S.; Hennkens, H.M.; Sisay, N.; Huclier-Markai, S.; Jurisson, S.S. Radiometals for combined imaging and therapy. *Chem. Rev.* **2013**, *113*, 858–883. [CrossRef] [PubMed]
42. Cal-Gonzalez, J.; Vaquero, J.J.; Herraiz, J.L.; Pérez-Liva, M.; Soto-Montenegro, M.L.; Peña-Zalbidea, S.; Desco, M.; Udías, J.M. Improving PET Quantification of Small Animal [68Ga]DOTA-Labeled PET/CT Studies by Using a CT-Based Positron Range Correction. *Mol. Imaging Biol.* **2018**, *20*, 584Y593. [CrossRef] [PubMed]
43. Edem, P.E.; Czorny, S.; Valliant, J.F. Synthesis and evaluation of radioiodinated acyloxymethyl ketones as activity-based probes for cathepsin B. *J. Med. Chem.* **2014**, *57*, 9564–9577. [CrossRef] [PubMed]

Sample Availability: Samples of the compounds are not available.

© 2020 by the authors. Licensee MDPI, Basel, Switzerland. This article is an open access article distributed under the terms and conditions of the Creative Commons Attribution (CC BY) license (http://creativecommons.org/licenses/by/4.0/).

Article

Kinetic Modelling of [^{68}Ga]Ga-DOTA-Siglec-9 in Porcine Osteomyelitis and Soft Tissue Infections

Lars Jødal [1,*], Anne Roivainen [2,3], Vesa Oikonen [2,3], Sirpa Jalkanen [4], Søren B. Hansen [5], Pia Afzelius [6], Aage K. O. Alstrup [5], Ole L. Nielsen [7] and Svend B. Jensen [1,8]

1. Department of Nuclear Medicine, Aalborg University Hospital, DK-9000 Aalborg, Denmark; svbj@rn.dk
2. Turku PET Centre, Turku University Hospital, FI-20520 Turku, Finland; aroivan@utu.fi (A.R.); vesa.oikonen@utu.fi (V.O.)
3. Turku PET Centre, University of Turku, FI-20520 Turku, Finland
4. MediCity Research Laboratory and Institute of Biomedicine, University of Turku, FI-20520 Turku, Finland; sirjal@utu.fi
5. Department of Nuclear Medicine and PET, Aarhus University Hospital, DK-8200 Aarhus, Denmark; soerehse@rm.dk (S.B.H.); aagealst@rm.dk (A.K.O.A.)
6. North Zealand Hospital, Hillerød, Copenhagen University Hospital, DK-3400 Hillerød, Denmark; pia.maria.tullia.afzelius@regionh.dk
7. Department of Veterinary and Animal Sciences, University of Copenhagen, DK-1870 Copenhagen, Denmark; ole.lerberg.nielsen@gmail.com
8. Department of Chemistry and Biosciences, Aalborg University, DK-9100 Aalborg, Denmark
* Correspondence: lajo@rn.dk; Tel.: +45-9766-5500

Academic Editor: Derek J. McPhee
Received: 2 October 2019; Accepted: 10 November 2019; Published: 13 November 2019

Abstract: Background: [^{68}Ga]Ga-DOTA-Siglec-9 is a positron emission tomography (PET) radioligand for vascular adhesion protein 1 (VAP-1), a protein involved in leukocyte trafficking. The tracer facilitates the imaging of inflammation and infection. Here, we studied the pharmacokinetic modelling of [^{68}Ga]Ga-DOTA-Siglec-9 in osteomyelitis and soft tissue infections in pigs. Methods: Eight pigs with osteomyelitis and soft tissue infections in the right hind limb were dynamically PET scanned for 60 min along with arterial blood sampling. The fraction of radioactivity in the blood accounted for by the parent tracer was evaluated with radio-high-performance liquid chromatography. One- and two-tissue compartment models were used for pharmacokinetic evaluation. Post-mortem soft tissue samples from one pig were analysed with anti-VAP-1 immunofluorescence. In each analysis, the animal's non-infected left hind limb was used as a control. Results: Tracer uptake was elevated in soft tissue infections but remained low in osteomyelitis. The kinetics of [^{68}Ga]Ga-DOTA-Siglec-9 followed a reversible 2-tissue compartment model. The tracer metabolized quickly; however, taking this into account, produced more ambiguous results. Infected soft tissue samples showed endothelial cell surface expression of the Siglec-9 receptor VAP-1. Conclusion: The kinetics of [^{68}Ga]Ga-DOTA-Siglec-9 uptake in porcine soft tissue infections are best described by the 2-tissue compartment model.

Keywords: kinetic analysis; Siglec-9; gallium-68; vascular adhesion protein; VAP-1; infection; inflammation; osteomyelitis; animal model; *Staphylococcus aureus*

1. Introduction

Infections causing localized lesions in the body, e.g., osteomyelitis, cannot always be successfully treated by the systemic administration of antibiotics; instead, surgical intervention may be necessary, which requires knowledge of the site(s) of infection. Assuming the availability of a suitable tracer, positron emission tomography (PET) is well suited for the task. However, it has proven difficult to find a tracer with both high and specific uptake in infected tissue.

The current standard tracer used for PET imaging of infectious and inflammatory diseases is the fluorine-18 labelled glucose analogue 2-deoxy-2-[^{18}F]-fluoro-D-glucose ([^{18}F]FDG) [1]. Generally, [^{18}F]FDG shows high uptake in sites of infection and inflammation, but the uptake is not specific: [^{18}F]FDG accumulates in all cells that use glucose as an energy source, including not only activated immune cells but also rapidly proliferating cancer cells, brain tissue and working muscle tissue.

Leukocytes (white blood cells, WBC) are natural seekers of infection. Leukocytes labelled with ^{111}In or ^{99m}Tc have been widely used for scintigraphy imaging of infection, but the labelling procedure is time consuming and requires withdrawing 30–40 mL of blood from the patient [2]. A simpler alternative is anti-granulocyte monoclonal antibodies (anti-G-mAbs) labelled with ^{99m}Tc. However, while the isotopes ^{111}In and ^{99m}Tc are suitable for gamma camera imaging, they cannot be used for PET imaging. This is a technical disadvantage, as the counting efficiency is typically an order of magnitude higher for the PET scanner than for the gamma camera, and the spatial resolution is superior. Labelling of leukocytes with [^{18}F]FDG for PET imaging has been investigated, but the labelling efficiency is lower, and [^{18}F]FDG elutes from the leukocytes over time [3]. Furthermore, the molar activity of labelled leukocytes will be many orders of magnitude higher than in the patient as a whole, for which reason the short-range β^+ radiation (positrons before annihilation) can result in damaging self-irradiation to [^{18}F]FDG-labelled leukocytes [4].

The relationship between time and physical decay must also be noted. Injected leukocytes take hours or more to accumulate at the infection site, and imaging can be performed 3–4 h after injection at the earliest [2]. These delays are acceptable with ^{111}In ($T_{\frac{1}{2}}$ = 2.8 days) and ^{99m}Tc ($T_{\frac{1}{2}}$ = 6 h), but in relation to PET imaging, the shorter physical half-lives of the two most widely used isotopes, ^{18}F ($T_{\frac{1}{2}}$ = 110 min) and ^{68}Ga ($T_{\frac{1}{2}}$ = 68 min), make long uptake times problematic.

Instead of labelled leukocytes, a tracer directly related to the processes causing leukocyte accumulation in infected tissues is being sought. Such a tracer would potentially enable faster imaging, and at the same time, the handling of blood products from the patient would be avoided. A relevant target molecule is vascular adhesion protein 1 (VAP-1), currently known as a primary amine oxidase (AOC3, EC 1.4.3.21), which is acting both as an adhesion molecule and as a regulatory enzyme in the process of leukocyte binding to the endothelium of blood vessels in infected and inflamed tissue [5,6]. Under normal conditions, VAP-1 is stored in intracellular granules. Upon an inflammatory stimulus, VAP-1 is rapidly translocated to the endothelial cell surface, where it is readily accessible to intravenously administered PET ligands.

As reviewed in [7], ^{68}Ga-labelled synthetic peptides that bind to VAP-1 have been investigated in different experimental infection and inflammation models, and some have shown a better distinction than [^{18}F]FDG between cancer and inflammation. Overall, the review considers VAP-1 to be an optimal target for imaging of inflammation. One of the reviewed tracers, [^{68}Ga]Ga-DOTA-Siglec-9, is currently in a phase I clinical trial (trial no. NCT03755245).

It may be noted that VAP-1 imaging will not distinguish between infection and inflammation, but because VAP-1 is involved in leukocyte extravasation, it is directly linked to the body's natural response to infection. Imaging of VAP-1 expression may even be used to study this process, especially if the kinetics of the VAP-1 imaging tracer are known.

Sialic acid-binding immunoglobulin-like lectin 9 (Siglec-9) is a natural ligand for VAP-1 and is expressed on monocytes and neutrophils [8,9]. The PET tracer [^{68}Ga]Ga-DOTA-Siglec-9 contains a modified fraction of the full Siglec-9 protein and has been found to accumulate in infected tissues [10,11] as well as inflammation sites [8,11–13] and certain tumours [8]. However, the kinetics of [^{68}Ga]Ga-DOTA-Siglec-9 have been investigated only in a single study that addressed inflammation [12]; thus studies on the kinetics of this tracer in infections are lacking.

In earlier studies, we investigated the PET imaging of infection in a porcine osteomyelitis model. In this model, *Staphylococcus aureus* (*S. aureus*) bacteria were injected into the right femoral artery of juvenile pigs to haematogenously induce osteomyelitis in the right hind limb without trauma to the bone [14–16]. In a number of animals, both osteomyelitis and soft tissue infections were induced in the limb, allowing us to compare the uptake in these two types of tissues. Each animal was scanned with a series of radioactive tracers within the same day, with planned delays to allow one tracer to physically decay before the next was injected [17].

The present study investigates the kinetics of [^{68}Ga]Ga-DOTA-Siglec-9 in this porcine osteomyelitis model, including a comparison of the uptake in osteomyelitis and in soft tissue infections.

2. Results

The characteristics of the eight pigs in the study are summarized in Table 1.

Table 1. Characteristics of the pigs in the study.

Pig No.*	Body Weight (kg)	Days from Inoculation to Scan	Injected Radioactivity (MBq)
6	22.0	7	178
9	23.0	8	175
10	22.5	7	191
22	19.0	9	208
23	19.0	9	150
24	20.0	8	126
25	20.5	8	253
26 †	24.5	8	227
			360

* The numbering follows the order in the overall osteomyelitis project described in the main text. Pigs no. 6, 9, and 10 have been described regarding other tracers in previous publications: [18,19] (same numbers) and [20] (where A–E corresponds to numbers 6–10). All pigs in the table except no. 24 were also included in [21] (with different numbering). Blood sample data on [^{68}Ga]Ga-DOTA-Siglec-9 in pig no. 10 have been described in [22]. † Pig no. 26 was scanned twice with [^{68}Ga]Ga-DOTA-Siglec-9. The second scan was performed 4.6 h (approximately 4 physical half-lives of ^{68}Ga) after the first scan and with a higher level of injected radioactivity. Prior to the second scan, 5 mg of unlabelled ("cold") DOTA-Siglec-9 peptide in 10 mL had been slowly injected intravenously (i.v.) into the pig to block receptors—see main text.

2.1. Visual Uptake

Visually, [^{68}Ga]Ga-DOTA-Siglec-9 showed elevated uptake in inflamed and infected soft tissues, while not being prominent in infected bone (osteomyelitis). As an example, a coronal view of the uptake around the femur in pig no. 25 is shown in Figure 1.

At late time points, the decay-corrected PET signal was markedly reduced in all tissues outside the bladder, indicating that uptake of the tracer is reversible rather than irreversible (see Figure 2).

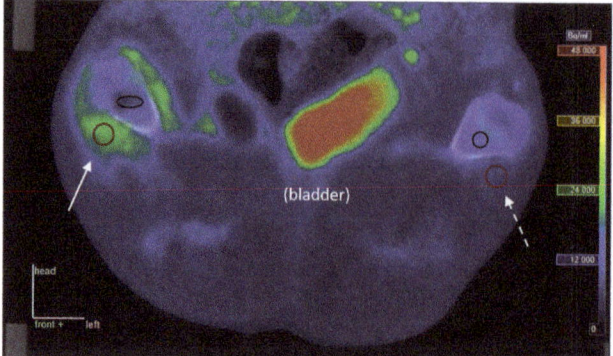

Figure 1. Representative PET/CT image from pig no. 25. The pig was lying supine with roughly upward-pointing limbs, and the view is coronal, seen from the ventral side of the animal ("front" = opposite of back is in the direction of the reader); the top of the image is in the direction of the animal's head. Full arrow: Uptake in phlegmon/early abscess at the right distal femur. Dotted arrow: Similar non-infected tissue for comparison. Red circles show sections of volumes of interest (VOIs) in soft tissue, black ellipse/circle show VOIs in bone tissue. The PET image represents the summed data from 15 to 30 min post-injection (p.i.). The tissue samples presented in Figure 3 were taken from positions approximately corresponding to the soft tissue VOIs.

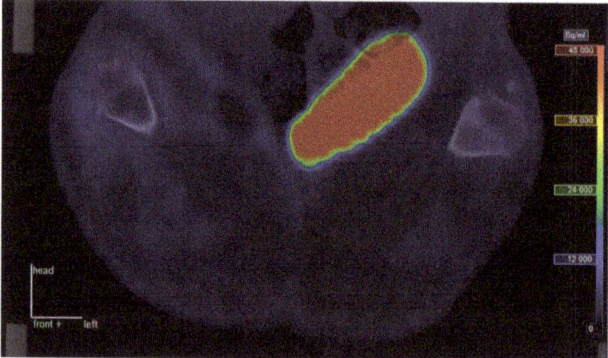

Figure 2. Late PET/CT image from pig no. 25. Static scan with [^{68}Ga]Ga-DOTA-Siglec-9 at approximately 70 min p.i.; position and colour scale correspond to those in Figure 1 (radioactivity is corrected for physical decay).

2.2. VAP-1 Expression Studied by Immunofluorescence (IF)

To analyse the expression of VAP-1 in infectious/inflammatory foci and to determine whether VAP-1 is translocated from intracellular granules to the endothelial cell surface, pig no. 25 was administered anti-VAP-1 antibody shortly before euthanasia, as described in the Methods section. The tissue samples were stained either with anti-VAP-1 + secondary antibody or with secondary antibody alone. Together, this pair of stainings allowed us to detect both the total VAP-1 pool and the part of the pool reached by the intravenously injected antibody; presumably only the latter pool is accessible to the imaging peptide. It can be noted that even though Siglec-9 and anti-VAP-1 both bind to VAP-1, their binding sites are different [9]; thus, they do not compete for binding.

VAP-1 was highly expressed in abscess-associated vasculature and a portion of this pool was detected with the intravenously injected antibody (Figure 3A,B). While VAP-1 is intracellularly expressed in certain vessels of normal soft tissue, no signal was detected with the intravenously administered antibody (Figure 3C,D). It thus appears that VAP-1 is accessible to blood-circulating antibodies only in the infected/inflammatory tissue, which can be seen as an indication that the translocation of VAP-1 to the cell surface is inflammation-induced.

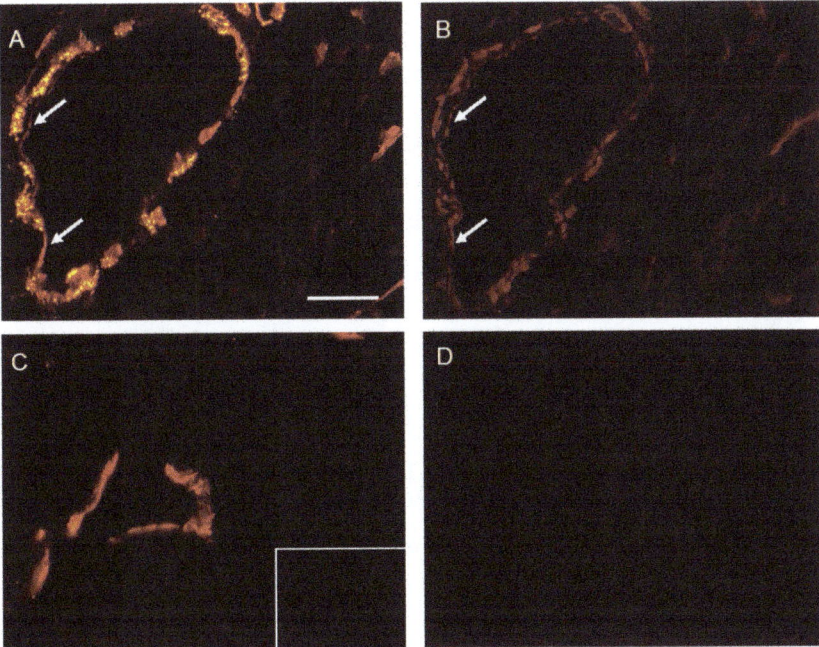

Figure 3. IF staining of infected and non-infected tissue. Stained samples from pig no. 25. **A** and **B** are stains of serial sections from phlegmon/early abscess at the right distal femur. For comparison, **C** and **D** are stains of a non-infected soft tissue sample from an anatomically corresponding position in the left limb of the same animal. (**A**) Staining with anti-VAP-1 and the secondary antibody showed high intracellular and surface expression of VAP-1 in vessels. The scale bar is 50 µm. (**B**) Staining only with the secondary antibody showed VAP-1 surface expression alone and thus indicated that a part of the VAP-1 pool is translocated to the endothelial cell surface (arrow). A signal is also detected on the abluminal side, indicating that the antibody gained access via inter-endothelial junctions. (**C**) Anti-VAP-1 staining followed by the secondary antibody. Negative control staining is shown in the inset. (**D**) Staining only with the secondary antibody showed no VAP-1 surface expression in the non-inflamed tissue.

2.3. Protein Binding of Radioactivity

The plasma protein binding results are shown in Figure 4. Overall, protein binding seemed to vary relatively little over the time of scanning (the late measurement in pig no. 10 deviates, but may also have the largest uncertainty due to the low level of radioactivity remaining at the late time point). A constant fraction of protein binding reflects an equilibrium between the protein-bound and non-bound tracer, allowing the plasma to be treated as a single compartment. Partly for this reason, and partly because we lack protein data for the rest of the pigs, we did not include protein binding in further analysis.

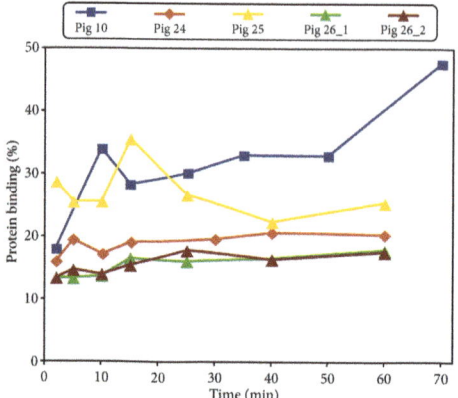

Figure 4. Plasma protein binding of [^{68}Ga]Ga-DOTA-Siglec-9 in the pigs. Percentage of binding as a function of minutes after tracer injection. For pig no. 10, the data are a summary of previously published results [22].

2.4. Analysis of the Parent Tracer Fraction

The parent tracer fraction (fraction of radioactivity originating from [^{68}Ga]Ga-DOTA-Siglec-9) rapidly decreased during the investigated time. A sample fraction curve for the parent tracer is shown in Figure 5. All fraction curves are shown in Supplementary Figure S1.

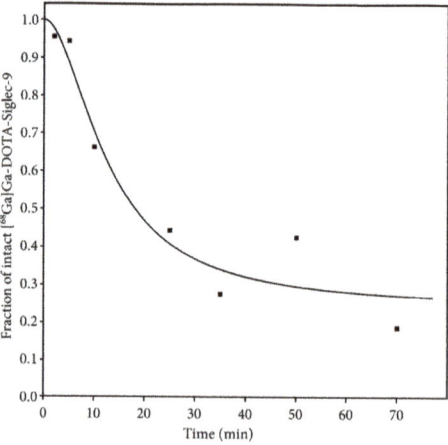

Figure 5. Fraction of radioactivity from intact [^{68}Ga]Ga-DOTA-Siglec-9 (the parent tracer fraction) in arterial plasma as a function of time. Data points and curve fits (Equation (2)) are from pig no. 9 as a representative example.

2.5. VOIs and Model Fits

In the 9 scans, a total of 24 VOI pairs were drawn: 10 VOI pairs were drawn in bone and 14 VOI pairs were drawn in soft tissue positions (counting the VOIs in pig no. 26 twice, as this pig was scanned twice). In pig no. 24, no infectious lesions were identified. In several pigs, bone lesions in the small pedal bones had to be excluded from VOI drawing and analysis, as the resolution of the PET scans would result in a partial volume effect that would be too high for the results to be robust.

The studied lesions are summarized in Tables 2 and 3. The individual VOI volumes (cm^3) can be found in Supplementary Table S1.

Table 2. Investigated osteomyelitis lesions, right hind limb.

Region	Pigs No.
Distal femur	6, 9, 22, 23, 25
Proximal tibia	6, 9, 10, 23
Distal tibia	10

Table 3. Investigated soft tissue infections sites, right hind limb.

Region	Pigs No.
Soft tissue at distal femur	22, 23, 25[†]
Soft tissue at distal tibia	10
Abscess at calcaneus	9
Soft tissue infection/abscess in the tissue plantar to lateral intermediary phalanxes	10, 25, 26*
Abscess at inoculation site (capsule)	22, 23
Abscess at inoculation site (centre)	22
Abscess in superficial popliteus lymph node	26*

* Pig no. 26 was scanned twice; between the two scans, 5 mg of unlabelled ("cold") DOTA-Siglec-9 peptide was injected into the animal. [†] See also Figure 3A,B.

For each VOI, the PET data were fitted with the models shown in Figure 6. Using pig no. 9 as an example, the model fits based on uncorrected and corrected input function are shown in Figures 7 and 8, respectively. All model fits are shown in Supplementary Figures S2–S9.

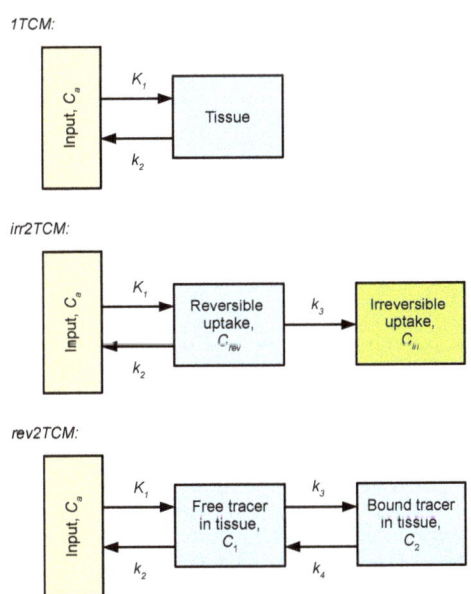

Figure 6. Kinetic models applied for the quantification of [^{68}Ga]Ga-DOTA-Siglec-9 uptake. From top to bottom: 1-tissue compartment model (1TCM), 2-tissue compartment model with irreversible uptake (irr2TCM), and 2TCM with reversible uptake (rev2TCM). The rate constants were fitted as K_1 (unit mL/min/cm^3 or mL/min/100 cm^3), the ratio K_1/k_2 (unit cm^3/mL), k_3 (unit min^{-1}), and the ratio k_3/k_4 (no unit). Additionally, the blood fraction in tissue V_b was fitted.

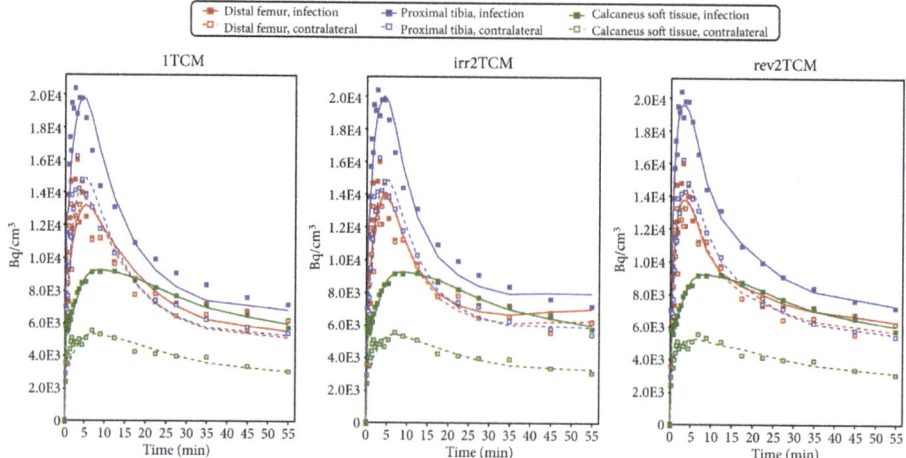

Figure 7. Model fits for pig no. 9 with the *uncorrected* input function. Infection data are from osteomyelitis lesions in the distal femur and proximal tibia and from a soft tissue infection at the calcaneus, all in the animal's right hind limb. Control data are from anatomically corresponding positions in the noninfected left hind limb. All data were modelled with each of the three kinetic models shown in Figure 6.

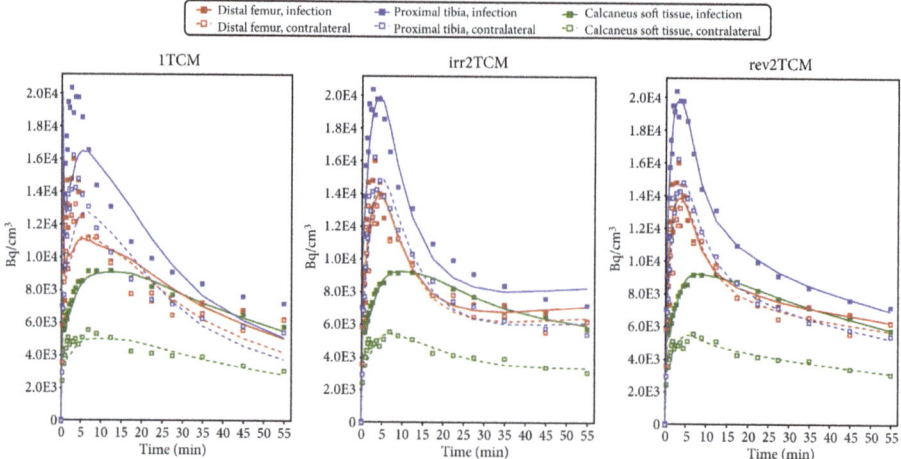

Figure 8. Model fits for pig no. 9 with the *corrected* input function. Same PET data as in Figure 7, different input function. The plots are very similar to those in Figure 7, but not identical; visual differences from Figure 7 are most notable for the early part of 1TCM.

Many of the Patlak plots showed signs of non-linearity and/or horizontal fits, indicating that uptake was reversible rather than irreversible. The Patlak plots for pig no. 9 are shown in Figure 9, and all Patlak plots are shown in Supplementary Figures S2–S9.

Plotting the fitted parameters from the corrected vs. uncorrected input function (plots not shown) revealed that the main difference was a highly elevated ratio k_3/k_4 if the corrected input function was used. A high value of k_3/k_4 indicates that the tracer leaves the second compartment only slowly, i.e., a nearly irreversible uptake. Accordingly, the Patlak plots based on the corrected input function were closer to linearity than the Patlak plots based on the uncorrected input function (cf. Figure 9 and Supplementary Figures S2–S9).

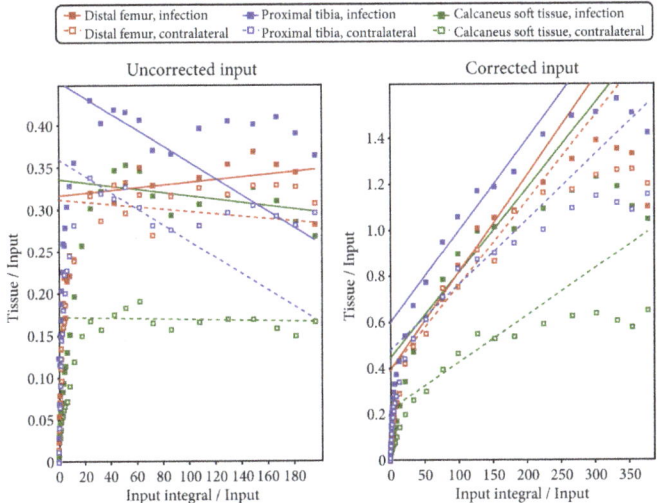

Figure 9. Patlak plots for pig no. 9. **Left**: Based on the uncorrected input function (data as in Figure 7). **Right**: Based on the corrected input function (data as in Figure 8). After the early points, irreversible uptake will be characterized by a linear Patlak plot with a positive slope, while nonlinearity or horizontal slope are signs of reversible uptake.

2.6. The Corrected Akaike Information Criterion (AIC$_c$)

The AIC$_c$ values favoured the rev2TCM model, whether the input function was the uncorrected or the corrected one. Specifically, when the uncorrected input function was used, rev2TCM gave the lowest AIC$_c$ in 18 of the 24 investigated (18/24) infection VOIs and 17/24 control VOIs. Using the corrected input function, the corresponding numbers were 17/24 and 20/24, respectively. See Supplementary Tables S2 and S3.

2.7. The Volume of Distribution in Tissue (V_T)

Based on the AIC$_c$ results, only the V_T results from the rev2TCM were considered.

The volume of distribution in tissue (V_T) from the VOIs in the infected (right hind limb) and the control (left hind limb) positions are compared in Figure 10. Generally, the V_T results for bone infections (osteomyelitis) were close to the identity line, i.e., similar uptake in infection and control tissue. For the soft tissue infections, most points were above the identity line, indicating higher uptake in the infected soft tissues than in the corresponding control tissues, although this result was less clear when the corrected input function was used.

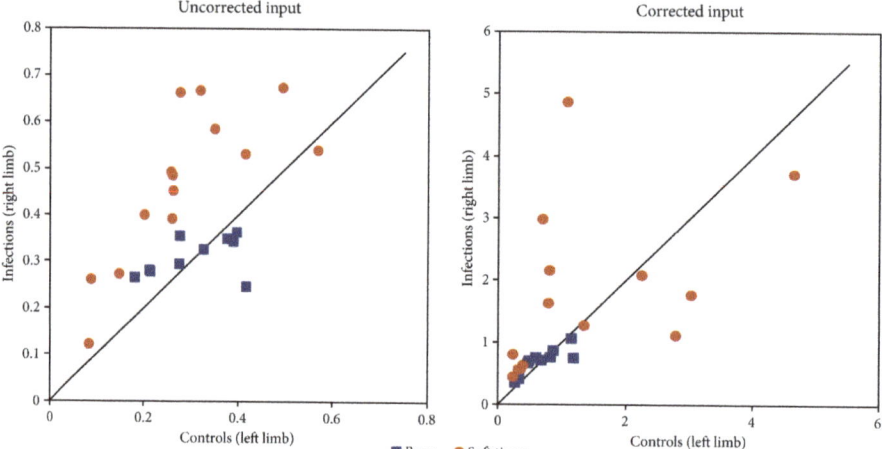

Figure 10. Volume of distribution in infected versus control tissues. Data are the V_T from Equation (6) in infection VOIs and control VOIs. Blue squares are bone VOIs, and red circles are soft tissue VOIs. **Left:** V_T from fit with the uncorrected input function. **Right:** V_T from fit with the corrected input function. Positions on the line correspond to equal values for the infection and control positions. The underlying data of the plots can be found in Supplementary Table S4. See Discussion for a comparison of the scales.

Based on the modelling results from uncorrected input function, the V_T differences between infection VOIs and control VOIs were statistically significant in the soft tissue infections ($p < 0.0001$), but not in the bone infections ($p = 0.83$). More precisely, for soft tissue VOI pairs ($n = 14$), the mean difference $\Delta V_T = V_{T,\text{infection}} - V_{T,\text{control}}$ was 0.18 mL/cm^3, with 95% limits of agreement (LOA) from 0.12 to 0.25 mL/cm^3. For bone VOI pairs ($n = 10$), the mean difference was 0.005 mL/cm^3, with 95% LOA from -0.050 to $+0.062$ mL/cm^3. No signs of non-normal distributions were found ($p > 0.1$ for both soft tissue and bone infections).

2.8. Correlation Between Standardized Uptake Value (SUV) and V_T

Only V_T values based on the uncorrected input functions were considered. The Pearson correlation coefficient was $r = 0.54$ ($r^2 = 0.29$, $n = 48$, $p < 0.0001$), indicating a moderate correlation. A plot is shown in Supplementary Figure S10.

Often, SUV is not evaluated as an absolute value but relative to a reference tissue. If relative rather than absolute values of V_T and SUV were plotted, then correlation rose to $r = 0.66$ ($r^2 = 0.43$, $n = 24$ pairs, $p = 0.0005$). A plot is shown in Figure 11.

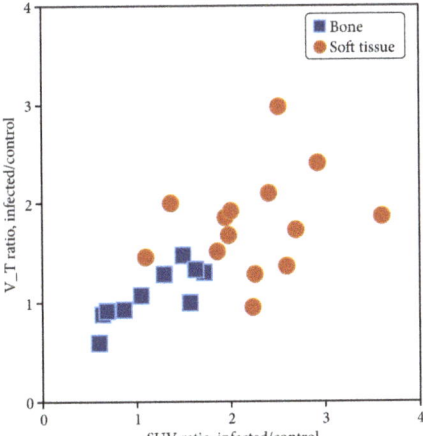

Figure 11. Correlation between relative distribution volume and relative SUV. In the plot $V_{T\text{-ratio}} = V_{T,\text{infected}} / V_{T,\text{control}}$ and $\text{SUV}_{\text{ratio}} = \text{SUV}_{\text{infected}} / \text{SUV}_{\text{control}}$, where for each infection VOI, the control VOI is at the corresponding location in the non-infected limb of the same animal VOI (see main text).

3. Discussion

In this study, we investigated the uptake of [^{68}Ga]Ga-DOTA-Siglec-9 in juvenile pigs with localized *S. aureus* infections, including both bone infection (osteomyelitis) and soft tissue infection.

As in previous studies [10,13], the tracer was found to have an affinity for infected tissues. Siglec-9 is known to bind to VAP-1, and IF staining was consistent with the expression of VAP-1 on cell surfaces in the infected soft tissue.

Kinetic modelling showed that [^{68}Ga]Ga-DOTA-Siglec-9 had reversible uptake kinetics, which could be described with a two-tissue compartment model (i.e., rev2TCM from Figure 6). Visually, the models with fewer parameters provided good fits in many cases (see the 1TCM and irr2TCM curves in Figures 7 and 8); however, the 1TCM curves provided poor fits for the initial part of many curves, and the Patlak plots revealed that models with irreversible uptake such as irr2TCM could not be generally applied (cf. Figure 9). Likewise, the analysis by the AIC_c values (cf. Supplementary Tables S2 and S3) favoured rev2TCM in the large majority of cases.

Based on these results, the following discussion will assume rev2TCM as the chosen model. Physiologically, rev2TCM indicates that the tracer is taken up by the tissue (first compartment), and then the tracer binds to receptors or otherwise changes status in the tissue (second compartment), but with the possibility of unbinding/changing back (reversible model).

3.1. Using the Corrected or the Uncorrected Input Function

The radio-HPLC analysis of blood samples revealed a rapid decrease in the parent tracer fraction (Figure 5). However, modelling showed no clear advantage of using a corrected input function. In most fits, the visual difference between using the uncorrected or corrected input function was only minor (compare Figures 7 and 8).

Physiologically, fitting with the uncorrected input function assumes that any radioactive metabolites have the same uptake kinetics as the parent tracer. In contrast, fitting with the corrected input function assumes that the radioactive metabolites have no uptake at all. The real situation is probably somewhere in between.

Mathematically, the inclusion of such a correction has the disadvantage of including a source of uncertainty, which becomes especially important in a case such as the present where the correction in the late part of the scans was considerable (cf. Figure 5).

The volume of distribution (V_T) measures how concentrated the tracer is in tissue relative to plasma (input), cf. Equation (4) in the Material and Methods section. As shown by comparing the scales in Figure 10, the calculated values of V_T depend markedly on the selection of the uncorrected or corrected input function. This dependency is a simple consequence of the math. The corrected input function is by definition only some fraction of the total, uncorrected input function (cf. Figure 5). A given radioactivity concentration in the tissue will be relatively higher when compared to a low number (the corrected input function) than when compared to a higher number (the uncorrected input function).

In summary, our data showed a rapid decrease in the parent tracer (cf. Figure 5), but unfortunately the modelling did not allow us to distinguish between the PET signal from the parent tracer (i.e., [^{68}Ga]Ga-DOTA-Siglec-9) and the PET signal from other radioactive species formed in vivo (i.e., all metabolites containing the ^{68}Ga isotope, possibly including free ^{68}Ga).

Consequently we reason that use of the corrected input function will lead to increased uncertainty of V_T due to possible errors in the measurement of parent tracer fraction. When the uncorrected input function is used, this pitfall is avoided, although at the cost of risking over-simplification. Pragmatically, the following discussion will therefore focus on the results found with the uncorrected input functions, bearing in mind that the physiological reality is likely more complicated than the model. Further studies investigating the nature of the radioactive products of [^{68}Ga]Ga-DOTA-Siglec-9 formed in vivo are warranted.

3.2. The Volume of Distribution V_T

The V_T for [^{68}Ga]Ga-DOTA-Siglec-9 in soft tissue infections was found to be higher than that of the corresponding control sites, (red circles in Figure 10, left part). In bone infections (osteomyelitis), however, V_T was not significantly different between the infection and control sites (blue squares in Figure 10, left part).

Retamal et al. [12] used the same tracer to study lung inflammation, also performing kinetic modelling with the model called rev2TCM in our terminology (with no mention of correction for parent tracer fraction), and found that the model described the time–activity curves well. The study also included protein binding, which was found to be constant over time, at a level of approximately 20% in healthy pigs and approximately 50% in the inflamed pigs. Our results (cf. Figure 4) are mostly in accordance with the first of these numbers, which could reflect that from a systemic perspective, a local infection in a single limb is more similar to a healthy pig than to a pig with severe inflammation in both lungs. Retamal et al. found increased uptake of [^{68}Ga]Ga-DOTA-Siglec-9 in inflamed lungs, which is consistent with our results on uptake in soft tissue infection.

In summary, [^{68}Ga]Ga-DOTA-Siglec-9 shows increased uptake in infected (and inflamed) soft tissue compared with control tissue; however, this study fails to demonstrate elevated uptake in infected bone (osteomyelitis). These quantitative results correspond to the visual impression of the sample image in Figure 1, where increased uptake is clear in infected soft tissue but not in the infected bone (cf. with pig no. 25 in Tables 2 and 3). Note, however, that this comparison is partly qualitative. The determination of V_T on a reliable absolute scale will depend on improved knowledge on the nature of the radioactive metabolite products of [^{68}Ga]Ga-DOTA-Siglec-9. The moderate correlation between SUV and V_T indicates that the volume of distribution gives information that is not just a complicated version of the SUV.

In a previous study [18] (in part performed on the same animals), we found only a small increase of blood perfusion in osteomyelitis lesions, while blood perfusion was considerably increased in soft tissue infections. As speculated in that study, an ineffective vascular response to infection may lead to too few leukocytes reaching the infected bone, in part explaining why osteomyelitis is difficult for the body to fight. Similarly, despite the uptake of [^{68}Ga]Ga-DOTA-Siglec-9 in infected tissue, imperfect perfusion can impair effective uptake, which may explain the difference in results for soft tissue infections and osteomyelitis.

3.3. Scan after the Injection of "Cold" DOTA-Siglec-9

Pig no. 26 was scanned twice, and the second scan was performed after the injection of 5 mg of unlabelled DOTA-Siglec-9 peptide, which was intended to block the VAP-1 receptors. The uptake curves for both scans are shown in Supplementary Figures S8 and S9. For the infected lymph node, the bolus passage and initial uptake show differences between the first and the second scan, but otherwise the two sets of curves are very similar. Quantitative distribution volumes are listed in Supplementary Table S4. Rather than the expected decrease from receptor blocking, both lesions in pig no. 26 show an approximately 20% increase in V_T from the first scan to the second when the uncorrected input function is used. With only two lesions in one pig, however, it is difficult to draw conclusions.

Using the corrected input function, Supplementary Table S4 shows pronounced V_T differences (still increases) for the two lesions in pig no. 26, but we are hesitant to draw conclusions from these results, as they may reflect the sensitivity of V_T to the correction (cf. Section 3.1). In addition, pig no. 26 unfortunately showed the lowest parent tracer fractions (cf. Supplementary Figure S1) and therefore had the largest sensitivity to possible errors in the correction of the input function.

3.4. Limitations

A porcine model has the advantage over, e.g., a murine model that the physical sizes involved in both surgery and scans are larger, but the disadvantage is that the cost per animal is higher. Accordingly, this study is limited by a relatively small number of animals.

The non-traumatic osteomyelitis protocol has the advantage that it is a very realistic model of haematogenous osteomyelitis (in humans and animals) and infections were reasonably limited to the right hind limb, but the numbers and locations of infections varied among the animals. However, within the individual pig, the other hind limb could be used as a control, and for soft tissue, the uptake measured as the volume of distribution (V_T) in infected versus control sites showed quite clear differences (Figure 10). As already noted, the study does not claim to report V_T on a robust absolute scale, which would also require data on protein binding in all animals.

A scanner replacement resulted in the last three pigs (no. 24–26) being scanned on a PET/CT scanner of a different brand and a newer generation than the initial scanner, enhancing the spatial resolution of the PET images. This allowed refined VOI drawing in these pigs, as reflected in the generally smaller VOI sizes for these animals (cf. Supplementary Table S1). More precisely drawn VOIs could reduce a possible partial volume effect, in which case uptake would be expected to be more pronounced in pigs no. 24–26 than in the pigs scanned on the initial scanner. However, the results for V_T based on the uncorrected input function (cf. Supplementary Table S4) do not indicate any pronounced effect. Using the corrected input function, pigs no. 25 and 26 do have high V_T values, but these values reflect the pronounced correction for the (possibly artificially low) parent tracer fractions in these pigs (cf. Supplementary Figure S1), while an effect of voxel size should also be reflected in V_T calculated using the uncorrected input function.

Another potential source of error may be that the scans were performed on anaesthetized pigs, as the anaesthetic may affect blood flow and hence the kinetics of the tracer. However, it was not practically feasible to scan awake pigs. Propofol was chosen because it provides relatively uniform and safe anaesthesia over many hours.

The use of penicillin in the animals may have reduced the extent of the studied infections, but has previously been found to provide a better balance between the successful development of osteomyelitis and the avoidance of systemic infections requiring euthanasia of the animal [23]. The use of opioids can have immunosuppressant effects, but so can pain reactions, which are reduced by the pain-killer effect; buprenorphine was chosen because it has shown weaker immune effects than morphine and fentanyl [24]. As the target molecule VAP-1 for the tracer is not related to the infecting agent but is instead part of the natural immune system response, there is no reason to expect direct interference between [^{68}Ga]Ga-DOTA-Siglec-9 and the antibiotic pharmaceuticals (penicillin and buprenorphine).

4. Materials and Methods

4.1. Porcine Infection Protocol

The animal protocol was approved by the Danish Animal Experimental Board, journals no. 2012-15-2934-00123 (original approval) and 2017-15-0201-01239 (renewed approval, no substantial changes to the protocol), and all procedures followed the European Directive 2010/63/EU on the protection of animals used for scientific purposes. The protocol for the haematogenous induction of osteomyelitis in domestic pigs has been detailed elsewhere [23]; for general background, see [14–16].

Briefly, *S. aureus* was inoculated into the right hind limb of juvenile (19–25 kg) Danish Landrace–Yorkshire crossbred female pigs. The pigs were pre-acclimated for at least one week, during which they were housed in groups in boxes with restricted access to food (Dia plus FI, DLG, Copenhagen, Denmark) and *ad libitum* access to tap water. The temperature was 20–24 °C, the humidity was 45–65%, and there were 12 h of darkness and 12 h of light in the stables. The pigs came from a specific pathogen free (SPF) herd and were clinically examined by a veterinarian before inoculation. The pigs were fasted for approximately 16 h before premedication with midazolam and Stresnil and propofol anaesthesia. After inoculation, the pigs were individually housed. The *S. aureus* used was the porcine strain S54F9 [25], and 8000 to 30,000 CFU/kg (colony forming units per kg body weight) were inoculated. To selectively infect the right hind limb, bacteria were injected into the right femoral artery. To further reduce the possibility of systemic infection, the pigs were administered penicillin (10,000 EI/kg) at the onset of the first clinical signs of disease; for such juvenile pigs this dosage has previously been shown to allow the development of osteomyelitis while minimizing cases of systemic infection [23]. To reduce pain in the days until euthanasia, the animals were treated every 8 h with buprenorphine (Temgesic; 0.3 to 0.9 mg intramuscularly, depending on clinical symptoms). Osteomyelitis was allowed to develop for approximately one week, after which the pig was scanned with PET and computed tomography (PET/CT, see below) and euthanized. If a pig reached predefined humane endpoints [23] before this time, it was euthanized (and not scanned).

Although the protocol was originally designed for inducing osteomyelitis [17], several pigs also developed soft tissue infections in the inoculated hind limb, typically related to bone infection or the site of inoculation. This turned out to be an advantage, as it allowed us to compare osteomyelitis and soft tissue infections. Some of the pigs also developed lung abscesses. However, these abscesses were outside the field-of-view of the dynamic PET scans and are therefore not included in this kinetics study.

4.2. Animals and Lesions

This study includes eight pigs dynamically scanned with PET/CT using [^{68}Ga]Ga-DOTA-Siglec-9 (details below). The characteristics of these pigs are summarized in Table 1. These eight pigs are a subset of the animals from the overall osteomyelitis project but represent all pigs in the project scanned with this tracer.

Before euthanasia, each pig was also PET/CT scanned with [^{18}F]FDG; these scans have been reported earlier and found [^{18}F]FDG to be a sensitive (but unspecific) marker of *S. aureus* infection [17,19,26]. Infectious lesions were identified on [^{18}F]FDG PET/CT scans. As part of the post-euthanasia analysis, selected lesions were verified by necropsy to be suppurative and to be caused by the inoculated *S. aureus* strain, S54F9.

For the identified lesions, volumes of interest (VOIs) for PET data analysis were drawn on the [^{68}Ga]Ga-DOTA-Siglec-9 PET/CT scans using Carimas 2.9 software (Turku PET Centre, www.turkupetcentre.fi/carimas). For osteomyelitis lesions, the VOI drawing was based on the CT part of the scan, while for soft tissue lesions the VOI drawing was guided by the PET scan.

For each lesion, a reference VOI was drawn in the anatomically corresponding position in the left, non-infected limb.

4.3. Radiochemistry

The radioactive labelling of DOTA-Siglec-9 has previously been discussed and described in detail [22]. The [^{68}Ga]Ga-DOTA-Siglec-9 radiosynthesis method called method 3 in the reference was applied in this study.

In summary, ^{68}Ga was eluted from a ^{68}Ge/^{68}Ga generator (GalliaPharm, Eckert & Ziegler, Berlin, Germany), trapped on a cation-exchange cartridge (Strata-XC 33 u Polymeric Strong, Phenomenex, Værløse, Denmark), and eluted from the cartridge with an acidified acetone solution. The pH was adjusted using HCl (0.1 M in metal-free water), and acetone was removed by heating. A solution of DOTA-Siglec-9 peptide in metal-free water was added, and ^{68}Ga incorporation took place. Water was added to the mixture, which was then run through a preconditioned C-18 Sep-Pak cartridge to trap [^{68}Ga]Ga-DOTA-Siglec-9. The product [^{68}Ga]Ga-DOTA-Siglec-9 was released from the cartridge with ethanol and diluted with saline. After this process, the product was ready for injection.

After 25 min, a 62% non-decay-corrected (ndc) yield of the product was obtained. The [^{68}Ga]Ga-DOTA-Siglec-9 was found by a radio HPLC system to be more than 98% radiochemically pure, and the specific radioactivity was approximately 35 MBq/nmol. Representative radio-HPLC chromatographs are shown in Supplementary Figure S11.

4.4. Dynamic PET Scans

Before scanning, each pig was anaesthetized with propofol, and catheters were implanted in the jugular vein and carotid artery [17]. After an initial CT scan, the pig was dynamically PET scanned for 60 min in 23 time frames: 8×15 s, 4×30 s, 2×60 s, 2×120 s, 4×300 s, and 3×600 s. The [^{68}Ga]Ga-DOTA-Siglec-9 tracer was injected into the jugular vein at the start of the PET scan. The tracer activities are shown in Table 1. All of these scans were performed at the Department of Nuclear Medicine, Aalborg University Hospital.

Pigs no. 6–23 were scanned on a GE VCT Discovery 64 PET/CT scanner (GE Healthcare, Chicago, IL, USA). The scan field covered 15 cm in the axial direction and was positioned over the pelvis and the hind limbs. The images were reconstructed with an ordered subset expectation maximization (OSEM) algorithm (3D Vue Point, GE). The reconstruction parameters were 2 iterations, 28 subsets, a 128×128 matrix in 47 slices, a $5.5 \times 5.5 \times 3.3$ mm^3 voxel size, and a 6 mm Gaussian filter.

Due to scanner replacement, pigs no. 24–26 were scanned on a different scanner than the previous pigs. Pigs no. 24–26 were scanned on a Siemens Biograph mCT (Siemens, Erlangen, Germany) with time-of-flight (TOF) detection. The scan field covered 22 cm in the axial direction, positioned over the pelvis and the hind limbs. The images were reconstructed with an OSEM algorithm without using the resolution recovery option (setting "Iterative + TOF"). The reconstruction parameters were 3 iterations, 21 subsets, a 400×400 matrix in $1.02 \times 1.02 \times 2.03$ mm^3 voxels, and a 3 mm Gaussian filter.

On both scanners, image reconstruction included decay-correction to the start of scanning and attenuation-correction based on the CT scan.

4.5. Blood Samples

An arterial blood sample was drawn before the tracer was injected (zero-sample). During the dynamic PET scan, 27 blood samples were drawn. All blood samples were drawn manually from the carotid artery, and the precise time (seconds) of each sample was recorded. Time zero was the time of injection, which was also the scan start time.

In pigs no. 6–10, the blood samples were drawn every 5 s for 50 s (10 samples), at 60, 80, 100, 120, and 150 s post-injection (p.i., 5 samples), and at 3, 4, 5, 6, 8, 10, 15, 20, 30, 40, 55, and 70 min p.i. (12 samples), i.e., 27 blood samples. Samples for analysis of the fraction of radioactivity originating from the parent tracer (rather than from radioactive metabolite products or free gallium) were drawn at 2, 5, 10, 15, 25, 35, 50, and 70 min p.i.

In pigs no. 22–26, the blood sample timing was slightly optimized. Blood samples were drawn every 5 s for 40 s (8 samples), at 50, 60, 80, 100, 120, and 150 s p.i. (6 samples), and at 3, 4, 5, 6, 8, 10, 15, 20, 25, 30, 40, 50, and 60 min p.i. (13 samples), i.e., 27 blood samples. Blood samples for analysis of the parent tracer fraction were drawn at 2, 5, 10, 15, 25, 40, and 60 min p.i.

As noted in Table 1, pig no. 26 was scanned twice, and 5 mg of "cold" (unlabelled) DOTA-Siglec-9 was injected before the second scan. Blood sampling was independently performed for the two scans.

Plasma samples were obtained by collecting the supernatant after the centrifugation of whole blood samples. Aliquots of the samples were counted in a calibrated Wizard 2480 gamma counter (PerkinElmer, Turku, Finland) with an energy window from 450 to 1200 keV. The counting results were converted to decay-corrected radioactivity concentrations (Bq/mL at the time of injection).

The plasma samples for analyses of the parent tracer fraction were denatured by thoroughly mixing 0.5 mL plasma with 0.5 mL acetonitrile, after which the mix was centrifuged (approximately 1 min, 10,000 rpm) to accelerate the precipitation of the plasma proteins. An aliquot was collected for HPLC analysis (see below). The radioactivity of the precipitate was determined with the Wizard 2480 gamma counter. Protein binding was calculated as

$$\text{protein binding} = \frac{\text{precipitate radioactivity}/0.5 \text{ mL}}{\text{plasma radioactivity concentration}} \times 100\%, \qquad (1)$$

using decay-corrected activities.

For the determination of parent tracer fractions in a sample, 0.2 mL of the supernatant was diluted with 0.8 mm water; this dilution was run through an HPLC with a fractionation collector. The fraction collector was set up with a 20 s delay to compensate for delays in the system. Twenty fractions of 45 s each were collected and counted in the Wizard 2480 gamma counter.

4.6. Tissue Samples and Immunofluorescence (IF)

As a proof of concept, the surface expression of VAP-1 in infected tissue was tested by an augmented protocol in one pig.

Approximately 10 min before euthanasia, pig no. 25 was administered 10 mg of VAP-1-binding antibody as an intravenous (i.v.) injection (10 mL injected liquid). The antibody was 1B2, a mouse IgM against human VAP-1 that also recognizes porcine VAP-1 [27]. The i.v. injection allowed 1B2 to bind to the VAP-1 expressed on cell surfaces, but not to the VAP-1 within intact cells.

After euthanasia of the pig, soft tissue samples were collected from acutely inflamed areas (phlegmon/early abscess) located peripheral to the osteomyelitis in the distal right femur and from similarly positioned non-inflamed areas in the left hind limb. All samples were embedded in a cryo-compound and frozen in petroleum spirit (VWR, Søborg, Denmark, cat. no. 87125.320) cooled with dry ice. The tissues were stored at −80 °C until use.

Immunofluorescence (IF) analysis was performed on 5 μm thick frozen sections of these samples. The first of two serial sections from the inflamed area (right limb) were stained with anti-VAP-1 mAb (1B2) or a class-matched negative control antibody, 7C7 (10 μg/mL; 1 h at room temperature) and then with a secondary antibody (Alexa 555-goat-anti-mouse IgM 1:100, Southern Biotech 1020-32; 30 min at room temperature), followed by nuclear staining with Hoechst 1:10,000 in PBS for 5 min, Thermo Scientific 6249, Waltham, MA, USA). The combined IF signal thus represented the VAP-1 within the cytoplasm as well as the VAP-1 expressed on cell surfaces. The second section was stained only with the secondary antibody; thus, the IF signal represented only the VAP-1 accessible to the i.v. injected antibody. A pair of sections from the left limb was similarly stained to represent the corresponding signals from non-infected tissue.

4.7. Input Function and Metabolite Correction

For each dynamic PET/CT scan, the decay-corrected plasma samples were used as a basic input function.

For metabolite correction, the parent tracer fraction data were fitted with a Hill-type function:

$$f(t) = 1 - \frac{(1-a)t^b}{c + t^b}, \qquad (2)$$

where t is the sampling time (seconds since injection). The function starts at $f(0) = 1$ (thus assuming no metabolism before injection) and has an asymptotic value $f(\infty) = a$. The parameters a, b and c were fitted for each injection of [^{68}Ga]Ga-DOTA-Siglec-9.

In the following text, the *uncorrected input function* denotes the total radioactivity concentration (decay-corrected Bq/mL) from plasma samples, and the *corrected input function* denotes the fraction $f(t)$ multiplied by the uncorrected input function.

As blood samples were taken from the carotid artery (a short distance from the heart), while the PET data were acquired over the hind end of the pig (a longer distance from the heart), the PET data were delayed by some seconds relative to the blood plasma data. To correct for these delays, an offset to the plasma time stamps was determined for each pig using the method described in [19]. The largest of these corrections was 8 s.

4.8. Kinetic Modelling

Data were modelled using the three kinetic models shown in Figure 6. This was performed twice for each model: once with the uncorrected input function and once with the corrected input function. The modelling was performed using software available on the Turku PET Centre website (fit2k for 1TCM, fit3k for irr2TCM, fit4k for rev2TCM) [28].

Data points were weighted based on time frame length (L) and decay:

$$w = L \times \exp(-\lambda t) = L \times 0.5^{t/T} \qquad (3)$$

This weighting scheme mirrors the overall count statistics of the decay-corrected PET data. Unlike weights based on counts in a VOI, the weights from Equation (3) do not in themselves contain noise. See reference [19] for a more detailed discussion of this weighting scheme.

For in vivo imaging, the volume of distribution in tissue (V_T) is defined as the ratio of tissue concentration to input concentration at a time when a steady-state has been reached [29], i.e.,

$$V_T = \frac{\text{tissue concentration}}{\text{input concentration}} \text{ (at steady-state)} \qquad (4)$$

The measurement of tissue concentration in Bq/cm³ (from the PET scan) and input concentration in Bq/mL (from plasma samples in the gamma counter) results in mL/cm³ as the unit of V_T.

For a given model with reversible uptake, the relationship between V_T and the model parameters can be theoretically calculated. In the cases of the 1TCM and the rev2TCM models, these relationships are [29]:

$$\text{1TCM}: V_T = \frac{K_1}{k_2} \qquad (5)$$

$$\text{rev2CTM}: V_T = \frac{K_1}{k_2} \cdot \left(1 + \frac{k_3}{k_4}\right) \qquad (6)$$

For models with irreversible uptake, such as irr2TCM, a steady-state is never reached and V_T is not defined. Mathematically, the tracer input concentration will be continually decreasing, while the uptake in the irreversible compartment will monotonically increase, and over time, the fraction in Equation (4) will diverge instead of converging.

The parent tracer fraction appeared to decrease faster in pig no. 26 (both scans) than in the other pigs, and was close to zero after approximately 30 min post-injection (see Supplementary Figure S1). The corrected input function in this case would therefore be close to zero (expectedly with high percentage errors) at late time points, which could lead to unreliable estimation of V_T = tissue concentration/input concentration. For these reasons, modelling of pig no. 26 with the corrected input function was performed using only the data within the interval 0–30 min rather than the full interval 0–60 min.

4.9. Evaluation

In addition to visual inspections of the fits, the three models were compared using the corrected Akaike information criterion (AIC_c), which favours a good fit, but penalizes the use of excessive fitting parameters [30,31]. The absolute value of AIC_c depends on both the data and the model, but for a given data set, the lowest value of AIC_c indicates the most favourable model.

Furthermore, Patlak plots [32,33] were calculated to help determine whether uptake was reversible or irreversible. A linear Patlak plot (after an equilibration time) is a sign of irreversible uptake, while a system with only reversible uptake will result in a non-linear Patlak plot that eventually approaches a constant value. The Patlak plot is directly calculated from the data, without the assumption of any specific uptake model. The Patlak plots were based on the data from 10 to 60 min (10–30 min in pig no. 26 with corrected input function).

As a measure not requiring modelling, standardized uptake values (SUV) were also calculated, and the correlation between SUV and the volume of distribution V_T was determined. The SUV calculation was based on the time interval 10 to 30 min, chosen as a time interval after the passage of the bolus peak, but not so late that a considerable part of the tracer with reversible binding (i.e., tracer not remaining in the tissue indefinitely) would have left the tissue yet.

4.10. Statistics

To compare the infected and control VOIs (i.e., right vs. left), a two-tailed paired t-test was used, with a significance level of $p < 0.05$. The normality of the differences was tested with the Shapiro–Wilk W test. The correlation between the volume of distribution V_T and SUV was defined using Pearson analysis. Statistics were calculated with StatsDirect version 3.1.14 (www.statsdirect.com).

5. Conclusions

Using the VAP-1-targeted leukocyte ligand Siglec-9, the immunofluorescence analysis of infected tissue samples indicated that VAP-1 was expressed on the cell surfaces in infected tissue, while surface VAP-1 was not observed in non-infected tissue.

The uptake kinetics of [^{68}Ga]Ga-DOTA-Siglec-9 with localized infection in pigs were found to be well described with a reversible 2-tissue compartment model, similar to the model used by Retamal et al. [12] in a study of severe lung inflammation, also in anaesthetized pigs.

We found that the parent tracer fraction decreased relatively rapidly, but despite this finding, we were unable to demonstrate an advantage of taking tracer metabolism into account in the analysis. More detailed analyses of the radioactive species occurring after the i.v. injection of [^{68}Ga]Ga-DOTA-Siglec-9 in the body are warranted.

The [^{68}Ga]Ga-DOTA-Siglec-9 uptake, evaluated as the volume of distribution, showed affinity to infection in soft tissue; however, no increased uptake in bone infections (osteomyelitis) could be demonstrated. This difference may be related to a previous report that found infected soft tissue to be more highly perfused than infected bone tissue (osteomyelitis).

Supplementary Materials: The following are available online, Figure S1: Parent tracer fraction curves, Figures S2 to S9: Model fits and Patlak plots, Figure S10: Volume of distribution (V_T) as a function of SUV, Figure S11: Sample radio-HPLC graphs, Table S1: VOI volumes, Tables S2 and S3: Akaike information criterion (AIC_c) results using, respectively, the uncorrected and the corrected input function, Table S4: Volume of distribution (V_T) as determined by rev2TCM.

Author Contributions: Conceptualization, L.J., A.R., S.J., S.B.H., P.A., A.K.O.A., O.L.N., S.B.J.; Methodology, L.J., P.A, A.K.O.A., O.L.N., S.B.J.; Software, L.J. and V.O; Validation, L.J. and S.B.H.; Formal Analysis, L.J. and S.B.H.; Investigation, S.J., P.A., A.K.O.A., O.L.N. and S.B.J.; Resources, A.R., S.J., P.A., A.K.O.A. and S.B.J.; Data Curation, L.J. and V.O.; Writing—Original Draft Preparation, L.J.; Writing—Review & Editing, L.J., A.R., V.O., S.J., S.B.H., P.A., A.K.O.A., O.L.N., S.B.J.; Visualization, L.J. and A.R.; Supervision, P.A., O.L.N. and S.B.J.; Project Administration, P.A., A.K.O.A., O.L.N. and S.B.J.; Funding Acquisition, A.R., A.K.O.A., O.L.N. and S.B.J.

Funding: This work was supported by the Danish Council for Independent Research, Technology and Production Sciences, grant number 0602-01911B (11-107077), and by Jane and Aatos Erkko Foundation.

Acknowledgments: The authors thank Timo Kattelus for technical assistance on the graphics. Part of the material in this paper has been presented at the European Conference on Medical Physics, Copenhagen 2018 [34].

Conflicts of Interest: Sirpa Jalkanen owns stocks in Faron Pharmaceuticals Ltd. The remaining authors declare no conflicts of interest.

References

1. Jamar, F.; Buscombe, J.; Chiti, A.; Christian, P.E.; Delbeke, D.; Donohoe, K.J.; Israel, O.; Martin-Comin, J.; Signore, A. EANM/SNMMI Guideline for ^{18}F-FDG Use in Inflammation and Infection. *J. Nucl. Med.* **2013**, *54*, 647–658. [CrossRef] [PubMed]
2. Signore, A.; Jamar, F.; Israel, O.; Buscombe, J.; Martin-Comin, J.; Lazzeri, E. Clinical indications, image acquisition and data interpretation for white blood cells and anti-granulocyte monoclonal antibody scintigraphy: An EANM procedural guideline. *Eur. J. Nucl. Med. Mol. Imaging* **2018**, *45*, 1816–1831. [CrossRef] [PubMed]
3. Palestro, C.J. Radionuclide imaging of osteomyelitis. *Semin. Nucl. Med.* **2015**, *45*, 32–46. [CrossRef] [PubMed]
4. Miñana, E.; Roldán, M.; Chivato, T.; Martínez, T.; Fuente, T. Quantification of the chromosomal radiation damage induced by labelling of leukocytes with [18F]FDG. *Nucl. Med. Biol.* **2015**, *42*, 720–723. [CrossRef]
5. Salmi, M.; Jalkanen, S. VAP-1: An adhesin and an enzyme. *Trends Immunol.* **2001**, *22*, 211–216. [CrossRef]
6. Jalkanen, S.; Salmi, M. VAP-1 and CD73, endothelial cell surface enzymes in leukocyte extravasation. *Arterioscler. Thromb. Vasc. Biol.* **2008**, *28*, 18–26. [CrossRef]
7. Roivainen, A.; Jalkanen, S.; Nanni, C. Gallium-labelled peptides for imaging of inflammation. *Eur. J. Nucl. Med. Mol. Imaging* **2012**, *39*, 68–77. [CrossRef]
8. Aalto, K.; Autio, A.; Kiss, E.A.; Elima, K.; Nymalm, Y.; Veres, T.Z.; Marttila-Ichihara, F.; Elovaara, H.; Saanijoki, T.; Crocker, P.R.; et al. Siglec-9 is a novel leukocyte ligand for vascular adhesion protein-1 and can be used in PET imaging of inflammation and cancer. *Blood* **2011**, *118*, 3725–3733. [CrossRef]
9. Salmi, M.; Jalkanen, S. Vascular adhesion protein-1: A cell surface amine oxidase in translation. *Antioxid. Redox Signal.* **2019**, *30*, 314–332. [CrossRef]
10. Ahtinen, H.; Kulkova, J.; Lindholm, L.; Eerola, E.; Hakanen, A.J.; Moritz, N.; Söderström, M.; Saanijoki, T.; Jalkanen, S.; Roivainen, A.; et al. ^{68}Ga-DOTA-Siglec-9 PET/CT imaging of peri-implant tissue responses and staphylococcal infections. *EJNMMI Res.* **2014**, *4*, 45. [CrossRef]
11. Siitonen, R.; Pietikäinen, A.; Liljenbäck, H.; Käkelä, M.; Söderström, M.; Jalkanen, S.; Hytönen, J.; Roivainen, A. Targeting of vascular adhesion protein-1 by positron emission tomography visualizes sites of inflammation in *Borrelia burgdorferi*-infected mice. *Arthritis Res. Ther.* **2017**, *19*, 254. [CrossRef]
12. Retamal, J.; Sörensen, J.; Lubberink, M.; Suarez-Sipmann, F.; Borges, J.B.; Feinstein, R.; Jalkanen, S.; Antoni, G.; Hedenstierna, G.; Roivainen, A.; et al. Feasibility of ^{68}Ga-labeled Siglec-9 peptide for the imaging of acute lung inflammation: A pilot study in a porcine model of acute respiratory distress syndrome. *Am. J. Nucl. Med. Mol. Imaging* **2016**, *6*, 18–31. [PubMed]
13. Virtanen, H.; Silvola, J.M.U.; Autio, A.; Li, X.-G.; Liljenbäck, H.; Hellberg, S.; Siitonen, R.; Ståhle, M.; Käkelä, M.; Airaksinen, A.J.; et al. Comparison of ^{68}Ga-DOTA-Siglec-9 and ^{18}F-fluorodeoxyribose-Siglec-9: Inflammation imaging and radiation dosimetry. *Contrast Media Mol. Imaging* **2017**, *2017*, 7645070. [CrossRef] [PubMed]
14. Johansen, L.K.; Svalastoga, E.L.; Frees, D.; Aalbæk, B.; Koch, J.; Iburg, T.M.; Nielsen, O.L.; Leifsson, P.S.; Jensen, H.E. A new technique for modeling of hematogenous osteomyelitis in pigs: Inoculation into femoral artery. *J. Invest. Surg.* **2013**, *26*, 149–153. [CrossRef] [PubMed]

15. Jensen, H.E.; Nielsen, O.L.; Agerholm, J.S.; Iburg, T.; Johansen, L.K.; Johannesson, E.; Møller, M.; Jahn, L.; Munk, L.; Aalbaek, B.; et al. A non-traumatic *Staphylococcus aureus* osteomyelitis model in pigs. *In Vivo* **2010**, *24*, 257–264.

16. Johansen, L.K.; Koch, J.; Frees, D.; Aalbæk, B.; Nielsen, O.L.; Leifsson, P.S.; Iburg, T.M.; Svalastoga, E.; Buelund, L.E.; Bjarnsholt, T.; et al. Pathology and biofilm formation in a porcine model of staphylococcal osteomyelitis. *J. Comp. Pathol.* **2012**, *147*, 343–353. [CrossRef]

17. Nielsen, O.L.; Afzelius, P.; Bender, D.; Schønheyder, H.C.; Leifsson, P.S.; Nielsen, K.M.; Larsen, J.O.; Jensen, S.B.; Alstrup, A.K.O. Comparison of autologous ^{111}In-leukocytes, ^{18}F-FDG, ^{11}C-methionine, ^{11}C-PK11195 and ^{68}Ga-citrate for diagnostic nuclear imaging in a juvenile porcine haematogenous staphylococcus aureus osteomyelitis model. *Am. J. Nucl. Med. Mol. Imaging* **2015**, *5*, 169–182.

18. Jødal, L.; Nielsen, O.L.; Afzelius, P.; Alstrup, A.K.O.; Hansen, S.B. Blood perfusion in osteomyelitis studied with [^{15}O] water PET in a juvenile porcine model. *EJNMMI Res.* **2017**, *7*, 4. [CrossRef]

19. Jødal, L.; Jensen, S.B.; Nielsen, O.L.; Afzelius, P.; Borghammer, P.; Alstrup, A.K.O.; Hansen, S.B. Kinetic modelling of infection tracers [^{18}F]FDG, [^{68}Ga]Ga-citrate, [^{11}C]methionine, and [^{11}C]donepezil in a porcine osteomyelitis model. *Contrast Media Mol. Imaging* **2017**, *2017*, 9256858. [CrossRef]

20. Afzelius, P.; Alstrup, A.K.O.; Schønheyder, H.C.; Borghammer, P.; Bender, D.; Jensen, S.B.; Nielsen, O.L. Utility of ^{11}C-methionine and ^{11}C-donepezil for imaging of *Staphylococcus aureus* induced osteomyelitis in a juvenile porcine model: Comparison to autologous ^{111}In-labelled leukocytes, ^{99m}Tc-DPD, and ^{18}F-FDG. *Am. J. Nucl. Med. Mol. Imaging* **2016**, *6*, 286–300.

21. Afzelius, P.; Nielsen, O.L.; Schønheyder, H.C.; Alstrup, A.K.O.; Hansen, S.B. An untapped potential for imaging of peripheral osteomyelitis in paediatrics using [^{18}F] FDG PET/CT—The inference from a juvenile porcine model. *EJNMMI Res.* **2019**, *9*, 29. [CrossRef] [PubMed]

22. Jensen, S.B.; Käkelä, M.; Jødal, L.; Moisio, O.; Alstrup, A.K.O.; Jalkanen, S.; Roivainen, A. Exploring the radiosynthesis and *in vitro* characteristics of [^{68}Ga]Ga-DOTA-Siglec-9. *J. Label. Compd. Radiopharm.* **2017**, *60*, 439–449. [CrossRef] [PubMed]

23. Alstrup, A.K.O.; Nielsen, K.M.; Schønheyder, H.C.; Jensen, S.B.; Afzelius, P.; Leifsson, P.S.; Nielsen, O.L. Refinement of a hematogenous localized osteomyelitis model in pigs. *Scand. J. Lab. Anim. Sci.* **2016**, *42*, 1–4.

24. Sacerdote, P. Opioid-induced immunosuppression. *Curr. Opin. Support. Palliat. Care* **2008**, *2*, 14–18. [CrossRef]

25. Aalbæk, B.; Jensen, L.K.; Jensen, H.E.; Olsen, J.E.; Christensen, H. Whole-genome sequence of *Staphylococcus aureus* S54F9 isolated from a chronic disseminated porcine lung abscess and used in human infection models. *Genome Announc.* **2015**, *3*, e01207-15. [CrossRef]

26. Afzelius, P.; Nielsen, O.L.; Alstrup, A.K.O.; Bender, D.; Leifsson, P.S.; Jensen, S.B.; Schønheyder, H.C. Biodistribution of the radionuclides ^{18}F-FDG, ^{11}C-methionine, ^{11}C-PK11195, and ^{68}Ga-citrate in domestic juvenile female pigs and morphological and molecular imaging of the tracers in hematogenously disseminated *Staphylococcus aureus* lesions. *Am. J. Nucl. Med. Mol. Imaging* **2016**, *6*, 42–58.

27. Jaakkola, K.; Nikula, T.; Holopainen, R.; Vähäsilta, T.; Matikainen, M.-T.; Laukkanen, M.-L.; Huupponen, R.; Halkola, L.; Nieminen, L.; Hiltunen, J.; et al. In vivo detection of vascular adhesion protein-1 in experimental inflammation. *Am. J. Pathol.* **2000**, *157*, 463–471. [CrossRef]

28. TPC List of Applications in TPCCLIB. Available online: http://www.turkupetcentre.net/programs/doc/index.html (accessed on 19 February 2019).

29. Innis, R.B.; Cunningham, V.J.; Delforge, J.; Fujita, M.; Gjedde, A.; Gunn, R.N.; Holden, J.; Houle, S.; Huang, S.-C.; Ichise, M.; et al. Consensus nomenclature for in vivo imaging of reversibly binding radioligands. *J. Cereb. Blood Flow Metab.* **2007**, *27*, 1533–1539. [CrossRef]

30. Akaike, H. A new look at the statistical model identification. *IEEE Trans. Autom. Control* **1974**, *19*, 716–723. [CrossRef]

31. Burnham, K.P.; Anderson, D.R. Multimodel inference: Understanding AIC and BIC in model selection. *Sociol. Methods Res.* **2004**, *33*, 261–304. [CrossRef]

32. Patlak, C.S.; Blasberg, R.G.; Fenstermacher, J.D. Graphical evaluation of blood-to-brain transfer constants from multiple-time uptake data. *J. Cereb. Blood Flow Metab.* **1983**, *3*, 1–7. [CrossRef] [PubMed]

33. Patlak, C.S.; Blasberg, R.G. Graphical evaluation of blood-to-brain transfer constants from multiple-time uptake data. Generalizations. *J. Cereb. Blood Flow Metab.* **1985**, *5*, 584–590. [CrossRef] [PubMed]
34. Jødal, L.; Roivainen, A.; Oikonen, V.; Jalkanen, S.; Hansen, S.B.; Afzelius, P.; Alstrup, A.K.O.; Nielsen, O.L.; Jensen, S.B. [P083] Kinetic modelling of [68Ga]Ga-DOTA-Siglec-9 in a porcine infection model. *Phys. Med.* **2018**, *52*, 124–125. [CrossRef]

Sample Availability: Data are available from the corresponding author upon reasonable request.

© 2019 by the authors. Licensee MDPI, Basel, Switzerland. This article is an open access article distributed under the terms and conditions of the Creative Commons Attribution (CC BY) license (http://creativecommons.org/licenses/by/4.0/).

Review

Methods to Enhance the Metabolic Stability of Peptide-Based PET Radiopharmaceuticals

Brendan J. Evans [1], Andrew T. King [1], Andrew Katsifis [2], Lidia Matesic [3] and Joanne F. Jamie [1,*]

1. Department of Molecular Sciences, Macquarie University, Sydney, NSW 2109, Australia; brendan.evans@hdr.mq.edu.au (B.J.E.); andrew.king2@hdr.mq.edu.au (A.T.K.)
2. Department of Molecular Imaging, Royal Prince Alfred Hospital, Camperdown, NSW 2050, Australia; andrewk@nucmed.rpa.cs.nsw.gov.au
3. Australian Nuclear Science and Technology Organisation (ANSTO), Lucas Heights, NSW 2234, Australia; lidia.matesic@ansto.gov.au
* Correspondence: joanne.jamie@mq.edu.au; Tel.: +61-2-9850-8283

Academic Editor: Anne Roivainen
Received: 1 April 2020; Accepted: 13 May 2020; Published: 14 May 2020

Abstract: The high affinity and specificity of peptides towards biological targets, in addition to their favorable pharmacological properties, has encouraged the development of many peptide-based pharmaceuticals, including peptide-based positron emission tomography (PET) radiopharmaceuticals. However, the poor in vivo stability of unmodified peptides against proteolysis is a major challenge that must be overcome, as it can result in an impractically short in vivo biological half-life and a subsequently poor bioavailability when used in imaging and therapeutic applications. Consequently, many biologically and pharmacologically interesting peptide-based drugs may never see application. A potential way to overcome this is using peptide analogues designed to mimic the pharmacophore of a native peptide while also containing unnatural modifications that act to maintain or improve the pharmacological properties. This review explores strategies that have been developed to increase the metabolic stability of peptide-based pharmaceuticals. It includes modifications of the C- and/or N-termini, introduction of D- or other unnatural amino acids, backbone modification, PEGylation and alkyl chain incorporation, cyclization and peptide bond substitution, and where those strategies have been, or could be, applied to PET peptide-based radiopharmaceuticals.

Keywords: radiopharmaceuticals; peptides; positron emission tomography; proteolysis; metabolic stability

---

1. Introduction

Positron emission tomography (PET) is a nuclear medicine imaging technique for the non-invasive quantitative measurement of specific biochemical, physiological, and pharmacological processes in vivo [1]. PET is useful in the diagnosis and staging of neurological, cardiovascular, and various oncology-based diseases [2]. PET imaging is achieved by administering a patient with a PET radiopharmaceutical which will localize into organs and/or tissues that express the desired biological target. Once localized, the distribution of the PET radiopharmaceutical throughout the body can be imaged with a PET scanner and a diagnosis can be made. PET radiopharmaceuticals are biologically active molecules that are labeled with positron-emitting radionuclides such as fluorine-18, gallium-68, or copper-64. A key component of a PET radiopharmaceutical is the targeting entity, which is designed to possess a pharmacophore that has high affinity and specificity towards a desired biological target present in an organ and/or tissue that is associated with a specific disease or malignancy [3]. The targeting entities of radiopharmaceuticals were initially developed as biologically active small molecules, as is the case for the most widely used PET radiopharmaceutical [^{18}F]fluorodeoxyglucose [4].

However, in recent years there has been a rapid development in the use of targeting entities developed from biologics, such as peptides, proteins, antibodies, and antibody fragments for the use as PET radiopharmaceuticals [5–7].

Peptides as Radiopharmaceuticals

The structure of a peptide-based radiopharmaceutical (Figure 1) typically contains the following components: a peptide to act as the targeting entity, a linker, a radionuclide bearing moiety, and a PET radionuclide. The linker is sometimes an optional component of a radiopharmaceutical that is incorporated to facilitate the conjugation of the targeting peptide and the radionuclide bearing moiety, and/or improve its pharmacokinetics such as by increasing metabolic stability or manipulating biodistribution [8–11]. The linker can also be used as a spacer to distance bulky portions of a radiopharmaceutical, such as chelators from the bioactive portions, to reduce steric interference and maintain high binding affinity [12].

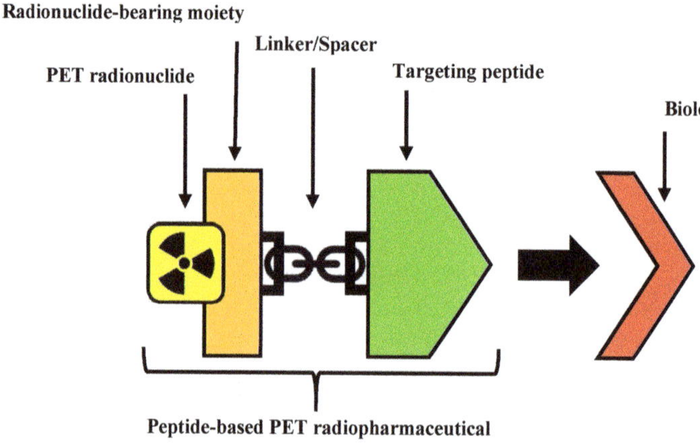

Figure 1. The structural components of a peptide-based positron emission tomography (PET) radiopharmaceuticals.

There are many advantages offered by utilizing peptides as the targeting entity in radiopharmaceuticals, especially in the field of oncology. These peptide-based radiopharmaceuticals can take the form of radiopharmaceuticals for the diagnosis of diseases by imaging the biological target associated with the disease in specific tissues, or as radiotherapeutics for the treatment of cancers by subjecting the tissue to localized ionizing radiation. Furthermore, the opportunity to exploit the same targeting agent with either an imaging or therapeutic radionuclide has given rise to the field of 'Theranostics' in nuclear medicine [13].

A key advantage of peptide-receptor targeting peptide-based radiopharmaceuticals is the higher density of peptide receptors expressed on tumor cells than in normal tissues, thus specific receptor-binding radiolabeled peptides can be designed to enable the efficient visualization and treatment of various tumors [14]. Due to the relatively small size of peptides compared to other biologics, such as proteins and antibodies, peptides often exhibit rapid pharmacokinetics, with the ability to efficiently penetrate tumors, fast clearance from the bloodstream and non-target tissues, and are not immunogenic [5,14–16]. Peptides can also be readily synthesized using conventional peptide synthesizers, and any desired modifications to the structure can be easily engineered by making the appropriate changes to the peptide sequence during synthesis and/or by adding other structural modifications after synthesis [9,17,18].

Consequently, the last 20 years have seen an explosive growth in the development of radiolabeled peptides for targeted diagnostic imaging and therapy. While radiolabeled peptides have been applied to various molecular imaging modalities that use nuclear probes, such as scintigraphy and single-photon emission computed tomography (SPECT), the superior image quality and quantitative data available from PET have resulted in a significant amount of research being devoted to the development of PET radiolabeled peptides [19]. Recently, [^{68}Ga]DOTATATE, also known as NETSPOT® (Figure 2), was the first radiolabeled peptide for PET imaging to receive regulatory approval from the Food and Drug Administration (FDA) [20].

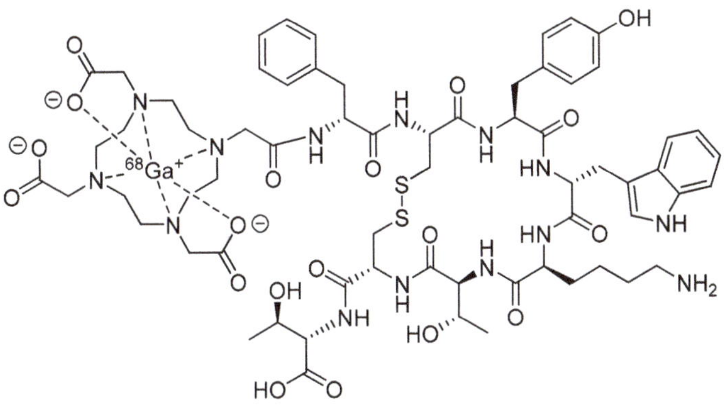

Figure 2. Structure of [^{68}Ga]DOTATATE (NETSPOT®).

Despite the advantages offered by peptides, there are several challenges inherent to the use of peptides in drug design and development, including for imaging and therapy. The most significant of these is that unmodified peptides often possess prohibitively short half-lives in vivo, primarily due to rapid proteolytic degradation in the blood, liver, and kidneys by endogenous proteases [21]. This liability results in a short duration of in vivo activity, poor bioavailability, and has significantly limited their application in drug development [22].

2. Challenges Faced by Peptide-Based Radiopharmaceuticals In Vivo

The rate at which a drug is metabolized and removed from the body determines its biological half-life [23]. In the case of radioactive drugs, such as PET radiopharmaceuticals, the physical half-life is determined by the incorporated radionuclide [23]. Thus, to achieve optimal results, the localization process of a radiopharmaceutical has to be fast relative to its biological and physical half-lives, such that the drug will localize to the target with adequate time for imaging or therapy before degrading below an effective activity and/or concentration [24]. In the case of peptide-based radiopharmaceuticals, the two most significant forms of degradation in vivo that impact the efficacy of the drug are loss of the radionuclide and degradation of the conjugated peptide. The premature degradation of peptide-based radiopharmaceuticals in vivo is of pressing concern since these drugs are typically administered in doses that are significantly smaller than conventional drugs and are therefore especially vulnerable to having their efficacy significantly disrupted by any amount of degradation in vivo.

2.1. Loss of the Radionuclide

Many radiopharmaceuticals are labeled with radiometals (e.g., technetium-99m, gallium-68, and copper-64). Ensuring that the radiometal is not lost from the radiopharmaceutical in vivo is of critical concern when developing these drugs as the free radiometal may exhibit high toxicity and cause significant damage to the body [25]. In the case of fluorine-18 labeled radiopharmaceuticals, in vivo radiodefluorination results in the release of free [^{18}F]fluoride ions that can readily accumulate into the calcium-rich fluorophilic bones of the body [26]. Radiodefluorination and in vivo metabolism of [^{18}F]radiopharmaceuticals also present major challenges to imaging studies as non-specific uptake of free [^{18}F]fluoride ions and [^{18}F]metabolites can lead to a degradation of the signal to noise ratio [27].

Further information on the radiodemetallation and radiodefluorination of radiopharmaceuticals, and the strategies that have been applied to mitigate these issues, are beyond the scope of this review but have been thoroughly reviewed elsewhere [28–30].

2.2. Degradation of the Peptide

The major challenge of using peptides in the active component of a pharmaceutical is that naturally occurring peptides are usually rapidly degraded in vivo [21,22]. Compared to other biologics, such as proteins and antibodies, peptides are generally more susceptible to enzymatic and/or chemical degradation. One of the key reasons a peptide sequence can also be susceptible to proteolytic degradation is due to its backbone containing a recognition motif for an endogenous protease [31]. In addition to proteolytic degradation, peptide bonds can also undergo spontaneous degradation under physiological conditions when particularly labile sequence motifs are present [32].

In peptide-based radiopharmaceuticals, degradation of the conjugated peptide will significantly disrupt its ability to localize to the target tissue, and the subsequent radioactive metabolites will undergo non-specific binding to other tissues and/or be rapidly cleared from the body. In the case of radiopharmaceuticals, this will reduce imaging sensitivity due to increasing background radiation [33]. In radiotherapeutics, this can result in insufficient irradiation of the target tissue while increasing the irradiation of non-target tissues [34,35].

Another point to consider is that amide bonds are often used to conjugate the radionuclide bearing moiety to the biomolecule or linker and these bonds are also susceptible to the same degradation pathways as peptide bonds via proteases and amidases [36]. As a result of these challenges, most peptides for use in radiopharmaceuticals are synthetically modified to minimize metabolic degradation in vivo [15,37].

3. Increasing the In Vivo Stability of Peptide-Based Radiopharmaceuticals

The peptide amide bond represents the central repeating structural element of peptides and proteins. This bond possesses partial double bond character, which is one of the key attributes that contributes to the rigidity of peptide chains. Its ability to form hydrogen bonds also makes the peptide bond play a crucial role in its recognition by and interactions with other proteins. Normally, the peptide amide bond is stable to hydrolysis, requiring harsh conditions involving concentrated acids or bases at increased temperatures [38,39]. However, the peptide bond can be readily cleaved under mild conditions at or even below room temperature in the presence of an appropriate protease or peptidase [39,40].

Peptidases can either be classified as exopeptidases, which specifically hydrolyze the C- or N-termini of a peptide, or as endopeptidases which are capable of hydrolyzing amide bonds within a peptide [41,42]. Mechanistically, hydrolysis of an amide bond with a peptidase occurs via a nucleophilic attack at the carbonyl carbon of the amide bond (Figure 3), with its pathway dependent on the amino acids present in the peptidase's active site [43]. Peptidases that have nucleophilic amino acids residues such as cysteine and serine in their active site, can attack the carbonyl carbon of the amide/peptide bond and form an acylated enzyme (as a thioester for cysteine or ester for serine), which is more vulnerable to attack by water than the original peptide bond (Figure 3A). Often these peptidases also have histidine and aspartic acid or glutamic acid within the active site, as a 'catalytic triad' [43,44]. With peptidases that have amino acid residues such as aspartate/aspartic acid or glutamate/glutamic acid (but no serine or cysteine) in their active site, these residues directly assist a water molecule in its nucleophilic attack of the carbonyl carbon of the amide bond, leading to direct hydrolysis of the amide bond (Figure 3B) [43].

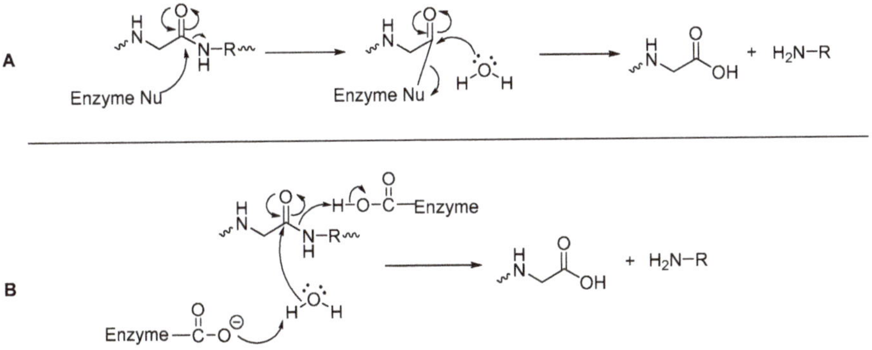

Figure 3. General mechanisms of hydrolysis via a peptidase (**A**) with nucleophilic amino acids; (**B**) with acidic amino acid residues [43,44].

Several strategies have been developed to synthesize peptide analogues in which vulnerable peptide bonds are either modified or obscured such that they are no longer targeted by proteolytic enzymes. The goal of this is to generate metabolically stable peptide analogues that maintain the biological activity of the original peptide. This review is focused on strategies to enhance metabolic stability of PET peptide-based radiopharmaceuticals. It includes modifications of the C- and/or N-termini, introduction of D- or other unnatural amino acids, backbone modification, PEGylation and alkyl chain incorporation, cyclization, and peptide bond substitution. While some of the examples discussed in this review have only been applied to SPECT radiopharmaceuticals, radiotherapeutics, or non-radioactive peptide-based pharmaceuticals, the strategies are applicable also to PET peptide-based radiopharmaceuticals.

3.1. D-Amino Acids

Apart from glycine, all amino acids possess chirality and can therefore exist in either levorotatory (L) or dextrorotatory (D) forms (Figure 4). However, nature has proven itself remarkably homochiral and most amino acids are found in their L-form in mammalian systems. D-Amino acids can be found in nature (e.g., in some species of frogs and bacteria), but these cases are exceedingly rare [45].

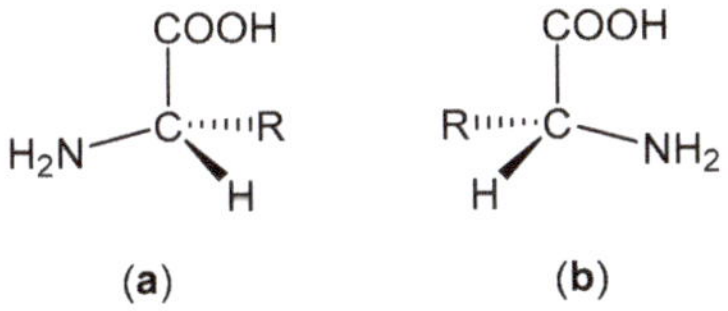

Figure 4. (**a**) Structure of an L-amino acid. (**b**) Structure of a D-amino acid.

The fundamental differences in chirality between L- and D-amino acids means that peptides built from D-amino acids are not recognized by many proteins, including proteases [46]. The result of this lack of recognition is that while most L-peptides are vulnerable to enzymatic degradation in vivo [22], analogous D-peptides are highly resistant to degradation by proteases and have low immunogenicity [47,48]. The simple substitution of all L-amino acids in a peptide with D-amino acids, however, is generally an ineffective strategy as the resulting changes in peptide conformation and side chain orientation can prevent the correct binding geometry and thus destroy target binding [49–52]. A common solution to this issue is retro-inversion, which constitutes reversing the D-peptide's sequence (Figure 5). This approach has proven successful in increasing the biological activity of some unstructured D-peptides by restoring the native L-amino acid side chain angles [53,54].

However, in structured peptides, retro-inversion is not enough to overcome the conformational changes caused by the introduction of D-amino acids [46]. For example, left-handed helices in D-peptides remain left-handed even after sequence reversal in retro-inversion, while helices in L-peptides are always right-handed; this difference results in a significant decrease in the binding efficiency of peptides to their biological target [51,55].

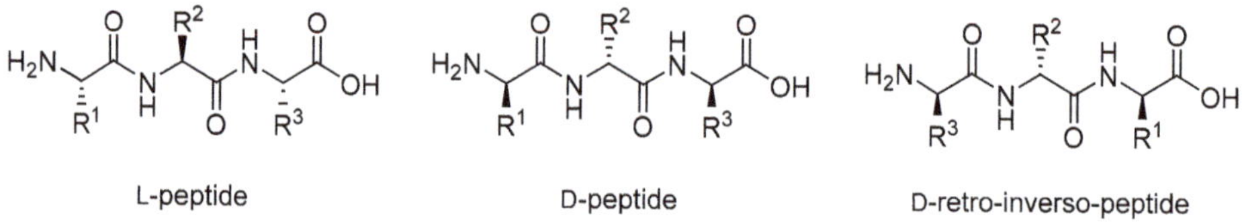

Figure 5. Structure of an L-peptide and its D-peptide analogue and D-retro-inverso-peptide analogue.

Within the Protein Data Bank (PDB), approximately 62% of protein–protein interactions involve helical structures [56]. Furthermore, approximately 80% of peptide drugs approved by the FDA contain helical structures [57]. Due to this, retro-inversion may not be able to be applied to most clinically interesting peptides to correct for the introduction of D-amino acids. However, the benefits offered by D-amino acids can also be conferred to a peptide without substituting every amino acid with its D-amino acid equivalent. For example, substituting the N-terminal L-amino acid of most proteins with the corresponding D-amino acid can significantly increase in vivo stability by preventing recognition of the N-terminus of the protein by proteases [58].

Research conducted by Donna et al. showed that the L-histidine and L-cysteine residues in the α5β1 integrin inhibitor peptide Ac-PHSCN-NH$_2$ (PHSCN) could be replaced with their D-stereoisomers to give the mixed chirality peptide Ac-PhScN-NH$_2$ (PhScN) (Figure 6) [59]. They found that the mixed chirality peptide PhScN showed significantly improved potency as an inhibitor of α5β1 integrin-mediated invasion of naturally occurring basement membranes in vitro, with IC$_{50}$ values of 0.097 pg/mL and 0.113 pg/mL for DU 145 and PC-3 cells, respectively, compared to 2600 pg/mL and 16,627 pg/mL for the unmodified analogue PHSCN [59]. Donna et al. proposed that the inclusion of the D-amino acids, D-histidine, and D-cysteine in the PhScN peptide could greatly increase systemic stability of the peptides compared to the unmodified analogue due to the resistance these unnatural amino acids show against endoproteinases [59].

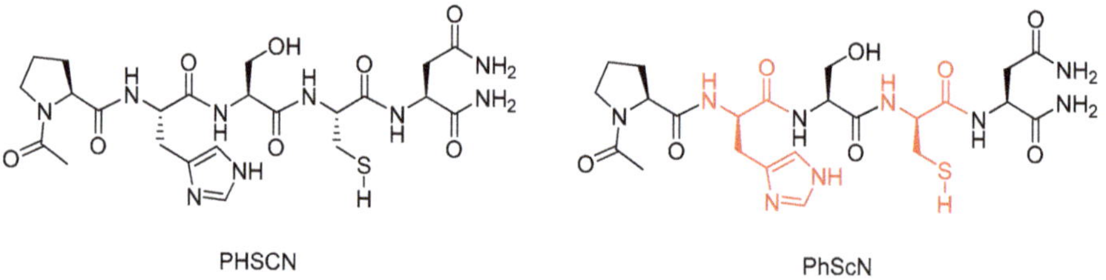

Figure 6. Structures of PHSCN and PhScN peptides, with the D-amino acids highlighted in red [59].

One of the best-known examples of targeted peptide modification for increasing the metabolic stability of a peptide is octreotide (Figure 7); a peptide-based drug developed from the endogenous hormone somatostatin [60]. Somatostatin (Figure 7) is a 14 amino acid long cyclic peptide that inhibits the secretion of a growth hormone. Somatostatin receptors are also found in high concentrations in various neuroendocrine tumors [60,61]. The clinical application of somatostatin is severely limited by its in vivo half-life of only 1–2 min in human plasma due to rapid enzymatic degradation [60]. This limitation spurred the development of somatostatin analogues with more useful properties, including octreotide, in which the somatostatin amino acid sequence was shortened from 14 to 8 amino acids, the L-tryptophan residue was replaced with D-tryptophan, and the N-terminal L-amino acid was replaced with a D-amino acid [60]. These modifications increased the in vivo half-life to 1.5 h in human plasma, while maintaining a high binding affinity for the somatostatin receptor subtype SSTR2, with an IC_{50} value of 0.56 nM for octreotide compared to 0.23 nM for the native somatostatin [60,62]. The improved half-life and maintained potency offered by octreotide and derivatives thereof led to their radiolabeling with a variety of radioisotopes for use in the diagnosis and treatment of various neuroendocrine tumors, including indium-111 for SPECT imaging [61,63], carbon-11 [64] and gallium-68 [65] for PET imaging, and yttrium-90 [61] and lutetium-177 [66] for radiotherapy.

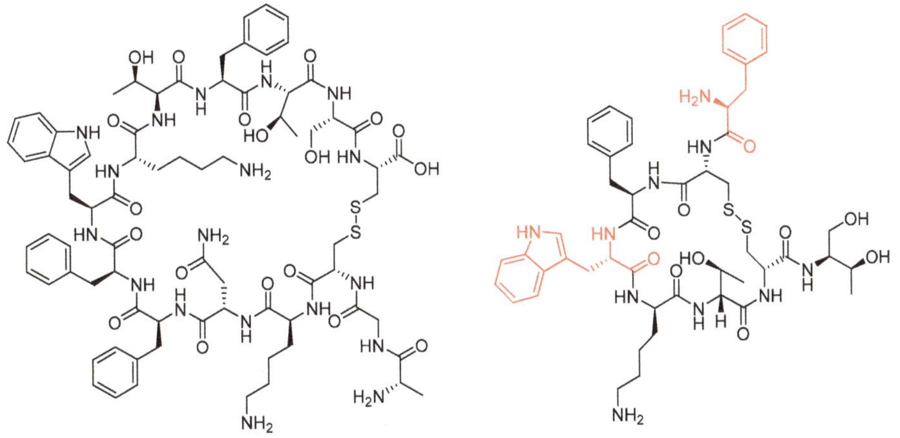

Figure 7. Structures of somatostatin and octreotide, with the D-amino acid modifications on octreotide highlighted in red.

Radiopharmaceuticals based upon the minigastrin peptide have been developed for the purposes of imaging and therapy to target cholecystokinin 2 receptors overexpressed in a variety of thyroid, lung, and ovarian tumors [67]. However, the clinical utility of early minigastrin radiopharmaceuticals was compromised due to high kidney retention, which can cause nephrotoxicity [67,68]. For example, the minigastrin radiopharmaceutical [^{111}In-DOTA]MG0 (Figure 8A) showed a kidney retention of approximately 127% ID/g [67]. To address the issue of kidney retention, new minigastrin analogues were developed with a decreased number of glutamic acid residues to reduce the negative charge on the peptide [68]. A minigastrin analogue with the five glutamic acid residues in the linker deleted from the sequence of [^{111}In-DOTA]MG0 had a significantly reduced kidney retention of approximately 0.3% ID/g [69], but it also had a decreased metabolic stability in human serum, with a mean half-life of only 2 h compared to 72.6 h for [^{111}In-DOTA]MG0, which ultimately reduced its suitability for clinical use [69]. Kolenic-Petial et al. found that the half-life of these minigastrin analogues could be improved by inserting a linker of non-ionic D-amino acids into the structure [68]. The authors further demonstrated the metabolic stability of the minigastrin analogue could be improved by increasing the length of the linker, with a linker comprised of six D-glutamine residues proving to be optimal and resulting in a half-life in human serum of approximately 495 h (Figure 8B) [68].

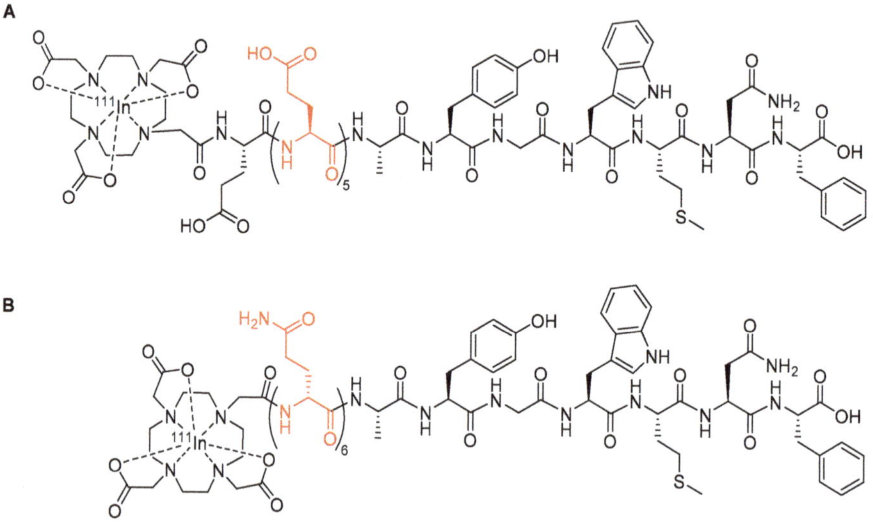

Figure 8. (**A**) Minigastrin analogue [^{111}In-DOTA]MG0 with the five L-glutamic acids linker highlighted in red [67]; (**B**) minigastrin peptide radiopharmaceutical developed by Kolenic-Petail et al., with the six D-amino acids linker highlighted in red [68].

While the research of Kolenic-Petail et al. [68] discussed above utilized indium-111 for SPECT imaging, the benefits achieved from utilizing D-amino acids could be easily carried over to PET studies, for example, by chelating a PET radiometal such as gallium-68 in place of indium-111 or by utilizing D-amino acids in the linker between a peptide and a radiofluorinated prosthetic group.

3.2. β-Amino Acids

β-Amino acids are widely found as biologically active natural products produced by plants [70], microorganisms [71], and marine organisms [72], but relatively few β-amino acids are found in mammalian systems [73]. The incorporation of single or multiple β-amino acids into peptides is of increasing interest in the pharmaceutical field as they can enhance in vivo metabolic stability [74–78] and potency [74,79]. This has been attributed to their different electronic environments and backbone/side chain configurations, compared to their α-amino acid analogues, decreasing recognition by proteases [80].

Garayoa et al. showed that the incorporation of a β-alanine-β-alanine (βAla-βAla) linker into a human tumor targeting bombesin peptide radiopharmaceutical (Figure 9) resulted in a two-fold increase in metabolic stability against proteolytic degradation and no decrease in receptor affinity when compared to the unmodified structure (sans linker) in studies performed using in vitro tumor cell cultures [81]. However, when the studies were performed in human plasma, the analogue with the βAla-βAla linker showed decreased metabolic stability with a half-life of 10 h compared to 16 h for the unmodified analogue [81].

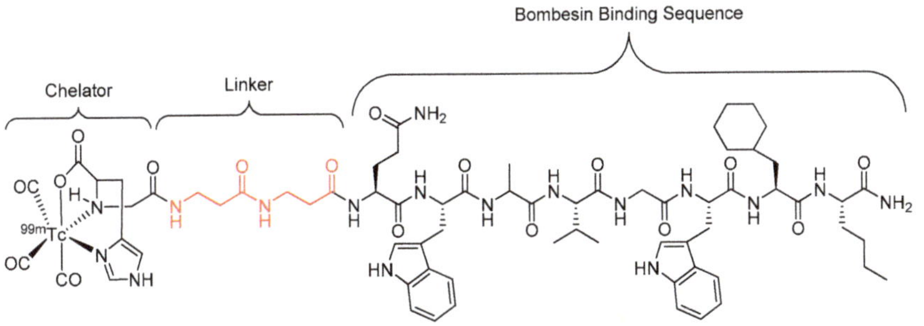

Figure 9. Bombesin-based peptide radiopharmaceutical investigated by Garayoa et al., with the βAla–βAla linker highlighted in red [81].

Further research by Garayoa et al. investigated alternative β-amino acid linkers (Figure 10) for the same bombesin peptide radiopharmaceutical as shown above (Figure 9), with a focus on β-amino acids that could hold a charge once incorporated into the final structure of the radiopharmaceutical [17]. The linkers were constructed from a combination of two or three of the selected β-amino acids in sequence. It was found that the bombesin analogue modified with $β^3$-homoglutamic acid (Figure 10d) in the linker, with one single negative charge, showed a significant increase in tumor uptake and tumor-to-tissue ratio, but did not increase metabolic stability compared to the previously developed radiopharmaceutical (Figure 9) [17].

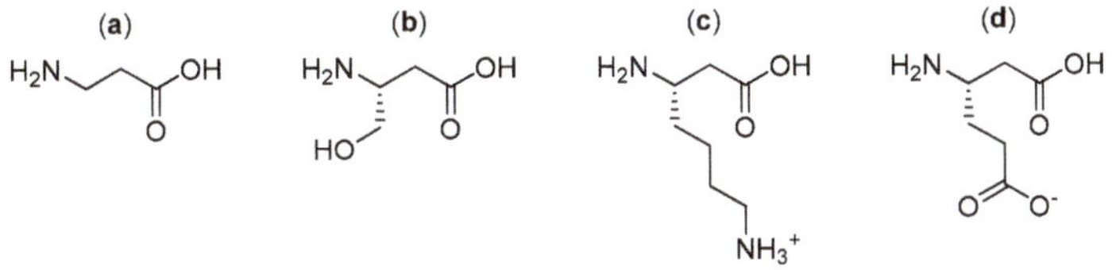

Figure 10. Alternate β-amino acids (a) β-alanine; (b) $β^3$-homoserine; (c) $β^3$-homolysine; and (d) $β^3$-homoglutamic acid investigated by Garayoa et al. for use as linkers [17].

In a similar study carried out by Popp et al., a single β-alanine amino acid, with a methylated nitrogen, was used as a linker in a statine-based GRPr-antagonist radiopharmaceutical (Figure 11) designed to target receptors on the surface of several human tumors [82]. The introduction of the N-methylated β-alanine did not disrupt the binding affinity and presented a similar in vivo stability in mice compared to the unmodified compound, with approximately 50% of the activity in the bloodstream representing intact compound 15 min post injection in the case of both compounds [82].

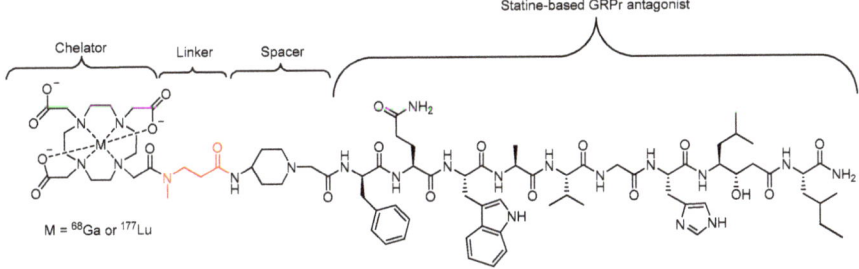

Figure 11. Statine-based GRPr antagonist radiopharmaceutical investigated by Popp et al., with the N-methylated β-alanine linker highlighted in red [82].

The research undertaken by Garayoa et al. [17,81] and Popp et al. [82] aimed to demonstrate that unnatural β-amino acids could be introduced into the structure of peptide-based radiopharmaceuticals to increase metabolic stability. While neither author achieved a significant increase in metabolic stability, they did produce compounds that maintained metabolic stability and binding affinity close to the unmodified analogue. This indicates that there is room for improvement for future peptide-based radiopharmaceuticals that incorporate unnatural β-amino acids.

While the research of Garayoa et al. only used the SPECT radionuclide technetium-99m, it should be apparent that the introduction of β-amino acids can easily be applied to PET studies through, for example, chelating PET radiometals or incorporating β-amino acids into peptide-based PET radiopharmaceuticals. Despite the potential benefits to pharmacological properties, there are currently no examples in the literature of β-amino acids being applied to improve the metabolic stability of peptide-based fluorine-18 labeled radiopharmaceuticals. Interestingly, the research by Schjoeth-Eskensen et al. successfully demonstrated the radiolabeling of the α-carbon of β-alanine with fluorine-18 (Figure 12), but no stability studies or subsequent conjugation to a peptide were performed [83].

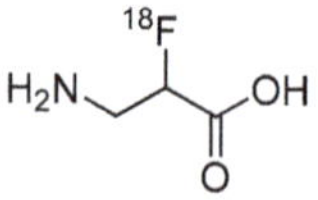

Figure 12. Fluorine-18 labeled β-alanine synthesized by Schjoeth-Eskensen et al. [83].

3.3. N-Methylation

Peptides can be modified through N-methylation (also known as N-alkylation). This method constitutes substituting one or more NH groups in a peptide backbone with N-methyl substituents (Figure 13). N-Methylation can confer several benefits compared to their unmodified analogues, including enhanced protease resistance, membrane permeability, and biological activity [84].

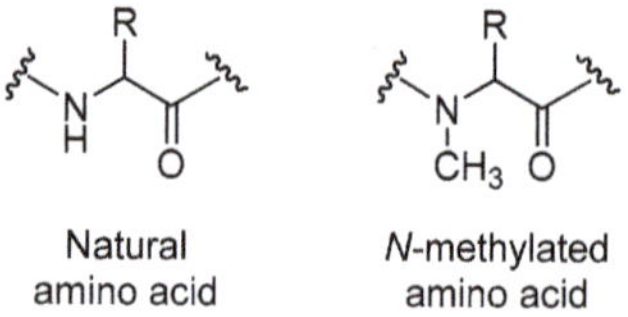

Figure 13. Structures of natural compared to N-methylated amino acids.

N-Methylation of a peptide backbone replaces a hydrogen bond donor (NH) with a potential steric clash (NCH$_3$), thus eliminating some inter- and intramolecular hydrogen bonds [85–87]. N-Methylation can also greatly alter the conformation of the entire peptide. This occurs because the N-methyl group will influence the conformational flexibility of both the peptide backbone and the side chains of the residues close to the N-methyl amino acids. Of particular note, the energy barrier between the *trans* and *cis* peptide bond conformations (Figure 14) is greatly reduced and consequently the *cis* peptide bond conformation becomes readily accessible [88].

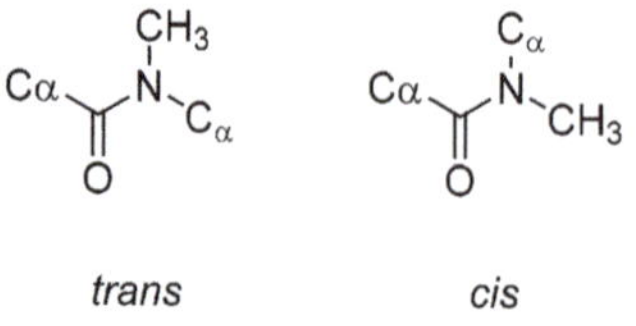

Figure 14. Comparison of the *trans* and *cis* conformations of N-methylated peptide bonds.

With the overall conformational change that can occur with N-methylated peptides, and the change in H-bonding capacity and steric features, N-methylation often leads to a decreased affinity of the peptide for the active site of proteases and therefore increased metabolic stability. It has also been found that N-methylation of an amide bond adjacent to the cleavage site can confer a greater resistance against enzymatic degradation than N-methylation of the amide bond at the cleavage site itself [89]. This behavior may result from N-methylation making the *cis* conformation readily accessible and then becoming the preferred conformation of the peptide in vivo. For example, the *cis* conformation may result in portions of the peptide being positioned such that they are now less accessible to proteolytic activity or simply no longer fit into the enzyme binding site, thus increasing the metabolic stability [88]. However, these structural changes may also disrupt intra- and intermolecular hydrogen bonds that may be important for the stabilization of biologically active conformations and for target receptor recognition [90]. Therefore, the use of N-methylation for increasing metabolic stability must be balanced against maintaining biological activity against the desired target receptor [91,92].

The endothelin peptides are potent vasoconstrictors. It has been found that highly potent antagonists of endothelin receptors can be developed from the C-terminal hexapeptide of endothelin [88]. However, these compounds are generally unstable towards enzymatic proteolysis and consequently have relatively short half-lives, which reduces their utility for clinical use [88]. Wayne et al. found that the N-methylation of a single isoleucine residue in the sequence of these previously developed endothelin receptor antagonists could significantly improve metabolic stability [88]. For example, N-methylation of the amide bond between the Ile^{19} and Ile^{20} residues increased the half-life in rat intestinal perfusate from 10.6 min to 538 min [88]. This modification also enhanced receptor binding affinity from an IC_{50} of 40 nM in the case of the unmodified compound down to 10 nM [88].

N-Methylation has also been effective when combined with other peptide modifications to further modulate the properties of a peptide-based radiopharmaceuticals. For example, as previously discussed in Section 3.2, Popp et al. successfully incorporated an N-methylated β-alanine residue as a linker in a statine-based GRPr antagonist radiopharmaceuticals [82].

3.4. PEGylation and Alkyl Linkers

PEGylation defines the process of linking one or more polyethylene glycol (PEG) polymer chains (Figure 15) to a peptide, protein, or non-peptide molecule. PEG possesses useful properties, including high solubility in water and many organic solvents, non-toxicity, and non-immunogenicity, and has been approved by the FDA for human use [93]. Many PEGylated peptide-based radiopharmaceuticals have been developed and shown to possess improved pharmacokinetic properties compared to their unmodified analogues, including increased receptor binding affinity, increased tumor uptake, and decreased kidney uptake [94–102]. Other potential improvements that have been achieved though PEGylation include longer circulatory times, increased aqueous solubility [103], and increased metabolic stability by creating steric hinderance that shields the molecule from proteases [104]. PEGylation also increases the molecular mass and size of peptides, which further increases body-residence time due to decreased kidney excretion [105]. The length and shape of PEGs (linear, branched, or dendritic) have been shown to influence the pharmacological properties of the PEGylated peptides and proteins, with branched PEG structures often most effective. This has been postulated to be due to both a greater degree of steric hinderance against proteases and possessing additional sites for conjugation with the target peptide or protein [106].

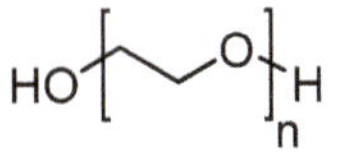

Figure 15. Polyethylene glycol (PEG) structure.

The advantages offered by PEGylation has seen it applied to a variety of peptide-based radiopharmaceuticals. For example, Hausner et al. found that the PEGylation of a radiopharmaceuticals derived from the $\alpha_v\beta_6$ integrin targeting peptide A20FMDV2, at both the C- and N-termini, greatly improves its pharmacokinetic properties [107]. When evaluated in mice, the PEGylated peptide showed good metabolic stability with approximately 80% of the compound remaining intact after incubation in mouse serum for 1 h. The modified compound also showed a higher uptake in $\alpha_v\beta_6$-expressing tumors compared to the unmodified peptide, with a tumor uptake in BxPC-3 cells of 4.7% ID/g compared to 0.69% ID/g at 1 h for the unmodified peptide (Figure 16) [107].

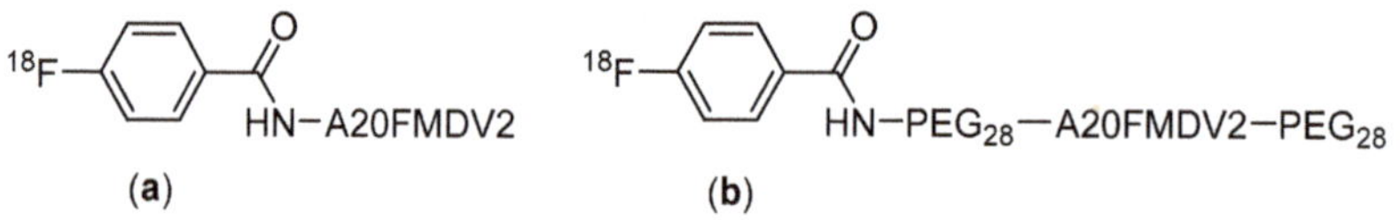

Figure 16. Structures of (a) [^{18}F]FBA-A20FMDV2 and (b) the bi-terminally PEGylated [^{18}F]FBA-PEG$_{28}$-A20FMDV2-PEG$_{28}$ [107].

The inclusion of large PEG groups has also been echoed by Dapp and co-workers [108,109], who used a 5 KDa PEG group in the linker of a bombesin peptide radiotherapeutic (Figure 17). The metabolic stability of the compound was assessed in vitro by incubating it in human serum for five days [109]. The inclusion of the PEG into the linker resulted in an increase in the amount of intact radiopharmaceuticals remaining after the five-day incubation, from 14% with the unmodified compound to 52% with the modified compound [109].

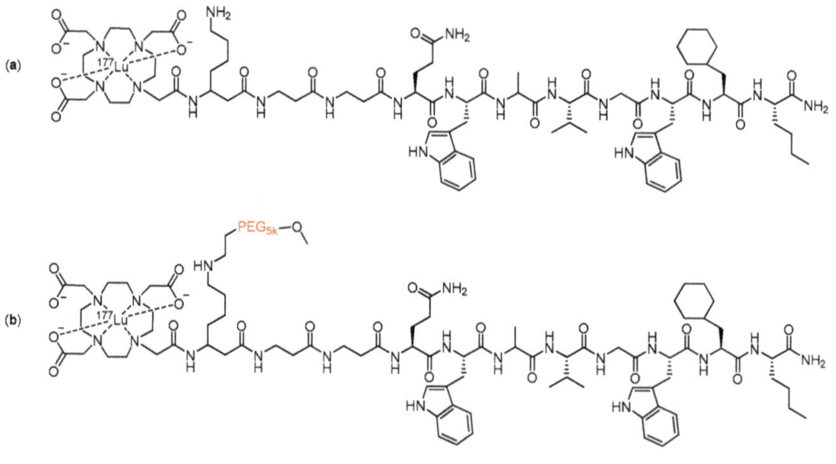

Figure 17. Structure of (a) unmodified and (b) modified bombesin peptide radiotherapeutics investigated by Dapp et al., with the PEG linker modification highlighted in red [109].

In another example, Wu et al. incorporated a mini-PEG (three ethylene oxide units) spacer into [^{18}F]FB-E[c(RGDyK)]$_2$ to produce [^{18}F]FB-mini-PEG-E[c(RGDyK)]$_2$ (Figure 18). This radiopharmaceutical showed greater radiolabeling yield, reduced renal excretion, and similar tumor uptake compared to the non-PEGylated analogue (Figure 18) [110]. This example also demonstrated that long PEG chains are not always necessary to significantly improve the properties of peptide-based radiopharmaceuticals.

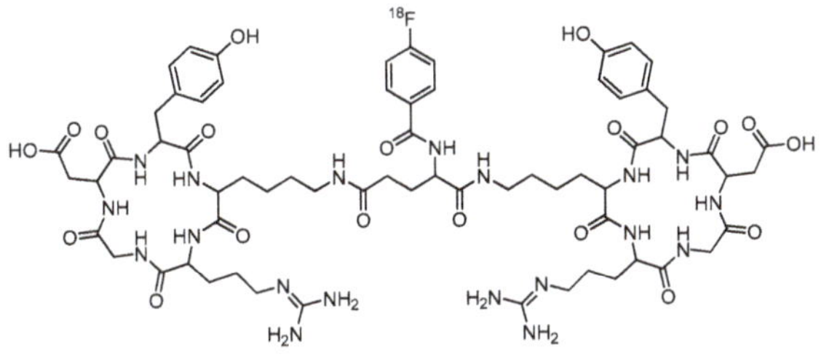

[^{18}F]FB-E[c(RGDyK)]$_2$

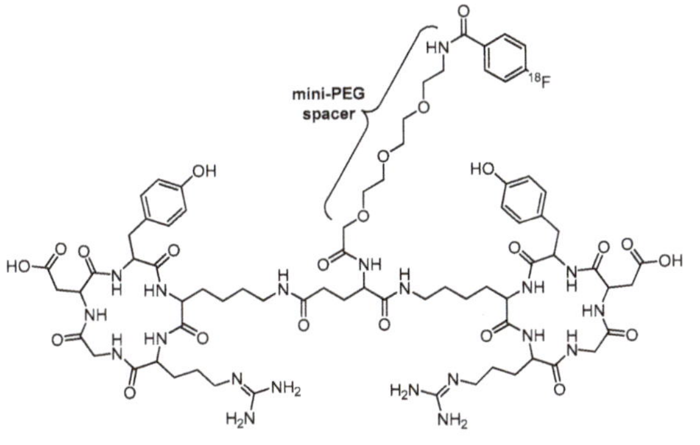

[^{18}F]FB-mini-PEG-E[c(RGDyK)]$_2$

Figure 18. Structures of [^{18}F]FB-E[c(RGDyK)]$_2$ and [^{18}F]FB-mini-PEG-E[c(RGDyK)]$_2$ [110].

Work conducted by the Maecke group [111] also investigated the effect of PEGylation on the previously developed bombesin analogue DOTA-Bombesin (7–14). Initial studies using a PEG_4 chain as a linker between the DOTA chelator and the peptide with gallium-68 or lutetium-177 as the radionuclide (Figure 19) found that the modified peptide-based radiopharmaceutical had superior pharmacokinetic properties in the PC-3 tumor-bearing mouse model than previously developed analogues [111]. In a subsequent study, the Maecke group synthesized a series of four lutetium-177 radiolabeled statine-based bombesin antagonist radiopharmaceuticals with PEG chains of different lengths (2, 4, 6, or 12 PEG units) as the linker between the DOTA chelator and the peptide analogue (Figure 19) [112]. The metabolic stability of these compounds was then assessed in human serum, and it was found that the half-life increased as PEG chain length increased up to the PEG_6 linker (half-life of PEG_2 = 246 h, PEG_4 = 407 h, and PEG_6 = 584 h), but began to decrease with the PEG_{12} linker (half-life of 407 h) [112].

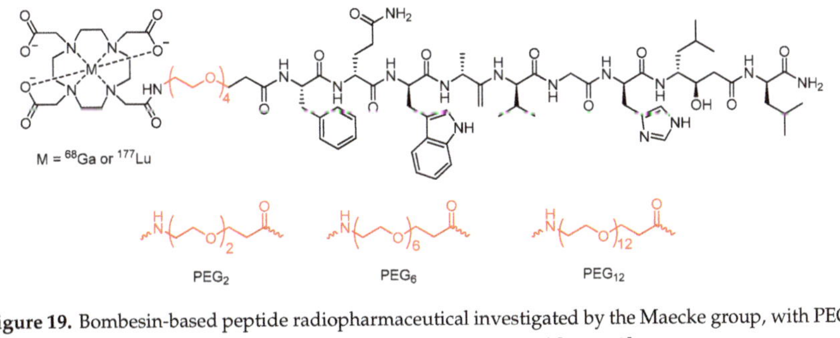

Figure 19. Bombesin-based peptide radiopharmaceutical investigated by the Maecke group, with PEG_4 chain linker and alternative PEG chain lengths highlighted in red [111,112].

In a similar study by Bacher et al., the incorporation of a PEG$_3$ linker into a different bombesin-based radiopharmaceutical was explored as a method to increase its metabolic stability (Figure 20) [113]. In vitro stability assays conducted in human serum found that incorporation of the PEG$_3$ linker led to a 9% increase in stability compared to the compound not containing a linker [113]. In the same study, Bacher et al. also explored the potential stabilizing effects of other alkyl chains as linkers, including 4-amino-1-carboxymethyl-piperidine and 6-aminohexanoic acid (Figure 20) [113]. However, these analogues were found to decrease metabolic stability by 20% and 7%, respectively, when compared to the radiopharmaceutical without any linker. In contrast to this, it was found that when the 4-amino-1-carboxylmethyl-piperdine linker was incorporated into a dimer of the bombesin-based radiopharmaceutical (Figure 21), there was a 52% increase in stability compared to the dimer without any linker [113]. The incorporation of the 6-aminohexanoic acid linker into the dimer resulted in decreased metabolic stability of the radiopharmaceutical [113].

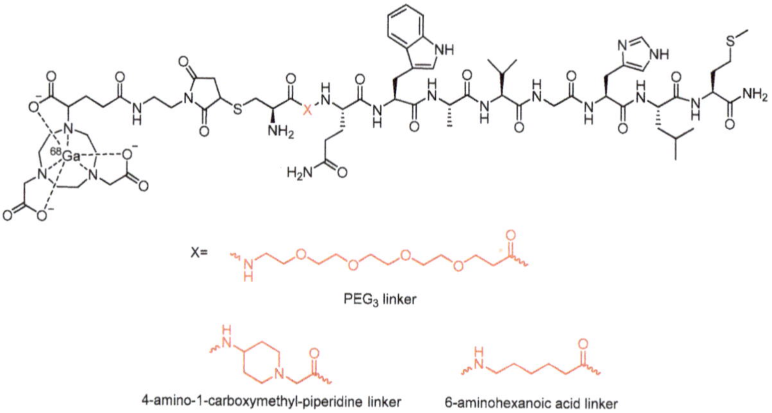

Figure 20. Bombesin-based radiopharmaceuticals investigated by Bacher et al., with the linker location and linkers highlighted in red [113].

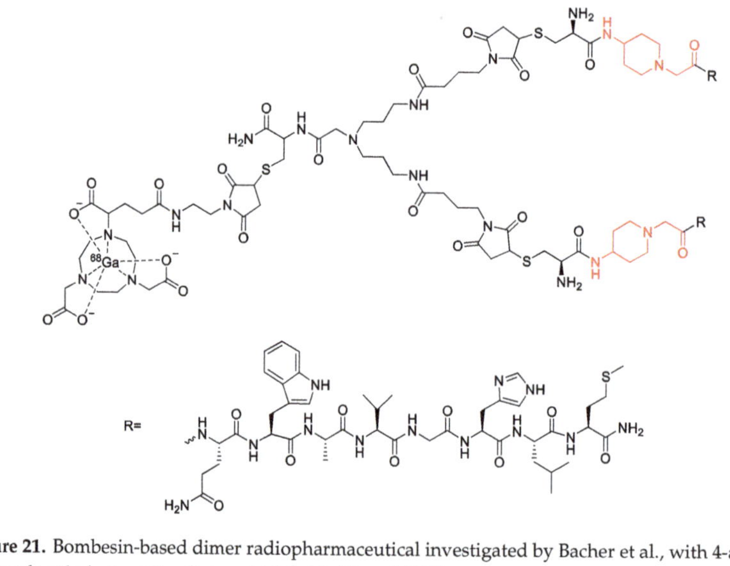

Figure 21. Bombesin-based dimer radiopharmaceutical investigated by Bacher et al., with 4-amino-1-carboxylmethyl-piperidine linkers highlighted in red [113].

While beyond the scope of this review, it is interesting to note that PEGylation has also been successfully applied to antibody fragments for the use as PET radiopharmaceuticals, with promising results [114,115].

3.5. Peptide Cyclization

Cyclization of a peptide sequence has been found to enhance stability against proteolytic degradation [116]. Cyclization is usually achieved by linking the C-terminus to the N-terminus of the peptide backbone [117]. However, it can also be advantageous to cyclize peptides by linking the C- or N-terminus to a side chain or linking one side chain to another side chain [118]. Depending on the desired cyclization site, cyclic peptides can be arranged in several ways, including head-to-tail, head-to-side chain, tail-to-side chain, and side chain-to-side chain (Figure 22) [119].

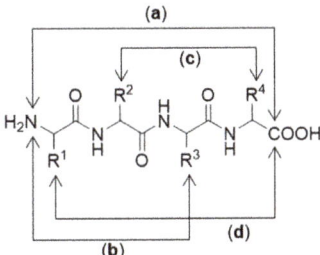

Figure 22. Different cyclization arrangements: (**a**) head-to-tail chain; (**b**) head-to-side chain; (**c**) side chain-to-side chain; (**d**) tail-to-side chain.

A distinct advantage of backbone cyclization over other cyclization methods is that cyclization is performed between backbone atoms, leaving the side chains that are usually essential for biological activity untouched [84].

Cyclization of a peptide through the formation of an amide bond between the C- and N-terminus can prevent degradation of the peptide by exopeptidases [60]. Cyclization of a linear peptide can also be used to increase the peptides' structural rigidity. This can increase metabolic stability by locking the peptide into a conformation that is less susceptible towards proteolytic enzymes (conformational constraints and/or selective molecular recognition). It can also be used to increase biological activity by locking the peptide into a more biologically active conformation [118,120,121].

Cyclization has been applied to the RGD (Arg-Gly-Asp) peptide sequence that can be used to target the $\alpha_v\beta_3$-integrin receptors, which are overexpressed on the surface of various tumor cells [33]. In cyclic RGD systems, the RGD peptide sequence is flanked by other amino acids to form a ring system that presents the RGD sequence in a specific conformation. Cyclic RGD systems are more potent, specific, and resistant to proteolysis than their linear analogues [122]. Consequently, cyclic RGD systems have been widely used in the development of various peptide-based PET radiopharmaceuticals [19,123].

3.6. Substitution of Amides with Sulfonamides

Another strategy that has been applied to increase the metabolic stability of peptide-based drugs is to substitute one or more amide groups in the backbone of a peptide with sulfonamide groups (Figure 23). Incorporation of the sulfonamide moiety into drugs is well known, as exemplified with the classical antibacterial sulfa drugs, and the sulfonamide group is inherently more stable than amides in mammalian systems [124].

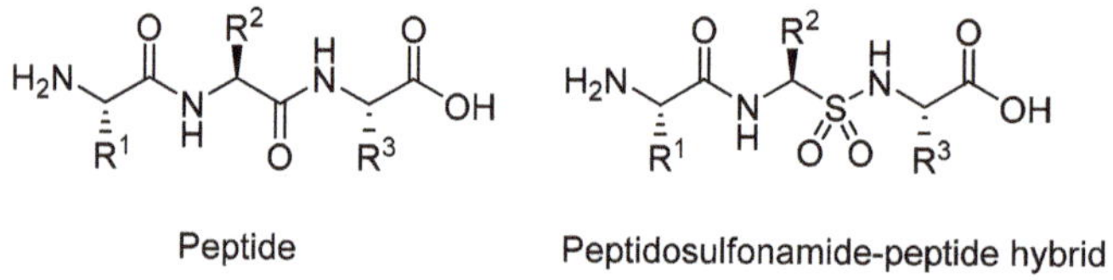

Figure 23. Structure of a regular peptides compared to sulfonamide analogue.

Sulfonamide groups contain a tetrahedral achiral sulfur atom directly bound to two electronegative oxygen atoms. These features with respect to geometry and electronic environment strongly resemble the transition state of peptide bond hydrolysis (Figure 24) [125]. Consequently, peptides containing sulfonamide substitutions have been investigated as transition state isosteres for the use as protease inhibitors [126].

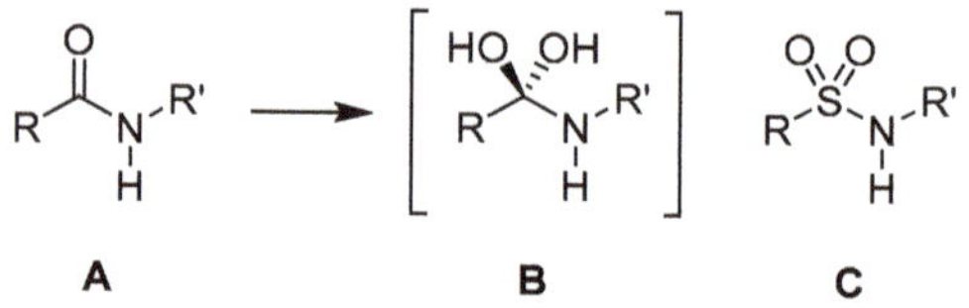

Figure 24. (**A**) Peptide bond structure. (**B**) Transition state for the hydrolysis of the peptide bond. (**C**) Sulfonamide bond as a suggested transition state isostere.

When compared to the amide moiety, the sulfonamide group is a stronger hydrogen bond donor [127] and the sulfonamide N-H is more acidic than the amide N-H, but a weaker hydrogen bond acceptor [127–129]. Furthermore, the hydrogen bond accepting character of the sulfonamide moiety is split between two accepting sites due to the two sulfonamide oxygens. These factors can impact the native hydrogen bonding network of the peptide and disrupt the formation of secondary structures. In contrast to the relatively rigid amide peptide bond, the sulfonamide bond is more freely rotatable and the *cis–trans* isomerism is not observed [127,130]. This greater rotational freedom allows for the sulfonamide oxygens to assume a variety of positions, where one oxygen occupies a *cis* or *trans* orientation with respect to the amide N-H, while the other oxygen is in neither a *cis* nor *trans* position. This can impede the formation of secondary structures by preventing the proper alignment of hydrogen bonds [127]. These potential disruptions to secondary structure formation have been found to have a greater effect on α-helices and a lesser effect on β-sheets [127].

The replacement of one or more amide bonds along a peptide backbone with sulfonamides has been successfully applied to develop peptidosulfonamide peptide analogues that display increased stability towards proteases compared to their unmodified analogues while also maintaining satisfactory biological activity [127,128,131]. The most common method of applying this strategy is to identify the preferred protease cleavage sites on a peptide and substitute the amides at those locations with sulfonamides. However, it has also been found that the substitution of amides close to cleavage sites can also increase metabolic stability [131]. This may be due to an effect similar to that seen in N-methylation where the substitution of the native amide bond with a more flexible bond, in this case a sulfonamide, allows the peptide to take a conformation that prevents proteases accessing the cleavage site [88,90].

The synthesis of a peptide in which all amides in the sequence are substituted with sulfonamides would lead to a peptidosulfonamide oligomer. However, this approach is not wise as α-amino sulfonamides are prone to fragmentation, releasing SO_2 [132]. This has been addressed by using β-aminosulfonamides, which are more stable than their α-amino analogues (Figure 25) [127].

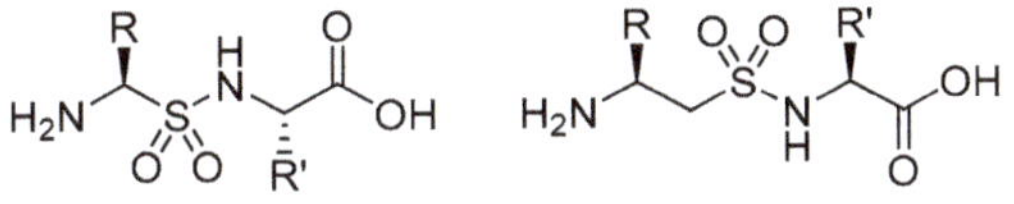

Figure 25. (a) Structure of α-peptidosulfonamide-α-peptide hybrid. (b) Structure of β-aminosulfonamide-α-peptide hybrid.

The substitution of the amide moiety with sulfonamides is starting to be explored in the development of peptide-based radiopharmaceuticals, including for linking of the peptide to the targeting moiety. For example, common amine-reactive prosthetic groups such as N-succinimidyl 4-[^{18}F]fluorobenzoate ([^{18}F]SFB) and 4-[^{18}F]fluorobenzoic acid ([^{18}F]FBA) are used to label peptides through the formation of amide bonds with primary amine residues (e.g., N-terminus or lysine) present in the peptide backbone [133,134]. While this method of labeling peptides has proven to be convenient, the susceptibility of the resulting amide bonds to hydrolysis in vivo is a potential vulnerability [36,135]. Löser et al. sought to explore this by comparing the metabolic stability of the fluorinated amide, N-(4-fluorophenyl)-fluoroacetanilide, and the fluorinated sulfonamide, N-(4-fluorophenyl)-3-fluoropropane-1-sulfonamide (Figure 26) [36]. The metabolic stability of both compounds were tested, and after 120 min of incubation in pig liver esterase (the porcine homologue of carboxylesterase), 95% of the N-(4-fluorophenyl)-3-fluoropropane-1-sulfonamide compared to only 20% of N-(4-fluorophenyl)-fluoroacetanilide remained intact [36]. While the compounds in this study were not complete structural analogues of each other, this research provides evidence of the potential benefits of substituting amide for sulfonamide bonds in radiopharmaceuticals.

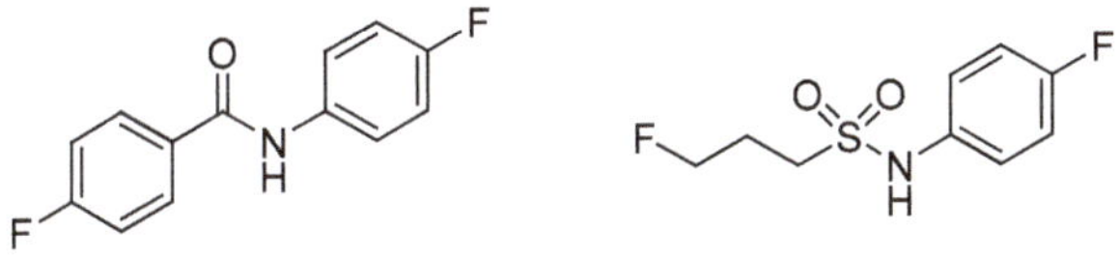

Figure 26. Structures of (a) N-(4-fluorophenyl)-fluoroacetanilide and (b) N-(4-fluorophenyl)-3-fluoropropane-1-sulfonamide [36].

4. Conclusions

The success of peptide-based PET radiopharmaceuticals, such as NETSPOT®, has sparked renewed interest in the development of new PET radiolabeled peptides for targeting diseases in the body. The applicability of new peptide-based radiopharmaceuticals will be influenced to a large extent by their in vivo stability as the inherently poor in vivo stability of natural peptides is one of the biggest challenges in the development of peptide-based radiopharmaceuticals, especially as degradation of the peptide can lead to non-specific binding. There have been several strategies developed to avoid this by modifying natural peptides to enhance their metabolic stability and sometimes other pharmacological properties such as receptor affinity. Effective strategies have included modification of the C- and/or N-termini, introduction of D- or other unnatural amino acids, backbone modification, PEGylation and alkyl chain incorporation, cyclization and peptide bond substitution. It has also been found that by applying more than one of these modifications in tandem on the same peptide, the different modifications can often work in concert to further enhance metabolic stability. However, no one approach fits all peptides and the decision of which strategy to apply must be made on a case-by-case basis. Consequently, it is rare to find individual studies where several different strategies have been applied to the same peptide to compare their efficacies. While some of the examples discussed in this review have only been applied to SPECT radiopharmaceuticals, radiotherapeutics, or non-radioactive peptide-based pharmaceuticals, the strategies could still be applied to PET radiopharmaceuticals. This could be achieved through simple substitution of a SPECT radiometal with a PET radiometal or by using the modification in the linker between a peptide and the radionuclide bearing moiety in a PET radiopharmaceutical. With the use of the discussed strategies to ensure in vivo stability, the number of successful peptide-based PET radiopharmaceuticals will continue to grow, and their clinical use will continue to expand.

Author Contributions: Conceptualization, B.J.E., A.T.K., A.K., L.M., and J.F.J.; writing—original draft preparation, B.J.E. and A.T.K.; writing—review and editing, B.J.E., A.T.K., A.K., L.M., and J.F.J.; supervision, L.M., A.K., and J.F.J.; project administration, J.F.J. All authors have read and agreed to the published version of the manuscript.

Funding: This research was funded by Macquarie University, including Macquarie University research training scholarships for B.J.E. and A.T.K.

Conflicts of Interest: The authors declare no conflict of interest.

References

1. Paans, A.M.J.; van Waarde, A.; Elsinga, P.H.; Willemsen, A.T.M.; Vaalburg, W. Positron emission tomography: The conceptual idea using a multidisciplinary approach. *Methods* **2002**, *27*, 195–207. [CrossRef]
2. Wood, K.A.; Hoskin, P.J.; Saunders, M.I. Positron emission tomography in oncology: A review. *Clin. Oncol.* **2007**, *19*, 237–255. [CrossRef] [PubMed]
3. Van de Bittner, G.C.; Ricq, E.L.; Hooker, J.M. A philosophy for CNS radiotracer design. *Acc. Chem. Res.* **2014**, *47*, 3127–3134. [CrossRef] [PubMed]
4. Coenen, H.H.; Elsinga, P.H.; Iwata, R.; Kilbourn, M.R.; Pillai, M.R.A.; Rajan, M.G.R.; Wagner, H.N.; Zaknun, J.J. Fluorine-18 radiopharmaceuticals beyond [18F]FDG for use in oncology and neurosciences. *Nucl. Med. Biol.* **2010**, *37*, 727–740. [CrossRef]
5. Olberg, D.E.; Hjelstuen, O.K. Labeling strategies of peptides with ^{18}F for positron emission tomography. *Curr. Top. Med. Chem.* **2010**, *10*, 1669–1679. [CrossRef]
6. Aluicio-Sarduy, E.; Ellison, P.A.; Barnhart, T.E.; Cai, W.; Nickles, R.J.; Engle, J.W. PET radiometals for antibody labeling. *J. Label. Compd. Radiopharm.* **2018**, *61*, 636–651. [CrossRef]
7. Fu, R.; Carroll, L.; Yahioglu, G.; Aboagye, E.O.; Miller, P.W. Antibody Fragment and Affibody ImmunoPET Imaging Agents: Radiolabelling Strategies and Applications. *ChemMedChem* **2018**, *13*, 2466–2478. [CrossRef]
8. Kolenc Peitl, P.; Tamma, M.; Kroselj, M.; Braun, F.; Waser, B.; Reubi, J.C.; Sollner Dolenc, M.; Maecke, H.R.; Mansi, R. Stereochemistry of Amino Acid Spacers Determines the Pharmacokinetics of 111In-DOTA-Minigastrin Analogues for Targeting the CCK2/Gastrin Receptor. *Bioconjug. Chem.* **2015**, *26*, 1113–1119. [CrossRef]
9. Antunes, P.; Ginj, M.; Walter, M.A.; Chen, J.; Reubi, J.-C.; Maecke, H.R. Influence of different spacers on the biological profile of a DOTA−somatostatin analogue. *Bioconjugate Chem.* **2007**, *18*, 84–92. [CrossRef]
10. Kapp, T.G.; Rechenmacher, F.; Neubauer, S.; Maltsev, O.V.; Cavalcanti-Adam, E.A.; Zarka, R.; Reuning, U.; Notni, J.; Wester, H.J.; Mas-Moruno, C.; et al. A comprehensive evaluation of the activity and selectivity profile of ligands for RGD-binding integrins. *Sci. Rep.* **2017**, *7*, 39805. [CrossRef]
11. Liese, S.; Netz, R.R. Influence of length and flexibility of spacers on the binding affinity of divalent ligands. *Beilstein J. Org. Chem.* **2015**, *11*, 804–816. [CrossRef] [PubMed]
12. Brandt, M.; Cardinale, J.; Aulsebrook, M.L.; Gasser, G.; Mindt, T.L. An overview of PET radiochemistry, part 2: Radiometals. *J. Nucl. Med.* **2018**, *59*, 1500–1506. [CrossRef] [PubMed]
13. Yordanova, A.; Eppard, E.; Kürpig, S.; Bundschuh, R.A.; Schönberger, S.; Gonzalez-Carmona, M.; Feldmann, G.; Ahmadzadehfar, H.; Essler, M. Theranostics in nuclear medicine practice. *Onco. Targets. Ther.* **2017**, *10*, 4821–4828. [CrossRef] [PubMed]
14. Reubi, J.C. Peptide receptors as molecular targets for cancer diagnosis and therapy. *Endocr. Rev.* **2003**, *24*, 389–427. [CrossRef]
15. Okarvi, S.M. Peptide-based radiopharmaceuticals: Future tools for diagnostic imaging of cancers and other diseases. *Med. Res. Rev.* **2004**, *24*, 357–397. [CrossRef]

16. Langer, M.; Beck-Sickinger, A. Peptides as carrier for tumor diagnosis and treatment. *Curr. Med. Chem. Agents* **2001**, *1*, 71–93. [CrossRef]
17. García Garayoa, E.; Schweinsberg, C.; Maes, V.; Brans, L.; Bläuenstein, P.; Tourwé, D.A.; Schibli, R.; Schubiger, P.A. Influence of the molecular charge on the biodistribution of bombesin analogues labeled with the [99m Tc(CO) 3]-core. *Bioconjug. Chem.* **2008**, *19*, 2409–2416. [CrossRef]
18. Eberle, A.N.; Mild, G.; Froidevaux, S. Receptor-mediated tumor targeting with radiopeptides. Part 1. General concepts and methods: Applications to somatostatin receptor-expressing tumors. *J. Recept. Signal Transduct.* **2004**, *24*, 319–455. [CrossRef]
19. Schottelius, M.; Wester, H.J. Molecular imaging targeting peptide receptors. *Methods* **2009**, *48*, 161–177. [CrossRef]
20. Mullard, A. FDA approvals for the first 6 months of 2016. *Nat. Rev. Drug Discov.* **2016**, *15*, 523. [CrossRef]
21. Diao, L.; Meibohm, B. Pharmacokinetics and pharmacokinetic–pharmacodynamic correlations of therapeutic peptides. *Clin. Pharmacokinet.* **2013**, *52*, 855–868. [CrossRef] [PubMed]
22. Vlieghe, P.; Lisowski, V.; Martinez, J.; Khrestchatisky, M. Synthetic therapeutic peptides: Science and market. *Drug Discov. Today* **2010**, *15*, 40–56. [CrossRef] [PubMed]
23. Van Grieken, R.; de Bruin, M. Nomenclature for radioanalytical chemistry (IUPAC Recommendations 1994). *Pure Appl. Chem.* **1994**, *66*, 2513–2526.
24. Elisa Crestoni, M. Radiopharmaceuticals for diagnosis and therapy. In *Reference Module in Chemistry, Molecular Sciences and Chemical Engineering*; Elsevier: Oxford, UK, 2018.
25. Fani, M.; Good, S.; Maecke, H.R. Radiometals (non-Tc, non-Re) and bifunctional labeling chemistry. In *Handbook of Nuclear Chemistry*; Springer: New York, NY, USA, 2011; pp. 2143–2178.
26. Van Dort, M.E.; Jung, Y.W.; Sherman, P.S.; Kilbourn, M.R.; Wieland, D.M. Fluorine for hydroxy substitution in biogenic amines: Asymmetric synthesis and biological evaluation of fluorine-18-labeled beta-fluorophenylalkylamines as model systems. *J. Med. Chem.* **1995**, *38*, 810–815. [CrossRef] [PubMed]
27. Magata, Y.; Lang, L.; Kiesewetter, D.O.; Jagoda, E.M.; Channing, M.A.; Eckelman, W.C. Biologically stable [(18)F]-labeled benzylfluoride derivatives. *Nucl. Med. Biol.* **2000**, *27*, 163–168. [CrossRef]
28. Carroll, V.; Demoin, D.W.; Hoffman, T.J.; Jurisson, S.S. Inorganic chemistry in nuclear imaging and radiotherapy: Current and future directions. *Radiochim. Acta* **2012**, *100*, 653–667. [CrossRef]
29. Price, E.W.; Orvig, C. Matching chelators to radiometals for radiopharmaceuticals. *Chem. Soc. Rev.* **2014**, *43*, 260–290. [CrossRef]
30. Kuchar, M.; Mamat, C. Methods to increase the metabolic stability of 18F-radiotracers. *Molecules* **2015**, *20*, 16186–16220. [CrossRef]
31. Werner, H.M.; Cabalteja, C.C.; Horne, W.S. Peptide backbone composition and protease susceptibility: Impact of modification type, position, and tandem substitution. *ChemBioChem* **2016**, *17*, 712–718. [CrossRef]
32. Furman, J.L.; Chiu, M.; Hunter, M.J. Early engineering approaches to improve peptide developability and manufacturability. *AAPS J.* **2015**, *17*, 111–120. [CrossRef]
33. Fani, M.; Maecke, H.R.; Okarvi, S.M. Radiolabeled peptides: Valuable tools for the detection and treatment of cancer. *Theranostics* **2012**, *2*, 481–501. [CrossRef]
34. Yeong, C.H.; Cheng, M.H.; Ng, K.H. Therapeutic radionuclides in nuclear medicine: Current and future prospects. *J. Zhejiang Univ. Sci. B* **2014**, *15*, 845–863. [CrossRef] [PubMed]
35. Liu, S.; Edwards, D.S. Bifunctional chelators for therapeutic lanthanide radiopharmaceuticals. *Bioconjug. Chem.* **2001**, *12*, 7–34. [CrossRef] [PubMed]
36. Löser, R.; Fischer, S.; Hiller, A.; Köckerling, M.; Funke, U.; Maisonial, A.; Brust, P.; Steinbach, J. Use of 3-[(18)F]fluoropropanesulfonyl chloride as a prosthetic agent for the radiolabelling of amines: Investigation of precursor molecules, labelling conditions and enzymatic stability of the corresponding sulfonamides. *Beilstein J. Org. Chem.* **2013**, *9*, 1002–1011. [CrossRef] [PubMed]
37. Behr, T.M.; Gotthardt, M.; Barth, A.; Béhé, M. Imaging tumors with peptide-based radioligands. *Q. J. Nucl. Med.* **2001**, *45*, 189–200. [PubMed]
38. Radzicka, A.; Wolfenden, R. Rates of uncatalyzed peptide bond hydrolysis in neutral solution and the transition state affinities of proteases. *J. Am. Chem. Soc.* **1996**, *26*, 6105–6109. [CrossRef]
39. Liskamp, R.M.J.; Rijkers, D.T.S.; Kruijtzer, J.A.W.; Kemmink, J. Peptides and proteins as a continuing exciting source of inspiration for peptidomimetics. *ChemBioChem* **2011**, *12*, 1626–1653. [CrossRef]

40. Mahesh, S.; Tang, K.-C.; Raj, M. Amide bond activation of biological molecules. *Molecules* **2018**, *23*, 2615. [CrossRef]
41. Haubner, R.; Decristoforo, C. Radiotracer II: Peptide-based radiopharmaceuticals. In *Small Animal Imaging: Basics and Practical Guide*; Springer: Berlin/Heidelberg, Germany, 2011; pp. 247–266. ISBN 9783642129445.
42. Charron, C.L.; Hickey, J.L.; Nsiama, T.K.; Cruickshank, D.R.; Turnbull, W.L.; Luyt, L.G. Molecular imaging probes derived from natural peptides. *Nat. Prod. Rep.* **2016**, *33*, 761–800. [CrossRef]
43. Brik, A.; Wong, C.H. HIV-1 protease: Mechanism and drug discovery. *Org. Biomol. Chem.* **2003**, *1*, 5–14. [CrossRef]
44. Hedstrom, L. Serine protease mechanism and specificity. *Chem. Rev.* **2002**, *102*, 4501–4523. [CrossRef] [PubMed]
45. Yamada, R.; Kera, Y. D-amino acid hydrolysing enzymes. In *D-amino Acids in Sequences of Secreted Peptides of Multicellular Organisms*; Birkhäuser Verlag: Basel, Switzerland, 1998; Volume 85.
46. Garton, M.; Nim, S.; Stone, T.A.; Wang, K.E.; Deber, C.M.; Kim, P.M. Method to generate highly stable D-amino acid analogs of bioactive helical peptides using a mirror image of the entire PDB. *Proc. Natl. Acad. Sci. USA* **2018**, *115*, 1505–1510. [CrossRef] [PubMed]
47. Miller, S.M.; Simon, R.J.; Ng, S.; Zuckermann, R.N.; Kerr, J.M.; Moos, W.H. Comparison of the proteolytic susceptibilities of homologous L-amino acid, D-amino acid, and N-substituted glycine peptide and peptoid oligomers. *Drug Dev. Res.* **1995**, *35*, 20–32. [CrossRef]
48. Uppalapati, M.; Lee, D.J.; Mandal, K.; Li, H.; Miranda, L.P.; Lowitz, J.; Kenney, J.; Adams, J.J.; Ault-Riché, D.; Kent, S.B.H.; et al. A potent d-protein antagonist of VEGF-A is nonimmunogenic, metabolically stable, and longer-circulating in vivo. *ACS Chem. Biol.* **2016**, *11*, 1058–1065. [CrossRef] [PubMed]
49. Gentilucci, L.; Cardillo, G.; Squassabia, F.; Tolomelli, A.; Spampinato, S.; Sparta, A.; Baiula, M. Inhibition of cancer cell adhesion by heterochiral Pro-containing RGD mimetics. *Bioorganic Med. Chem. Lett.* **2007**, *17*, 2329–2333. [CrossRef] [PubMed]
50. Durani, S. Protein design with L- and D-α-amino acid structures as the alphabet. *Acc. Chem. Res.* **2008**, *41*, 1301–1308. [CrossRef]
51. Li, C.; Zhan, C.; Zhao, L.; Chen, X.; Lu, W.Y.; Lu, W. Functional consequences of retro-inverso isomerization of a miniature protein inhibitor of the p53-MDM2 interaction. *Bioorganic Med. Chem.* **2013**, *21*, 4045–4050. [CrossRef]
52. Pallerla, S.; Naik, H.; Singh, S.; Gauthier, T.; Sable, R.; Jois, S.D. Design of cyclic and D-amino acids containing peptidomimetics for inhibition of protein-protein interactions of HER2-HER3. *J. Pept. Sci.* **2018**, *24*, 2070–2073. [CrossRef]
53. Li, H.; Kem, D.C.; Zhang, L.; Huang, B.; Liles, C.; Benbrook, A.; Gali, H.; Veitla, V.; Scherlag, B.J.; Cunningham, M.W.; et al. Novel retro-inverso peptide inhibitor reverses angiotensin receptor autoantibody-induced hypertension in the rabbit. *Hypertension* **2015**, *65*, 793–799. [CrossRef]
54. Ben-Yedidia, T.; Beignon, A.-S.; Partidos, C.D.; Muller, S.; Arnon, R. A retro-inverso peptide analogue of influenza virus hemagglutinin B-cell epitope 91-108 induces a strong mucosal and systemic immune response and confers protection in mice after intranasal immunization. *Mol. Immunol.* **2002**, *39*, 323–331. [CrossRef]
55. Li, C.; Pazgier, M.; Li, J.; Li, C.; Liu, M.; Zou, G.; Li, Z.; Chen, J.; Tarasov, S.G.; Lu, W.Y.; et al. Limitations of peptide retro-inverso isomerization in molecular mimicry. *J. Biol. Chem.* **2010**, *285*, 19572–19581. [CrossRef] [PubMed]
56. Jochim, A.L.; Arora, P.S. Systematic analysis of helical protein interfaces reveals targets for synthetic inhibitors. *ACS Chem. Biol.* **2010**, *5*, 919–923. [CrossRef] [PubMed]
57. Law, V.; Knox, C.; Djoumbou, Y.; Jewison, T.; Guo, A.C.; Liu, Y.; MacIejewski, A.; Arndt, D.; Wilson, M.; Neveu, V.; et al. DrugBank 4.0: Shedding new light on drug metabolism. *Nucleic Acids Res.* **2014**, *42*, 1091–1097. [CrossRef] [PubMed]
58. Rabideau, A.E.; Pentelute, B.L. A D-amino acid at the N-terminus of a protein abrogates its degradation by the N-end rule pathway. *ACS Cent. Sci.* **2015**, *1*, 423–430. [CrossRef] [PubMed]
59. Veine, D.M.; Yao, H.; Stafford, D.R.; Fay, K.S.; Livant, D.L. A D-amino acid containing peptide as a potent, noncovalent inhibitor of α5β1 integrin in human prostate cancer invasion and lung colonization. *Clin. Exp. Metastasis* **2014**, *31*, 379–393. [CrossRef] [PubMed]
60. Werle, M.; Bernkop-Schnürch, A. Strategies to improve plasma half life time of peptide and protein drugs. *Amino Acids* **2006**, *30*, 351–367. [CrossRef]

61. Paganelli, G.; Bodei, L.; Handkiewicz Junak, D.; Rocca, P.; Papi, S.; Lopera Sierra, M.; Gatti, M.; Chinol, M.; Bartolomei, M.; Fiorenza, M.; et al. 90Y-DOTA-D-Phe1-Try3-octreotide in therapy of neuroendocrine malignancies. *Pept. Sci. Orig. Res. Biomol.* **2002**, *66*, 339–398. [CrossRef]
62. Harris, A.G. Somatostatin and somatostatin analogues: Pharmacokinetics and pharmacodynamic effects. *Gut* **1994**, *35*, S1–S4. [CrossRef]
63. Olsen, J.O.; Pozderac, R.V.; Hinkle, G.; Hill, T.; O'Dorisio, T.M.; Schirmer, W.J.; Ellison, E.C.; O'Dorisio, M.S. Somatostatin receptor imaging of neuroendocrine tumors with indium-111 pentetreotide (Octreoscan). *Semin. Nucl. Med.* **1995**, *25*, 251–261. [CrossRef]
64. Chin, J.; Vesnaver, M.; Bernard-Gauthier, V.; Saucke-Lacelle, E.; Wängler, B.; Wängler, C.; Schirrmacher, R. Direct one-step labeling of cysteine residues on peptides with [11C]methyl triflate for the synthesis of PET radiopharmaceuticals. *Amino Acids* **2013**, *45*, 1097–1108. [CrossRef]
65. Deppen, S.A.; Blume, J.; Bobbey, A.J.; Shah, C.; Graham, M.M.; Lee, P.; Delbeke, D.; Walker, R.C. 68Ga-DOTATATE compared with 111In-DTPA-octreotide and conventional imaging for pulmonary and gastroenteropancreatic neuroendocrine tumors: A systematic review and meta-analysis. *J. Nucl. Med.* **2016**, *57*, 872–878. [CrossRef]
66. Maqsood, M.H.; Tameez Ud Din, A.; Khan, A.H. Neuroendocrine tumor therapy with lutetium-177: A literature review. *Cureus* **2019**. [CrossRef] [PubMed]
67. Kaloudi, A.; Nock, B.A.; Lymperis, E.; Krenning, E.P.; De Jong, M.; Maina, T. Improving the in vivo profile of minigastrin radiotracers: A comparative study involving the neutral endopeptidase inhibitor phosphoramidon. *Cancer Biother. Radiopharm.* **2016**, *31*, 20–28. [CrossRef] [PubMed]
68. Kolenc-Peitl, P.; Mansi, R.; Tamma, M.; Gmeiner-Stopar, T.; Sollner-Dolenc, M.; Waser, B.; Baum, R.P.; Reubi, J.C.; Maecke, H.R. Highly improved metabolic stability and pharmacokinetics of indium-111-DOTA-gastrin conjugates for targeting of the gastrin receptor. *J. Med. Chem.* **2011**, *54*, 2602–2609. [CrossRef] [PubMed]
69. Good, S.; Walter, M.A.; Waser, B.; Wang, X.; Müller-Brand, J.; Béhé, M.P.; Reubi, J.C.; Maecke, H.R. Macrocyclic chelator-coupled gastrin-based radiopharmaceuticals for targeting of gastrin receptor-expressing tumours. *Eur. J. Nucl. Med. Mol. Imaging* **2008**, *35*, 1868–1877. [CrossRef] [PubMed]
70. Nicolaou, K.C.; Dai, W.-M.; Guy, R.K. Chemistry and biology of taxol. *Angew. Chemie Int. Ed. English* **1994**, *33*, 15–44. [CrossRef]
71. Minami, Y.; Yoshida, K.I.; Azuma, R.; Saeki, M.; Otani, T. Structure of an aromatization product of C-1027 chromophore. *Tetrahedron Lett.* **1993**, *34*, 2633–2636. [CrossRef]
72. Cardillo, G.; Tomasini, C. Asymmetric synthesis of β-amino acids and α-substituted β-amino acids. *Chem. Soc. Rev.* **1996**, *25*, 117–128. [CrossRef]
73. Griffith, O.W. β-amino acids: Mammalian metabolism and utility as α-amino acid analogues. *Annu. Rev. Biochem.* **1986**, *55*, 855–878. [CrossRef]
74. Juaristi, E.; Soloshonok, V.A. *Enantioselective Synthesis of [Beta]-amino Acids*; John Wiley & Sons, Inc.: Hoboken, NJ, USA, 2005; ISBN 9780471467380.
75. Tan, C.Y.K.; Weaver, D.F. A one-pot synthesis of 3-amino-3-arylpropionic acids. *Tetrahedron* **2002**, *58*, 7449–7461. [CrossRef]
76. Frackenpohl, J.; Arvidsson, P.I.; Schreiber, J.V.; Seebach, D. The outstanding biological stability of - and -peptides toward proteolytic enzymes: An in vitro investigation with fifteen peptidases. *ChemBioChem* **2001**, *2*, 445–455. [CrossRef]
77. Hansen, T.; Moe, M.K.; Anderssen, T.; Strøm, M.B. Metabolism of small antimicrobial β2,2-amino acid derivatives by murine liver microsomes. *Eur. J. Drug Metab. Pharmacokinet.* **2012**, *37*, 191–201. [CrossRef] [PubMed]
78. Tørfoss, V.; Ausbacher, D.; Cavalcanti-Jacobsen, C. de A.; Hansen, T.; Brandsdal, B.-O.; Havelkova, M.; Strøm, M.B. Synthesis of anticancer heptapeptides containing a unique lipophilic β2,2-amino acid building block. *J. Pept. Sci.* **2012**, *18*, 170–176. [CrossRef] [PubMed]
79. Ali, S.M.; Hoemann, M.Z.; Aube, J.; Mitscher, L.A.; Georg, G.I.; McCall, R.; Jayasinghe, L.R. Novel cytotoxic 3'-(tert-Butyl) 3'-diphenyl analogs of paclitaxel and docetaxel. *J. Med. Chem.* **1995**, *38*, 3821–3828. [CrossRef]
80. Hook, D.F.; Bindschädler, P.; Mahajan, Y.R.; Šebesta, R.; Kast, P.; Seebach, D. The proteolytic stability of "designed" β-peptides containing α-peptide-bond mimics and of mixed α,β-peptides: Application to the construction of MHC-binding peptides. *Chem. Biodivers.* **2005**, *2*, 591–632. [CrossRef]

81. García Garayoa, E.; Rüegg, D.; Bläuenstein, P.; Zwimpfer, M.; Khan, I.U.; Maes, V.; Blanc, A.; Beck-Sickinger, A.G.; Tourwé, D.A.; Schubiger, P.A. Chemical and biological characterization of new Re(CO)3/[99mTc](CO)3 bombesin analogues. *Nucl. Med. Biol.* **2007**, *34*, 17–28. [CrossRef]
82. Popp, I.; Del Pozzo, L.; Waser, B.; Reubi, J.C.; Meyer, P.T.; Maecke, H.R.; Gourni, E. Approaches to improve metabolic stability of a statine-based GRP receptor antagonist. *Nucl. Med. Biol.* **2017**, *45*, 22–29. [CrossRef]
83. Schjoeth-Eskesen, C.; Hansen, P.R.; Kjaer, A.; Gillings, N. Efficient regioselective ring opening of activated aziridine-2-carboxylates with [18F]fluoride. *ChemistryOpen* **2015**, *4*, 65–71. [CrossRef]
84. Linde, Y.; Ovadia, O.; Safrai, E.; Xiang, Z.; Portillo, F.P.; Shalev, D.E.; Haskell-Luevano, C.; Hoffman, A.; Gilon, C. Structure-activity relationship and metabolic stability studies of Backbone cyclization and N-methylation of melanocortin peptides. *Wiley InterSci.* **2008**, *90*, 671–682. [CrossRef]
85. Antolíková, E.; Žáková, L.; Turkenburg, J.P.; Watson, C.J.; Hančlová, I.; Šanda, M.; Cooper, A.; Kraus, T.; Brzozowski, A.M.; Jiráček, J. Non-equivalent role of inter- and intramolecular hydrogen bonds in the insulin dimer interface. *J. Biol. Chem.* **2011**, *286*, 36968–36977. [CrossRef]
86. Wormser, U.; Laufer, R.; Hart, Y.; Chorev, M.; Gilon, C.; Selinger, Z. Highly selective agonists for substance P receptor subtypes. *EMBO J.* **1986**, *5*, 2805–2808. [CrossRef]
87. Haviv, F.; Fitzpatrick, T.D.; Swenson, R.E.; Nichols, C.J.; Mort, N.A.; Bush, E.N.; Diaz, G.; Bammert, G.; Nguyen, A.; Rhutasel, N.S. Effect of N-methyl substitution of the peptide bonds in luteinizing hormone-releasing hormone agonists. *J. Med. Chem.* **1993**, *36*, 363–369. [CrossRef]
88. Cody, W.L.; He, J.X.; Reily, M.D.; Haleen, S.J.; Walker, D.M.; Reyner, E.L.; Stewart, B.H.; Doherty, A.M. Design of a potent combined pseudopeptide endothelin-A/endothelin-B receptor antagonist, Ac- d Bhg 16 -Leu-Asp-Ile-[NMe]Ile-Trp 21 (PD 156252): Examination of its pharmacokinetic and spectral properties. *J. Med. Chem.* **1997**, *40*, 2228–2240. [CrossRef]
89. Tal-Gan, Y.; Freeman, N.S.; Klein, S.; Levitzki, A.; Gilon, C. Metabolic stability of peptidomimetics: N-methyl and aza heptapeptide analogs of a PKB/Akt inhibitor. *Chem. Biol. Drug Des.* **2011**, *78*, 887–892. [CrossRef]
90. Snyder, J.P. Probable unimportance of intramolecular hydrogen bonds for determining the secondary structure of cyclic hexapeptides. Roseotoxin B. *J. Am. Chem. Soc.* **1984**, *106*, 2393–2400. [CrossRef]
91. White, T.R.; Renzelman, C.M.; Rand, A.C.; Rezai, T.; McEwen, C.M.; Gelev, V.M.; Turner, R.A.; Linington, R.G.; Leung, S.S.F.; Kalgutkar, A.S.; et al. On-resin N-methylation of cyclic peptides for discovery of orally bioavailable scaffolds. *Nat. Chem. Biol.* **2011**, *7*, 810–817. [CrossRef]
92. Räder, A.F.B.; Reichart, F.; Weinmüller, M.; Kessler, H. Improving oral bioavailability of cyclic peptides by N-methylation. *Bioorg. Med. Chem.* **2018**, *26*, 2766–2773. [CrossRef]
93. Veronese, F.M.; Pasut, G. PEGylation, successful approach to drug delivery. *Drug Discov. Today* **2005**, *10*, 1451–1458. [CrossRef]
94. Autio, A.; Henttinen, T.; Sipilä, H.J.; Jalkanen, S.; Roivainen, A. Mini-PEG spacering of VAP-1-targeting 68Ga-DOTAVAP-P1 peptide improves PET imaging of inflammation. *EJNMMI Res.* **2011**, *1*, 1–7. [CrossRef]
95. Accardo, A.; Galli, F.; Mansi, R.; Del Pozzo, L.; Aurilio, M.; Morisco, A.; Ringhieri, P.; Signore, A.; Morelli, G.; Aloj, L. Pre-clinical evaluation of eight DOTA coupled gastrin-releasing peptide receptor (GRP-R) ligands for in vivo targeting of receptor-expressing tumors. *EJNMMI Res.* **2016**, *6*, 1–10. [CrossRef]
96. Liu, Z.; Niu, G.; Shi, J.; Liu, S.; Wang, F.; Liu, S.; Chen, X. 68Ga-labeled cyclic RGD dimers with Gly3 and PEG 4 linkers: Promising agents for tumor integrin $\alpha v \beta 3$ PET imaging. *Eur. J. Nucl. Med. Mol. Imaging* **2009**, *36*, 947–957. [CrossRef] [PubMed]
97. Colin, D.J.; Inkster, J.A.H.; Germain, S.; Seimbille, Y. Preclinical validations of [18F]FPyPEGCBT-c(RGDfK): A 18F-labelled RGD peptide prepared by ligation of 2-cyanobenzothiazole and 1,2-aminothiol to image angiogenesis. *EJNMMI Radiopharm. Chem.* **2017**, *1*, 16. [CrossRef] [PubMed]
98. Chen, X.; Hou, Y.; Tohme, M.; Park, R.; Khankaldyyan, V.; Gonzales-Gomez, I.; Bading, J.R.; Laug, W.E.; Conti, P.S. Pegylated Arg-Gly-Asp Peptide: 64Cu Labeling and PET Imaging of Brain Tumor $\alpha v \beta 3$-Integrin Expression. *J. Nucl. Med.* **2004**, *45*, 1776–1783.
99. Shi, J.; Kim, Y.S.; Zhai, S.; Liu, Z.; Chen, X.; Liu, S. Improving tumor uptake and pharmacokinetics of 64Cu-labeled cyclic RGD peptide dimers with Gly 3 and PEG 4 linkers. *Bioconjug. Chem.* **2009**, *20*, 750–759. [CrossRef]
100. Dijkgraaf, I.; Liu, S.; Kruijtzer, J.A.W.; Soede, A.C.; Oyen, W.J.G.; Liskamp, R.M.J.; Corstens, F.H.M.; Boerman, O.C. Effects of linker variation on the in vitro and in vivo characteristics of an 111In-labeled RGD peptide. *Nucl. Med. Biol.* **2007**, *34*, 29–35. [CrossRef]

101. Wang, L.; Shi, J.; Kim, Y.S.; Zhai, S.; Jia, B.; Zhao, H.; Liu, Z.; Wang, F.; Chen, X.; Liu, S. Improving tumor-targeting capability and pharmacokinetics of 99mtc-Labeled cyclic RGD dimers with PEG 4 linkers. *Mol. Pharm.* **2009**, *6*, 231–245. [CrossRef]
102. Varasteh, Z.; Rosenström, U.; Velikyan, I.; Mitran, B.; Altai, M.; Honarvar, H.; Rosestedt, M.; Lindeberg, G.; Sörensen, J.; Larhed, M.; et al. The Effect of Mini-PEG-Based Spacer Length on Binding and Pharmacokinetic Properties of a 68Ga-Labeled NOTA-Conjugated Antagonistic Analog of Bombesin. *Molecules* **2014**, *19*, 10455–10472. [CrossRef]
103. Zhang, X.; Wang, H.; Ma, Z.; Wu, B. Effects of pharmaceutical PEGylation on drug metabolism and its clinical concerns. *Expert Opin. Drug Metab. Toxicol.* **2014**, *10*, 1691–1702. [CrossRef]
104. Khandare, J.; Minko, T. Polymer–drug conjugates: Progress in polymeric prodrugs. *Prog. Polym. Sci.* **2006**, *31*, 359–397. [CrossRef]
105. Pasut, G.; Guiotto, A.; Veronese, F. Protein, peptide and non-peptide drug PEGylation for therapeutic application. *Expert Opin. Ther. Pat.* **2004**, *14*, 859–894. [CrossRef]
106. Mahato, R.; Tai, W.; Cheng, K. Prodrugs for improving tumor targetability and efficiency. *Adv. Drug Deliv. Rev.* **2011**, *63*, 659–670. [CrossRef]
107. Hausner, S.H.; Bauer, N.; Hu, L.Y.; Knight, L.M.; Sutcliffe, J.L. The effect of bi-terminal PEGylation of an integrin $\alpha v \beta_6$-targeted ^{18}F peptide on pharmacokinetics and tumor uptake. *J. Nucl. Med.* **2015**, *56*, 784–790. [CrossRef]
108. Däpp, S.; Garayoa, E.G.; Maes, V.; Brans, L.; Tourwé, D.A.; Müller, C.; Schibli, R. PEGylation of 99mTc-labeled bombesin analogues improves their pharmacokinetic properties. *Nucl. Med. Biol.* **2011**, *38*, 997–1009. [CrossRef]
109. Däpp, S.; Müller, C.; Garayoa, E.G.; Bläuenstein, P.; Maes, V.; Brans, L.; Tourwé, D.A.; Schibli, R. PEGylation, increasing specific activity and multiple dosing as strategies to improve the risk-benefit profile of targeted radionuclide therapy with 177Lu- DOTA-bombesin analogues. *EJNMMI Res.* **2012**, *2*, 1–12. [CrossRef]
110. Wu, Z.; Li, Z.-B.; Cai, W.; He, L.; Chin, F.T.; Li, F.; Chen, X. 18F-labeled mini-PEG spacered RGD dimer (18F-FPRGD2): Synthesis and microPET imaging of $\alpha v \beta 3$ integrin expression. *Eur. J. Nucl. Med. Mol. Imaging* **2007**, *34*, 1823. [CrossRef]
111. Zhang, H.; Schuhmacher, J.; Waser, B.; Wild, D.; Eisenhut, M.; Reubi, J.C.; Maecke, H.R. DOTA-PESIN, a DOTA-conjugated bombesin derivative designed for the imaging and targeted radionuclide treatment of bombesin receptor-positive tumours. *Eur. J. Nucl. Med. Mol. Imaging* **2007**, *34*, 1198–1208. [CrossRef]
112. Jamous, M.; Tamma, M.L.; Gourni, E.; Waser, B.; Reubi, J.C.; Maecke, H.R.; Mansi, R. PEG spacers of different length influence the biological profile of bombesin-based radiolabeled antagonists. *Nucl. Med. Biol.* **2014**, *41*, 464–470. [CrossRef]
113. Bacher, L.; Fischer, G.; Litau, S.; Schirrmacher, R.; Wängler, B.; Baller, M.; Wängler, C. Improving the stability of peptidic radiotracers by the introduction of artificial scaffolds: Which structure element is most useful? *J. Label. Compd. Radiopharm.* **2015**, *58*, 395–402. [CrossRef]
114. Li, L.; Turatti, F.; Crow, D.; Bading, J.R.; Anderson, A.L.; Poku, E.; Yazaki, P.J.; Williams, L.E.; Tamvakis, D.; Sanders, P.; et al. Monodispersed DOTA-PEG-conjugated anti-TAG-72 diabody has low kidney uptake and high tumor-to-blood ratios resulting in improved64Cu PET. *J. Nucl. Med.* **2010**, *51*, 1139–1146. [CrossRef]
115. Li, L.; Crow, D.; Turatti, F.; Bading, J.R.; Anderson, A.L.; Poku, E.; Yazaki, P.J.; Carmichael, J.; Leong, D.; Wheatcroft, M.P.; et al. Site-specific conjugation of monodispersed DOTA-PEGn to a thiolated diabody reveals the effect of increasing PEG size on kidney clearance and tumor uptake with improved 64-copper PET imaging. *Bioconjug. Chem.* **2011**, *22*, 709–716. [CrossRef]
116. Tugyi, R.; Mező, G.; Fellinger, E.; Andreu, D.; Hudecz, F. The effect of cyclization on the enzymatic degradation of herpes simplex virus glycoprotein D derived epitope peptide. *J. Pept. Sci.* **2005**, *11*, 642–649. [CrossRef] [PubMed]
117. Cardillo, G.; Gentilucci, L.; Tolomelli, A.; Spinosa, R.; Calienni, M.; Qasem, A.R.; Spampinato, S. Synthesis and evaluation of the affinity toward μ-opioid receptors of atypical, lipophilic ligands based on the sequence c[-Tyr-Pro-Trp-Phe-Gly-]. *J. Med. Chem.* **2004**, *21*, 5198–5203. [CrossRef] [PubMed]
118. Reichelt, A.; Martin, S.F. Synthesis and properties of cyclopropane-derived peptidomimetics. *Acc. Chem. Res.* **2006**, *39*, 433–442. [CrossRef] [PubMed]
119. Qvit, N.; Rubin, S.J.S.; Urban, T.J.; Mochly-Rosen, D.; Gross, E.R. Peptidomimetic therapeutics: Scientific approaches and opportunities. *Drug Discov. Today* **2017**, *22*, 454–462. [CrossRef] [PubMed]

120. Hruby, V.J. Conformational restrictions of biologically active peptides via amino acid side chain groups. *Life Sci.* **1982**, *31*, 189–199. [CrossRef]
121. Cantel, S.; Isaad, A.L.C.; Scrima, M.; Levy, J.J.; DiMarchi, R.D.; Rovero, P.; Halperin, J.A.; D'Ursi, A.M.; Papini, A.M.; Chorev, M. Synthesis and conformational analysis of a cyclic peptide obtained via i to i+4 intramolecular side-chain to side-chain azide-alkyne 1,3-dipolar cycloaddition. *J. Org. Chem.* **2008**, *73*, 5663–5674. [CrossRef]
122. Schaffner, P.; Dard, M.M. Structure and function of RGD peptides involved in bone biology. *Cell. Mol. Life Sci.* **2003**, *60*, 119–132. [CrossRef]
123. Kim, H.L.; Sachin, K.; Jeong, H.J.; Choi, W.; Lee, H.S.; Kim, D.W. F-18 labeled RGD probes based on bioorthogonal strain-promoted click reaction for PET imaging. *ACS Med. Chem. Lett.* **2015**, *6*, 402–407. [CrossRef]
124. Kalgutkar, A.S.; Jones, R.; Sawant, A. Sulfonamide as an Essential Functional Group in Drug Design. In *Metabolism, Pharmacokinetics and Toxicity of Functional Groups: Impact of Chemical Building Blocks on ADMET*; The Royal Society of Chemistry: Cambridge, UK, 2010; pp. 210–247. ISBN 978-1-84973-016-7.
125. Radkiewicz, J.L.; McAllister, M.A.; Goldstein, E.; Houk, K.N. A theoretical investigation of phosphonamidates and sulfonamides as protease transition state isosteres. *J. Org. Chem.* **1998**, *63*, 1419–1428. [CrossRef]
126. Moree, W.J.; van der Marel, G.A.; Liskamp, R.J. Synthesis of peptidosulfinamides and peptidosulfonamides: Peptidomimetics containing the sulfinamide or sulfonamide transition-state isostere. *J. Org. Chem.* **1995**, *60*, 5157–5169. [CrossRef]
127. Liskamp, R.M.J.; Kruijtzer, J.A.W. Peptide transformation leading to peptide-peptidosulfonamide hybrids and oligo peptidosulfonamides. *Mol. Divers.* **2004**, *8*, 79–87. [CrossRef] [PubMed]
128. Calcagni, A.; Gavuzzo, E.; Lucente, G.; Mazza, F.; Morera, E.; Paglialunga Paradisi, M.; Rossi, D. Peptides containing the sulfonamide junction. 2. Structure and conformation of Z-Tau-Pro-D-Phe-NHiPr. *Biopolymers* **2000**, *54*, 379–387. [CrossRef]
129. Brouwer, A.J.; Liskamp, R.M.J. Synthesis of cyclic peptidosulfonamides by ring-closing metathesis. *J. Org. Chem.* **2004**, *69*, 3662–3668. [CrossRef]
130. Moree, W.J.; Schouten, A.; Kroon, J.; Liskamp, R.M.J. Peptides containing the sulfonamide transition-state isostere: Synthesis and structure of N-acetyl-tauryl-l-proline methylamide. *Int. J. Pept. Protein Res.* **2009**, *45*, 501–507. [CrossRef] [PubMed]
131. De Bont, D.B.A.; Sliedregt-Bol, K.M.; Hofmeyer, L.J.F.; Liskamp, R.M.J. Increased stability of peptidesulfonamide peptidomimetics towards protease catalyzed degradation. *Bioorg. Med. Chem.* **1999**, *7*, 1043–1047. [CrossRef]
132. Moree, W.J.; van Gent, L.C.; van der Marel, G.A.; Liskamp, R.M.J. Synthesis of peptides containing a sulfinamide or a sulfonamide transition-state isostere. *Tetrahedron* **1993**, *49*, 1133–1150. [CrossRef]
133. Vaidyanathan, G.; Zalutsky, M.R. Labeling proteins with fluorine-18 using N-succinimidyl 4-[18F]fluorobenzoate. *Int. J. Rad. Appl. Instrum. B.* **1992**, *19*, 275–281. [CrossRef]
134. Sutcliffe-Goulden, J.L.; O'Doherty, M.J.; Marsden, P.K.; Hart, I.R.; Marshall, J.F.; Bansal, S.S. Rapid solid phase synthesis and biodistribution of 18F-labelled linear peptides. *Eur. J. Nucl. Med. Mol. Imaging* **2002**, *29*, 754–759. [CrossRef]
135. Briard, E.; Zoghbi, S.S.; Siméon, F.G.; Imaizumi, M.; Gourley, J.P.; Shetty, H.U.; Lu, S.; Fujita, M.; Innis, R.B.; Pike, V.W. Single-step high-yield radiosynthesis and evaluation of a sensitive 18F-labeled ligand for imaging brain peripheral benzodiazepine receptors with PET. *J. Med. Chem.* **2009**, *52*, 688–699. [CrossRef]

© 2020 by the authors. Licensee MDPI, Basel, Switzerland. This article is an open access article distributed under the terms and conditions of the Creative Commons Attribution (CC BY) license (http://creativecommons.org/licenses/by/4.0/).

Review

The Rise of Synaptic Density PET Imaging

Guillaume Becker *,†, Sylvestre Dammicco †, Mohamed Ali Bahri † and Eric Salmon †

GIGA-Cyclotron Research Center-in vivo imaging, University of Liège, Allée du 6 Août, B30, 4000 Liege, Belgium; sdammicco@uliege.be (S.D.); m.bahri@uliege.be (M.A.B.); eric.salmon@uliege.be (E.S.)
* Correspondence: g.becker@uliege.be; Tel.: +32-4366-2334
† These authors have equally contributed.

Academic Editors: Anne Roivainen and Xiang-Guo Li
Received: 24 April 2020; Accepted: 8 May 2020; Published: 14 May 2020

Abstract: Many neurological disorders are related to synaptic loss or pathologies. Before the boom of positrons emission tomography (PET) imaging of synapses, synaptic quantification could only be achieved in vitro on brain samples after autopsy or surgical resections. Until the mid-2010s, electron microscopy and immunohistochemical labelling of synaptic proteins were the gold-standard methods for such analyses. Over the last decade, several PET radiotracers for the synaptic vesicle 2A protein have been developed to achieve in vivo synapses visualization and quantification. Different strategies were used, namely radiolabelling with either ^{11}C or ^{18}F, preclinical development in rodent and non-human primates, and binding quantification with different kinetic modelling methods. This review provides an overview of these PET tracers and underlines their perspectives and limitations by focusing on radiochemical aspects, as well as preclinical proof-of-concept and the main clinical outcomes described so far.

Keywords: SV2A protein; PET radiotracers; synaptic loss; radiochemistry; preclinical development; clinical outcomes

1. Introduction

The synaptic vesicle glycoprotein 2A (SV2A) has been studied for three decades which were punctuated by important milestones. Among others, the discovery by Lynch et al. of SV2A being the binding site of the first-in-class antiepileptic drug levetiracetam is one of these significant steps [1]. The most recent milestone is undoubtedly the current possibility to study SV2A in vivo via PET imaging with specific radiopharmaceuticals. The protein is ubiquitously expressed throughout the brain and is known to be essential for proper brain development and functioning [2,3]. Numerous functions have been hypothesized for SV2A and were recently reviewed [4]. It is noteworthy that pharmacological modulation of SV2A protein may have an impact on Alzheimer's disease (AD) progression [5,6]. Besides the obvious importance of SV2A in epilepsy, AD is a perfect illustration of the clinical interest of synaptic density quantification [7,8]. Before the onset of synaptic density PET imaging, the quantification of synaptic density in brain tissue was performed using immunohistochemistry of key proteins located in the pre- or postsynaptic neurons, such as synaptophysin, or using electron microscopy. Terry et al. used the immunohistochemistry of synaptophysin and demonstrated a lower synaptic density in the midfrontal and inferior parietal cortices of AD patients. Moreover, they reported a positive correlation between the synaptophysin optical density (as an index of synaptic density) and the mini-mental state examination (MMSE) score of the patients [7]. More than a decade later, Scheff et al. counted the synapses in the hippocampus of AD patients using electron microscopy and found a decreased synaptic density in early AD compare to MCI and healthy controls. Consistently with previously mentioned results, Scheff et al. described a positive correlation between the number of synapses in the hippocampus and the MMSE score [8]. Another pertinent example consists in

Parkinson's disease of which the synaptopathy in mainly consists in dysfunction of dopaminergic nigrostriatal terminals and corticostriatal synapses, probably due to the toxic effect of misfolded α-Synuclein. In that case, the majority of genes implicated in PD have a critical role in presynaptic function and thus presumably lead to synaptic dysfunctions. With this in mind, the underlying concept is that PET imaging of synaptic density could be achieved by targeting proteins embedded in synaptic vesicles thanks to appropriate specific radiotracers. Strongly supported by the experience of UCB Pharma in the development of SV2A ligands, several PET research centres embarked on the development of SV2A PET radiotracers. In this review, we return to this "success story" with an emphasis on updated radiochemical considerations of available PET tracers for SV2A. In a second part, we summarize the preclinical developments that were made with these radiotracers, and finally, we report the major clinical outcomes achieved by the existing PET radiotracers.

2. Radiochemistry of SV2A PET Radiotracers

Over the last decades UCB has developed several effective antiepileptic drugs sharing the same mode of action for binding SV2A: levetiracetam (Keppra®) and brivaracetam (Briviact®) (Figure 1) [9–12].

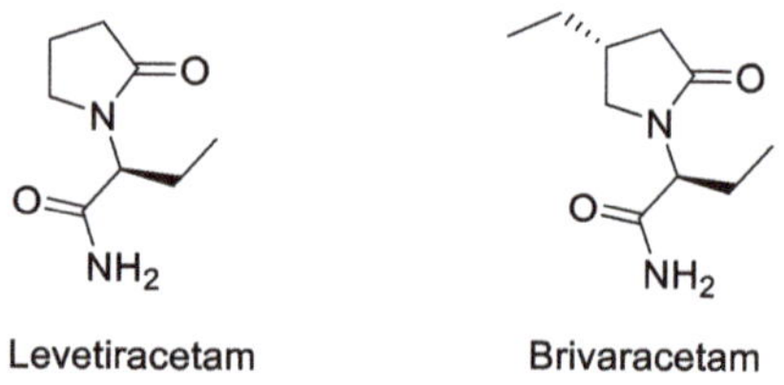

Figure 1. Two antiepileptic drugs: Levetiracetam and Brivaracetam.

After further exploration, the acetamide motif was replaced by different heteroaryl groups as imidazole and pyridine derivatives mimicking the acetamide motif [13]. The modification of the substitution of the alky group on the lactam function was also studied and revealed that (poly)-fluorinated aryls provided the best affinities at SV2A obtained so far.

Quantification of SV2A-PET signal aims to provide a measure of brain synaptic density. Specific requirements for designing CNS PET ligands have been recently reviewed extensively [14–17]. Several parameters are required to obtain a good PET tracer. The radioligand must have a high affinity for the target (nanomolar or subnanomolar) and a B_{max}/K_d ratio in a suitable range to measure a specific signal in vivo (above 10). It has to be able to cross the BBB to reach its target in the CNS. Typically, the LogP of the compound should be between 1 and 3. Since the target density is usually low, the molar activity of the tracer has to be as high as possible. Finally, the radiosynthesis of the tracer with a PET radionuclide as fluorine-18 or carbon-11 must be fast and easily implemented.

Initially, in 2014, the radiochemical synthesis of levetiracetam was performed with carbon-11 (Table 1, entry 1) by Cai et al. but the affinity for SV2A of levetiracetam (Ki = 2.5 µM) was too low for in vivo imaging application [18]. Several other candidates presenting good affinity for SVA were also developed by UCB and were labelled with fluorine-18 and carbon-11: [^{18}F]UCB-H, [^{11}C]UCB-A and [^{11}C]UCB-J (Table 1, entry 2–5). All these compounds were synthesized in partnership with PET centers, in the University of Liège for [^{18}F]UCB-H, in Uppsala University in Sweden for [^{11}C]UCB-A and finally, in Yale University in New Haven for [^{11}C]UCB-J. These compounds have two enantiomeric forms due to the chiral carbon on the lactam ring. In every case, the R form presents best affinity for the SV2A target.

The first published radiotracer was [^{18}F]UCBH. A multi-step synthesis was described in 2013 by Aerts et al. involving nucleophilic ^{18}F-labelling (entry 2) of a pyridine derivative precursor followed by reductive amination, reduction and ring closure reaction [19,20]. This approach afforded UCB-H with a no decay corrected radiochemical yield (ndc RCY) of 30%. The duration of the synthesis including the purification and formulation steps was of 150 min and the molar activity of 96.2 GBq·µmol^{-1}. This process was fully automated on a FASTlab synthesizer and was until recently used in the Liège laboratory for cGMP routine production of the radiopharmaceutical.

In 2014, an alternative at this time consuming multi step synthesis was proposed by the same laboratory and greatly simplified the automation process [21]. This late stage approach (entry 3) requires the labelling of a Pyridyl(4-methoxypehnyl)iodonium triflate salt precursor according to the Pike methodology [22]. In order to improve the yield of the labelling (< 1%), TEMPO was added as a radical scavenger to stabilize the diaryliodonium salt precursor. Indeed, these types of compounds are known to decompose into radical intermediates upon heating. Although TEMPO is known to be genotoxic, the QC study shows that the presence of traces of this radical in the final injectable solution was far below the specification (500–1000 times). This process, fully automated on an AIO from Trasis, proceeds at the Curie level with an RCY of 35% (dc). The duration of the synthesis is only of 50 min and the molar activity is very high (815 GBq·µmol^{-1}). However, even if the affinity of UCB-H is slightly lower compared to UCB-J, leading to more difficulties in detecting the variation of SV2A availability, this radiotracer has a more convenient half live of 110 min and its precursor is commercial available as the dedicated cassette for automated synthesis.

In 2016, [^{11}C]UCB-A was labelled with [^{11}C]methyl triflate reagent (entry 4) reacting with the triphenylmethyl UCB-A precursor [23]. After an acidic treatment and about 40 min of synthesis, the formulated radiotracer was obtained with a RCY of 14% (dc) and a molar activity of 62.9 GBq·µmol^{-1}. Unfortunately, [^{11}C]UCB-A showed slow kinetic and slow metabolism with a brain Tmax of 20 min with a maximum of 5 min for the two other radiotracers leading to quantification issue.

The same year, [^{11}C]UCB-J was produced by labelling the trifluoroborate precursor with [^{11}C]methyl iodide (entry 5) under Suzuki-Miyaura coupling conditions with a palladium catalyst [24]. In a first step, [^{11}C]methyl iodide was reacted with the palladium catalyst followed by in situ hydrolysis of the trifluorobrate precursor to activate it. This coupling reaction lead to a 35% decay corrected radiochemical yield and good molar activity (215 GBq·0µmol^{-1}). The authors noticed that the efficiency of the labelling by methylation depended from the purity of the precursor. The presence of traces amounts (a few percent) of the corresponding boronic acid was required to obtain good yield. Recently, Rokka et al. changed the solvent of the reaction (THF-H$_2$O instead of DMF-H$_2$O) and performed the reaction in a single step by mixing all the reagents before the addition of [^{11}C] methyl iodide [25]. The RCY of the coupling reaction between [^{11}C]methyl iodide and the precursor was 39 ± 5% decay corrected within 40 min with a molar activity of 390 GBq·µmol^{-1} at the end of synthesis. More recently, Sephton et al. proposed a completely automated radiosynthesis of [^{11}C]UCB-J with cGMP compliant conditions [26]. The precursor was preactivated with HCl to form the more reactive boronic acid derivative. The RCY of the coupling was 35 ± 4% dc and 11 ± 1 ndc with a molar activity around 30 GBq·µmol^{-1}. UCB-J presents the best in class SV2A radiotracer with fast kinetic, high brain uptake and rapid metabolization [24]. However, the low half-life of the carbone-11 (20.4 min) requires on-site cyclotron production resulting in some issue for commercial distribution to PET centers.

To circumvent this half-life issue, Li et al. from Yale University have investigated the feasibility of ^{18}F-labelling one of the three fluorine atoms presents on the aromatic ring of the UCB-J molecule [27]. First, the authors tried the traditional ^{18}F-fluorine nucleophilic substitution of a halogen leaving group and the isotopic exchange but none of these methods were successful. They managed to obtain [^{18}F]UCB-J by ^{18}F-labelling an iodonium salt and a iodonium ylide (entry 6). Both reactions generated a racemization of the compound due to the high temperature of the labelling reaction (above 170 °C) requiring a chiral HPLC purification. Those syntheses provided very low RCY (1–2%) and the molar activity was moderate (59 ± 36 GBq·µmol^{-1}). Other radiofluorination methods with arylstannanes, phenofluor derivatives and boronic ester were tested without further success. Although [^{18}F]UCB-J presented similar imaging properties (clearance, metabolism, kinetic) than [^{11}C]UCB-J due to equivalent chemical structure, the radiosynthesis issue led to a need for new SV2A tracers.

In 2018, and a few months apart, the Yale University group and Invicro published a new UCB-J analogue with two fluorine atoms instead of three on the aromatic cycle [28,29]. This compound was initially named [^{18}F]SDM-8 by the Yale University group and [^{18}F]MNI-1126 by Invicro. The two groups agreed to call it [^{18}F]SynVesT-1 (entry 7) for later publication and this name will be used in this paper. This tracer has an affinity comparable to UCB-J (8.4 vs 8.2 respectively) and offers the advantage of a fluorine-18 labelling. Different synthesis route for the preparation of desired tracer were investigated by Li et al. They tried iodonium ylide, boronic ester and arylstannane precursors. The ^{18}F-fluorination of the iodonium precursor afforded very low yield (< 1%) partially due to the chiral HPLC purification of the racemate resulting from the high temperature use for the labelling. Note that the authors did not tested the addition of a radical scavenger as TEMPO in the previous study [21]. The boronic ester derivative provided better yields (11% at 150 °C and 6% at 110 °C) but the stannic precursor is the most promising with a RCY of 19% (ndc) after 95min of radiosynthesis and a significant molar activity of 242 MBq·µmol^{-1}. The radiolabelling occurred at 110 °C which is lower than the critical racemization temperature of 120 °C. Only the synthesis of the R-enantiomer is reported in this publication. Constantinescu et al. (Invicro) worked also on the tin precursor but they synthesized all the enantiomeric forms of the compound (SynVesT-1 (R), MNI-1128 (S) and the racemate) and compared their PET imaging properties. They managed to obtain RCY between 10 and 20% decay corrected with high specific activities (100–370 GBq·µmol^{-1}).

The synthesis of another analogue of UCB-J was also published by the Yale University group. This compound ([^{18}F]SMD-2) with only one fluorine atom on the aromatic ring was renamed later in [^{18}F]SynVesT-2 [30,31]. Like SynVest-1, SynVesT-2 is also an interesting candidate for SV2A imaging probe. It has been synthesized via an iodonium ylide and an arylstannane precursor. The iodonium approach did not provide high radiochemical yield (1%) but radiosynthesis from the tin precursor provided higher RCY (7%, dc). An average molar activity of 141 Gb·µmol^{-1} was obtained after a 90 min synthesis including the purification and formation processes.

In 2019, a screening of a large panel of potential radioligands for SV2A imaging, was performed to identify compounds with similar binding affinities to UCB-J. From these in vitro homogenate binding, autoradiography and in vivo micro-PET studies [32], it appears that SynVesT-1 and SynVesT-2 are promising candidates for SV2A PET imaging, confirming the previous studies (19–22).

In 2019, the synthesis of a ^{18}F-difluoromethyl analogue of UCB-J was reported by Trump et al. [33]. As the substitution of the methyl group presents on the pyridine ring of UCB-J by a fluorine atom slightly reduced the affinity of the molecule for SV2A, these authors investigated the possibility to have a CHF^{18}F function instead of the methyl. The radiosynthesis of this difluoromethylated compound [^{18}F]1 (entry 9) implied a late-stage ^{18}F-difluoromethylation step. This method proceeds by C-H activation and does not require the synthesis of a specific precursor. The catalyzed addition of the CHF^{18}F moiety is generated by UV from a previously synthesized [^{18}F]difluorobenzothiazole sulfone. The reaction is not regiospecific and several isomers can be obtained. Even if the RCY (dc) was of only 1.5%, the fully automated process affords sufficient amount of the labelled molecule, with relatively high molar activity (40–80 GBq·µmol^{-1}) for subsequent preliminary µPET animal studies [34]. This ^{18}F-fluoromethyl labelling approach of N-heteroaromatic compounds by C-H activation represents an interesting way to rapidly evaluate new SV2A PET tracers. However, for future PET clinical imaging applications, the radiochemical synthesis of the [^{18}F]difluorobenzothiazole sulfone should be improved.

Table 1. Affinity, molar activity and radiochemical yield (RCY) of published SV2A tracers.

Entry	Tracer	Ref	Synthesis of the Radiotracer	pIC$_{50}$ for Human SV2A	Ki (nM) for Human SV2A	Molar Activity (GBq. µmol^{-1})	RCY (%)
1	[^{11}C]Levetir-acetam	[18]		5.7 [24]	2500	17	8.3 (dc)
2	[^{18}F]UCB-H	[19]		7.8	9.0	518	15 (ndc)
3	[^{18}F]UCB-H	[21]				815 ± 185	35 (ndc)

167

Table 1. Cont.

Entry	Tracer	Ref	Synthesis of the Radiotracer	pIC$_{50}$ for Human SV2A	Ki (nM) for Human SV2A	Molar Activity (GBq. μmol^{-1})	RCY (%)
4	[^{18}F]UCB-A	[23]		7.9 [24]	ND [a]	65	14 (dc)
5	[^{11}C]UCB-J	[24]		8.2	1.5	215	35 (dc)
6	[^{18}F]UCB-J	[27]		Similar to [^{11}C]UCB-J	Similar to [^{11}C]UCB-J	59 ± 36	1–2 (ndc)

Table 1. *Cont.*

Entry	Tracer	Ref	Synthesis of the Radiotracer	pIC$_{50}$ for Human SV2A	Ki (nM) for Human SV2A	Molar Activity (GBq·µmol^{-1})	RCY (%)
7	[^{18}F]SynVesT-1	[28,29]		8.4	2.2–4.7 [b]	242	19 (ndc)
8	[^{18}F]SynVesT-2	[31]		ND [a]	12	141	7 (dc)
9	[^{18}F]1	[33]		8.3	ND [a]	40–80	1.5 (dc)

[a] Unpublished information. [b] Respectively the value obtained by Invicro and Yale University.

[¹¹C]UCB-J and [¹⁸F]UCB-H are currently the two most widely used radio tracers for SV2A imaging. [¹¹C]UCB-J has a higher affinity for the target but the half-life of ¹¹C drastically limits its use in PET centers without cyclotron. Conversely, the synthesis of [¹⁸F]UCB-H is commercially available with good yield and high specific activity. However, due to the limited imaging properties of [¹⁸F]UCB-H, a new generation of radiotracer has recently been released. [¹⁸F]SynVesT-1 and [¹⁸F]SynVesT-2 were produced via labelling of an arylstannane precursor. The disadvantage of this chemistry is the toxicity of organotin compounds as well as copper used as a catalyst which could lead to complications for GMP applications. Indeed, an ICP-MS analysis seems essential to quality control for injection to the patient. On the other hand, the iodonium precursors radiolabelling did not show as interesting yields as for the [¹⁸F]UCB-H. The use of radical scavenger has not yet been considered, which has however proved its usefulness for the labelling of [¹⁸F]UCB-H.

3. Preclinical Developments of SV2A PET Radiotracers

3.1. Drug Metabolism and Pharmacokinetic (DMPK)

The first generation of SV2A PET radiotracers, namely UCB-A, UCB-H and UCB-J shared the same chemical backbone and were profiled with DMPK studies in male Wistar rats (single intravenous injection at 0.1 mg·kg^{-1} for each compound) [13]. We report here the main DMPK features for these three radiotracers.

Among the three compounds, UCB-A displayed good affinity for SV2A protein (Table 1, entry 4) and a good logD value of 1.4 (measured from octanol/water partition coefficient at 25 °C and pH 7.4). The A > B apparent permeability (P_{app}) was measured on CACO-2 cells and the Efflux ratio (ER) was defined as the ratio of apparent permeabilities (B > A / A > B) with values of 383 nm·s^{-1} and 1.2 respectively. The intrinsic clearance (Clint) measured in human microsomes was 20 µL·min^{-1}·mg·protein^{-1}. Finally, the fraction unbound in rat brain tissue (Fu% brain) is 12% with a ratio of free brain concentration versus free plasma concentration at 20 min post-administration (Free B/P ratio) is 0.6.

As a comparison, UCB-H and UCB-J displayed a good affinity as well (Table 1, entry 2,3,5). The reported logD values were 2.3 for UCB-H and 2.5 for UCB-J. The P_{app}/ER were 707 nm·s^{-1} /0.7 for UCB-H and 323 nm·s^{-1} /0.8 for UCB-J. The Clint were 12 µL·min^{-1}·mg·protein^{-1} for UCB-H and 16 µL·min^{-1}·mg·protein^{-1} for UCB-J. Finally, the Fu% brain and the Free B/P ratio were 8 and 1.6, and 4.5 and 1.6 for UCB-H and UCB-J respectively.

Overall, these data (summarized in Table 2) highlighted that the UCB compounds family have a high permeability and suitable PKPD profile as PET radiotracers. The LogD, a key feature for brain PET radiotracers, is nearly equal for UCB-H and UCB-J, and even lower for UCB-A. In all cases, it is in the suitable range for blood brain barrier crossing. However, another important feature, the free B/P ratio clearly highlights that UCB-H and UCB-J possess the greatest potential (free B/P ratio higher than 1 for both, whereas this ratio is less than 1 for UCB-A).

Table 2. Main PK/PD characteristics of the first-generation of SV2A radiotracers derived from IUCB Pharma library. LogP measured from octanol/water partition coefficient at 258C and pH 7.4, Efflux ratio (ER) derived from apparent permeabilities, the intrinsic clearance (Clint) measured in human microsomes, the fraction unbound in rat brain tissue (Fu% brain) and the ratio of free brain concentration versus free plasma concentration (Free B/P ratio).

Tracers	LogD	ER	Clint µL·min^{-1}·mg·Protein^{-1}	Fu% Brain	Free B/P Ratio
[¹¹C]UCB-A	1.4	1.2	20	12	0.6
[¹⁸F]UCB-H	2.3	0.7	12	8	1.6
[¹¹C]UCB-J	2.5	0.8	16	4.5	1.6

Concerning the pharmacological profile, UCB-J exhibited a greater than 10-fold and greater than 100-fold selectivity for SV2A over SV2C and SV2B protein, respectively [24]. Regarding UCB-H, the in vivo selectivity for SV2A over the two other isoforms (B and C) was assessed by microPET imaging with pharmacological blocking experiment [35]. The authors performed pharmacological competitions with either vehicle, SV2A competitor (levetiracetam at 10 mg·kg^{-1}), SV2B competitor (UCB5203 at 3 mg·kg^{-1}), and SV2C competitor (UCB0949 at 3 mg·kg^{-1}). Statistical analysis revealed differences between levetiracetam pre-treated group and all the other groups of the study, thereby highlighting the in vivo selectivity of UCB-H for SV2A.

To summarized, all of the three radiotracers initially developed by UCB Pharma displayed very good features for central nervous system PET imaging, with excellent brain penetration and the important requirement that none of them is substrate of P-glycoprotein (P-gp) efflux.

As previously mentioned, the second generation of SV2A PET radiotracers includes mainly the two compounds SynVesT-1 and SynVesT-2. SynVesT-1 seems the most promising compound with an IC50 of 3.52 nM and a logP value measured at 2.32. Plasma free fraction of SynVesT-1 was high, at 43 ± 2% [28]. SynVesT-2 hold a high affinity toward SV2A assessed by Ki values of 7.6 and 12 nM for rat and human SV2A, respectively. Finally, SynVesT-2 possesses a Log D of 2.17 ± 0.02 and a plasma free fraction similar of 41% ± 2% [31].

3.2. Preclinical PET Imaging with SV2A Radiotracers

The first in vivo description of SV2A PET radiotracer has been achieved in mice with UCB-H for dosimetric analysis purpose [36]. This study revealed by ex vivo tissue distribution that the brain is one of the mice organs most exposed to radioactive doses, along with the urinary bladder wall and the liver, all the three organs receiving a resulting effective dose of $1.88 \cdot 10^{-2}$ mSv·MBq^{-1}. This study concluded that UCB-H tracer met the standard criteria for radiation exposure in clinical studies with an estimated effective dose of 2.8 mSv for an injected dose of 150 MBq and a maximum injectable dose of approximately 325 MBq per participant. Moreover, this study highlighted the high brain penetration of the tracer in mice.

Afterward, preclinical investigations with UCB-H were conducted in rats (Sprague Dawley) considering that their size better suits microPET brain imaging, as well as images quantification with invasive protocols [19]. Warnock et al. proceeded to a full quantification using an arterial input function (AIF). The AIF was measured thanks to an arteriovenous shunt and a beta-probe. To compute the full kinetic model, they used AIF corrected for plasma-to-whole blood ratio and in vivo metabolism. The metabolism was highly reproducible and fitted a bi-exponential curve. This parent fraction curve revealed that 40% of the UCB-H was metabolized 5 min after the injection, while it remains 20% of the parent compound 30 min after injection. The authors showed that the Logan graphic analysis (start time ranged from 7.5 to 15 min.) afforded a highly reproducible distribution volume (V_T) measurement in a test-retest experimental design. Moreover, they validated the pharmacological blocking by increasing doses of levetiracetam, which ranged from 0.1 to 100 mg·kg^{-1} leading respectively to −9.0 ± 4.6% and −43.8 ± 4.7% decreases in the V_T value for the whole brain.

This in vivo evaluation of UCB-H as a PET tracer for SV2A protein was followed by several studies that aimed at simplifying the binding quantification, AIF in rats being highly invasive and time consuming. The first step for a non-invasive quantification of UCB-H in rat brain used a population-based input function (PBIF) [37]. The authors averaged 8 individuals AIF to compute the PBIF and validate the used of the PBIF on each individual rat. They reported a high correlation of the V_T measured either by individual AIF or by the PBIF ($R^2 = 0.99$ for all fits, the reported V_T values for both methods being almost equals for each rat). Beyond the quantification part, the authors used HPLC-MS/MS to identify the N-oxide as the main plasmatic metabolite of UCB-H (the UCB-H-N-oxide represented 90.2% of the formed metabolites). Then, they proceed to the radiolabelling of this UCB-H-N-oxide and performed PET imaging with [^{18}F]UCB-H-N-oxide thereby establishing the complete absence of UCB-H radiometabolite in the rat brain.

The second simplification step for preclinical studies with UCB-H was the validation of a static PET acquisition and standardized uptake value assessment [38]. The authors compared the two parameters: the V_T obtained from Logan graphical analysis using the previously developed PBIF, and the standardized uptake value (SUV). The authors used pharmacological competition design with 1 and 10 mg·kg^{-1} of levetiracetam in rats to study the impact of the length of the dynamic acquisition (90 versus 60 min) and the correlations between V_T and SUV over 20 min consecutive timeframes. They reported no bias between the V_T values obtained for both dynamic acquisition times, and a high correlation between the V_T derived from 90 min acquisition time and the SUV computed from 20 to 40 min static acquisition. The authors identified the 20–40 min timeframe as the best situation for a static PET acquisition with UCB-H in rat. This method of static acquisition and SUV assessment was further used to investigate the in vivo modification in SV2A protein expression in a preclinical rat model of temporal lobe epilepsy. Although pathophysiological findings are beyond the scope of this review, this study highlights the efficiency of static PET analysis with UCB-H to detect variations in SV2A expression in the brain of an epileptic rat model, as well as physiological modifications related to brain development and maturation [39].

The UCB-J was initially developed in rhesus monkey [24]. As it has been shown with UCB-H, UCB-J displayed a fast metabolism, the parent fraction in the plasma accounted for approximately 40% and about 25% of the radioactivity at 30 and 90 min after injection, respectively. PET images revealed a high brain uptake throughout the grey matter, consistent with the ubiquitous distribution of SV2A. The pre-injection of levetiracetam (10 mg·kg^{-1}) substantially blocked [^{11}C]UCB-J binding and co-injection of cold UCB-J (150 mg·kg^{-1}) drastically decreased specific UCB-J binding. [^{11}C]UCB-J displayed a high uptake and rapid kinetics, with an SUV peak of 5–8 in grey matter areas and peak uptake times ranging from 10 to 50 min. Nabulsi et al. tested several kinetic models to quantify UCB-J binding and chose the 1-Tissu compartmental model (1T model) as it produced reliable V_T estimates with low variability (compared to the other kinetic analyses tested, the multilinear analysis and the 2-tissues compartmental model). Then, using the 1T model, the author reported a significant reduction in regional V_T due to levetiracetam pre-injection (10 and 30 mg·kg^{-1}). The resulting receptor occupancy analysis revealed that 59% and 89% of SV2A in rhesus monkey brain were occupied by 10 and 30 mg·kg^{-1} of levetiracetam respectively. The blocking experiment using 17, 50, and 150 mg·kg^{-1} of UCB-J showed respectively an occupancy of occupancies of 46%, 68%, and 87%. Thanks to these results, the authors were able to compute the Kd value for UCB-J estimated 3.4 ± 0.2 nM and the B_{max} for several brain regions (highest value in cingulate cortex 350 nmol·L^{-1}, and lowest value in the pons and the brain stem 123 and 124 nmol·L^{-1} respectively).

Regarding the dosimetry, they performed a whole-body distribution studies showing that the liver was receiving the largest doses for males (0.0199 mGy·MBq^{-1}) whereas the brain received the largest dose for females (0.0181 mGy·MBq^{-1}). They estimated the effective dose equivalent value of approximately 4.5 mSv·MBq^{-1} and conclude that the maximum effective dose equivalent from a single 740 MBq (20 mCi) administration of [^{11}C]UCB-J is equivalent to 3.4 mSv, thus being fully compliant with regulation for human PET imaging.

The same group published in 2016 the only complete validation of SV2A PET as marker of synaptic density [40]. This well-designed study was realized in vivo with [^{11}C]UCB-J PET imaging in an olive baboon (Papio anubis), after which the animal was sacrificed, and the brain was dissected for post-mortem tissue studies. The post-mortem analyses (involving 12 brain regions) consisted in western blotting, SV2A homogenate binding assays and immunohistochemical staining for SV2A and synaptophysin protein as reference marker for synaptic density measures. To established that SV2A can be used as marker of synaptic density, they compared regional densities of SV2A and the "gold standard" synaptic marker synaptophysin using selective antibodies. SV2A and synaptophysin signals were strong and specific in all grey matter regions but absent or weak in the centrum semiovale (CS). There was an excellent linear correlation between SV2A and synaptophysin across all grey matter regions. They were able to establish that SV2A can be used as an alternative to Synaptophysin for accurate synapse quantification. In addition, the authors established a strong correlation between the in vitro regional distribution of SV2A and the [^{11}C]UCB-J V_T measured in vivo by PET. Afterward, to further evaluate the relationship between in vivo [^{11}C]UCB-J binding and SV2A density, homogenate binding studies were performed to determine affinity (Kd) and regional SV2A densities B_{max}. The authors reported a clear correlation between the in vitro B_{max} values measured in the 12 brain regions and the in vivo V_T values. As a cross validation, there was an excellent correlation between the Kd derived from homogenate binding and the regional SV2A Western blot measurements.

Recently, UCB-J was investigated in mice and more specifically in the amyloid precursor protein and presenilin 1 double-transgenic (APPswe/PS1DE9 [APP/PS1]) mouse model of Alzheimer disease [41]. The authors intended to perform a longitudinal [^{11}C]UCB-J PET on these AD mice to measure the treatment effects of saracatinib, an inhibitor of the tyrosine kinase Fyn which is believed to be useful in AD treatment. The experimental design consisted in the testing of two groups: the control group with wild type (WT) mice and the test group with APPswe/PS1DE9 mice. Both groups underwent three [^{11}C]UCB-J PET measurements: at baseline, after treatment, and during drug washout (more than 27 d after the end of treatment). The quantification was achieved with several parameters, the first to be used was the BP_{ND}, computed by the simplified reference tissue model with the brain stem as reference region. Then the authors compared this BP_{ND} with the SUVratio using the brain stem signal for normalization ($SUVR_{BS}$, static acquisition time of 30 min between 30 and 60 min post-injection). The authors also tested another SUVratio normalized by the whole brain ($SUVR_{WB}$). The authors found that BP_{ND} and $SUVR_{BS}$ demonstrate excellent agreement with correlation coefficient $R^2 = 0.85$ for the whole brain region. However, for treatment response assessment, the authors claimed that $SUVR_{WB}$ gave less variability mainly due to the small size of the used regions of interest. Then, Toyonaga et al. stated that [^{11}C]UCB-J PET allows to differentiate hippocampal $SUVR_{WB}$ in APP/PS1 mice and in WT mice at baseline and to detect a significant increase in hippocampal $SUVR_{WB}$ after AD treatment with saracatinib.

In 2016, UCB Pharma, in collaboration with Yale University, reported preclinical data on their newly developed antiepileptic drug named Brivaracetam (acting on SV2A, like its predecessor compound levetiracetam). They used PET imaging to prove a faster SV2A occupancy by brivaracetam compared with levetiracetam [42]. Displacement experiments using [^{11}C]UCB-J were performed in rhesus monkeys to estimate the time course of tracer exit from the brain after single IV dosing of brivaracetam (5 mg·kg^{-1}) or levetiracetam (30 mg·kg^{-1}). The estimated displacement half-times were 10 min for brivaracetam and 30 min for levetiracetam. Further, using kinetic modelling, they were able to predict drug entry half-times, which were 3 min for brivaracetam and 23 min for levetiracetam.

The third SV2A radiotracer belonging to the first generation is the UCB-A [23]. Unlike the two other UCB tracers, UCB-A displayed a relatively slow metabolism, with 93% and 42% intact tracer present at 5 and 40 min post-injection. Surprisingly, the [^{11}C]UCB-A showed a slow kinetic profile with an accumulation during the first 60 min, followed by a slow wash out from the brain. The time activity curves of the whole brain clearly demonstrated a successful blocking of [^{11}C]UCB-A binding by the pre-injection of an acetamide-based ligand of SV2A (named compound 4 in the study, 10 mg·kg^{-1}). Further, the reversible binding of [^{11}C]UCB-A was demonstrated by the administration of brivaracetam (21 mg·kg^{-1}), at 45 min post-injection, which fully and rapidly displaced [^{11}C]UCB-A. Afterward, the [^{11}C]UCB-A was further developed in pig where PET images were quantified by whole blood time-activity measures and metabolite-corrected arterial input curves for kinetic modelling. The authors used the Akaike criteria to determine that [^{11}C]UCB-A kinetics were best described by a 1T model. With this model, they could estimate both V_T and k_1 parameters. However, the 1T model was unable to properly estimate V_T in the blocking condition. The authors found that Logan graphical analysis could reliably fit baseline and blocking conditions and therefore decided to use it in the occupancy analysis of their blocking conditions. Regarding the dosimetry in rats, the organ receiving the highest absorbed dose was the liver, at 12.5 and 26.3 µGy·MBq^{-1} in males and females, respectively. Effective doses were 2.9 µSv·MBq^{-1} and 4.6 µSv·MBq^{-1}, for males and females, respectively. The authors claimed that given the effective doses observed in rats, an effective dose of no more than 1.8 mSv per typical administration of 400 MBq would be achievable.

Currently, UCB-J has been fluorinated and tested in vivo in non-human primates [27]. As the chemical structure is the same as UCB-J, all features of UCB-J radiotracer remain the same and the authors stated that the tracer possesses the same in vivo behavior has its parent compounds UCB-J.

The second generation of radiotracer is leaded by two tracers: SynVesT-1 and SynVesT-2. Both tracers display a very high affinity for SV2A (reported previously in this review). Both of them were evaluated in non-human primates [28,31]. SynVesT-1 and SynVesT-2 share a very similar metabolism rate in plasma. The first one displayed 42 ± 13% of parent radiotracer remaining at 30 min after injection, which further decreased to 27 ± 5% and 23 ± 5%, respectively, at 60 and 90 min. post-injection. For the second one, the authors reported that 34% ± 0.1% of intact parent radiotracer remains at 30 min post-injection, which further decreased to 24% ± 3% and 22% ± 4%, at 60 and 90 min, respectively. In rhesus monkeys, both tracer with high tracer uptake in gray matter and very low uptake in white matter (the lower value being found in the CS, as in the case of UCB-J PET imaging). SynVesT-1 has been drastically displaced by the administration of levetiracetam (30 mg·kg^{-1}) at 90 min post-injection. The binding specificity was emphasized by blocking studies, either by pretreatment with levetiracetam (30 mg·kg^{-1}) or UCB-J (0.15 mg·kg^{-1}). Both competitors dramatically decreased regional uptake levels across all brain regions and resulted in earlier peak uptake. Regarding SynVesT-2 preblocking study with unlabelled UCB-J (150 μg·kg^{-1}) completely reduced the specific binding of the tracer. Although both tracers look very similar, it is worth mentioning that their brain kinetics largely differ. Indeed, SynVesT-1 displayed an SUV peak within 30 min post-injection in all brain regions followed by a moderate rate of clearance over time, whereas SynVesT-2 showed a much faster brain kinetic with a SUV peak reached at 10 min post-injection. On the contrary, both tracers display a very similar uptake pattern with the highest tissue uptake levels found in the frontal cortex and putamen (SUV > 8 for both), and lowest in CS (SUV < 2 for both). Concerning the quantification of PET images, both tracers were analyzed with either the 1T model or the simplified reference tissue model (SRTM) using the CS as reference tissue. The 1T model allows to compute the V_T parameter which has proven to be more than two times higher for SynVesT-1 than for SynVesT-2 across all brain regions. In addition, the V_T values for SynVesT-1 are much closer to those obtained with UCB-J in rhesus monkeys. Similarly, with the BP_{ND} obtained with SRTM, values for SynVesT-1 were higher for all brain regions, but this time BP_{ND} values for SynVesT-2 are closer to those obtained with UCB-J.

In a nutshell, SynVesT-1 and SynVesT-2 share lots of common features in terms of metabolism and in vivo behavior, although SynVesT-1 seems the most promising one with close outcomes to UCB-J PET imaging.

4. Clinical Studies with SV2A PET Radiotracers

4.1. Quantification of SV2A PET Radiotracers Binding

4.1.1. UCB-H in Human Brain

The distribution of UCB-H in the human brain was studied using the full kinetic modelling approach [43,44]. Authors have motivated their orientation to the full kinetic modelling by the ubiquitous distribution of SV2A in the brain, which makes very unlikely the identification of a "reference region" with all its necessary characteristics for modelling the radiotracer distribution [45]. Moreover, the white matter usually used as reference region for several tracers and studies was also discarded because of the presence of white matter lesions in several elderly and AD participants [46]. In order to overcome the problem of patient discomfort during the arterial input function (AIF) sampling, an alternative method using carotid artery image-derived input function was developed for UCB-H quantification [43]. Even with this image-derived input function, blood sampling was not completely discarded, and authors collected venous blood samples in a group-representative number of subjects at six time-points post injection in order to determine the plasmatic parent fraction needed for image-derived input function correction.

The input function was derived from dynamic PET images [43]. Briefly, the method extracts time series of radiotracer activity in the carotid arteries [47]. The identification of voxels belonging to the carotids is based on the computation of the Pearson product-moment correlation coefficient between a "seeding region" and voxels in a mask containing the carotid. As, during the first 2min, radioactivity is mainly localized in the vessels, inducing a large spill-out effect [47], the signal was corrected for this spill-out effect using the geometric transfer matrix approach [48]. The extracted time-series signal was then corrected for the mean unchanged plasma fraction and used for kinetic modelling.

As image analysis and kinetic modelling, UCB-H PET dynamic data frames underwent a series of processing steps before voxel-wise parametric map extraction. Indeed, data frames were corrected for participant's motion during the acquisition and co-registered to the corresponding MRI structural image (the sum of frames between 2 and 30 min as source image), and then partial volume corrected. The PET scanner low resolutions as well the small structures of interest (e.g., hippocampus and the basal forebrain that are atrophied in early Alzheimer's disease are the main raisons for partial volume correction [49]. The iterative Yang (PETPVC toolbox, [50]) voxel-wise PVC method was used, with grey matter, white matter, CSF as ROI mask. Kinetic modelling using PVE-corrected dynamic PET data and image-derived input function was done with PMOD software (PMOD Technologies, Zurich, Switzerland). Logan graphical analysis method was used to calculate the V_T map of UCB-H in the brain with t* of 25 min. The obtained V_T map was then spatially normalized into the MNI space using the transformation parameters obtained during structural MRI spatial normalization. Spatial normalization is usually done through the structural MRI image as the last one has a higher resolution. Mean regional V_T values were then extracted from V_T map using the AAL atlas [51].

4.1.2. UCB-J in Human Brain

UCB-J in vivo quantification in human brain was presented in the publication of Finnema 2016 where the authors proposed the CS as a reference region for non-displaceable binding [40]. Regional non-displaceable binding potential (BP_{ND}) was calculated as the ratio between the distribution volume in region of interest and the distribution volume in the CS minus one ($V_{T,region} / V_{T,centrum\_semiovale}$ −1). The volume of distribution was calculated using a full kinetic modelling through the 1T model with measured arterial input function and regional time activity curves (TACs). The extraction of the regional time activity curves was done on the subject PET space using AAL atlas. Briefly, the inverse transformation parameters obtained during the spatial normalization of the motion corrected UCB-J dynamic data were applied to the AAL atlas to bring it into the PET space. Then, regional TACs were extracted end expressed in SUV by normalization of the regional activity concentration by the ratio of injected dose to subject weight.

Parametric BPND maps were also generated with the simplified reference tissue model 2 using the CS as a reference region [52]. The author showed that CS has some degree of specific binding and further follow-up studies and analysis were still needed to justify its use as a reference region.

In a recent study aiming at the validation of the CS, Rossano et al. investigated the region of interest definition and the reconstruction parameters (number of iteration of the ordered subset expectation minimization "OS-EM" [53]. Indeed, optimization of the CS ROI definition aims to minimize the inclusion of grey matter and partial volume effects that would lead to erroneous CS V_T values. This was done using an improved ROI definition strategy based on the use of the CS AAL to mask the individual white matter and the definition of a group CS ROIs with increasing volume (2, 4, and 6 mL). Additionally, convergence using the OS-EM algorithm was investigated by increasing the iterations used to reconstruct the PET image, which would minimize any positive bias that may be observed in the low-intensity white matter, surrounded by the high-intensity grey matter. Moreover, baseline and blocking (levetiracetam and brivaracetam) UCB-J scans were used to estimate the V_{ND} of the grey matter, and the validity of using CS V_T to estimate V_{ND} in calculating BP_{ND} was evaluated. Even if the results showed a specific uptake in the white matter lower than 10% of that in the grey matter and the CS uptake was predominantly due to non-displaceable binding, the authors stressed the importance of assessing differences in white matter uptake when investigating differences across disease groups, especially in diseases with white matter pathology. Koole et al. also investigated CS as reference region for UCB-J quantification showing considerable agreement with Rossano conclusions [54]. Compared to a full kinetic analysis, with 1T compartment method with measured AIF, simplified reference tissue model approach using a 90 min acquisition interval and the CS as reference tissue provided a negligible bias of less than 6%. Moreover, the authors suggested shortening the dynamic acquisition time to 70 min instead of 120 min. Alternately to dynamic tracer uptake modelling, SUV ratio relative to the CS for 30 min acquisition starting 60 min after tracer injection was proposed as a quantitative approximation for UCB-J brain PET imaging. The use of CS white matter as reference region for [^{11}C]UCB-J quantification has been further supported by in vitro autoradiographic observations of low binding for [^{11}C]UCB-J in this region, in either human non-human primates [55].

4.2. Clinical Outcomes with SV2A PET Imaging

The first in-human study with SV2A specific PET tracer was published by Bretin et al. and concerned biodistribution description and radiation dosimetry assessment [56]. The authors reported the results of dynamic whole-body PET with UCB-H on five healthy males with an injected activity between 139.1–156.5 MBq. The radiation dosimetry was calculated using OLINDA/EXM. Regarding the brain, as main organ of interest, the absorbed dose was $1.89 \times 10^{-2} \pm 2.32 \times 10^{-3}$ mGy·MBq^{-1}. The methodological aspects of UCB-H brain images quantification have been described above in this review [43]. Thanks to the implementation of the image-derived blood input function, the same group reported the outcomes of UCB-H PET imaging in twenty-four patients with mild cognitive impairment or Alzheimer's disease (with positive [^{18}F]Flutemetamol amyloid-PET) compared to 19 healthy controls (Figure 2., [44]). Using voxel-wise analysis on V_T maps after partial volume effect correction (amyloid positive patients vs control group), the authors described a significant reduction of synaptic density in the right anterior hippocampus extending to the entorhinal cortex (26.9% decrease in right hippocampal [^{18}F]UCB-H V_T). A sub-analysis focusing only on mild AD patient (n = 14) showed reduced synaptic density in the right superior temporal gyrus. Finally, the authors computed Pearson correlations between [^{18}F]UCB-H values in selected ROIs and cognitive measures. When taking into account the correction for multiple comparisons, the authors showed correlations between awareness of memory problems and MMSE scores on the one hand and [^{18}F]UCB-H values in the hippocampus on the other hand (Pearson correlation coefficient of −0.75 and 0.57 respectively, at $p < 0.005$). Thus, Bastin et al. clearly confirmed that SV2A-PET imaging with [^{18}F]UCB-H, allows to image in vivo synaptic changes in Aβ-positive patients with neurocognitive disorder and to relate them to cognitive impairment with regional specificity according to the cognitive domain.

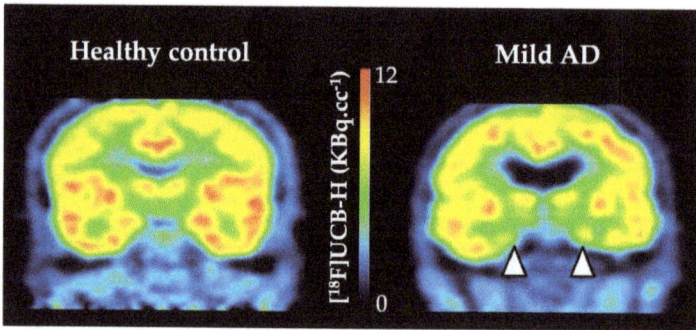

Figure 2. Individual example of [^{18}F]UCB-H in healthy control and in mild AD patient. The white arrow heads highlight the reduced [^{18}F]UCB-H uptake in temporal lobe areas of the mild AD patient.

The first study reporting SV2A PET assessment in AD was published by the University of Yale and used UCB-J [57]. The outcomes parameters used as imaging outcomes were the V_T, the delivery rate constant K_1 and the BP_{ND} (obtained with the 1T and the SRTM 2 models). The study included twenty-one participants, with 10 amnestic MCI due to AD or mild AD dementia participants and 11 healthy controls. The results were consistent with the hypothesis derived from AD Braak staging, showing that [^{11}C]UCB-J binding was significantly reduced in the hippocampus of participants with AD compared to the control group. In the hippocampus of AD participants, the V_T was 28% lower and the BP_{ND} was reduced by 44%. Parametric map of K1 value, reflecting the cerebral blood flow clearly highlighted that the pattern of regional k_1 reduction in the participant with AD was similar to that of hypometabolism observed with FDG-PET in AD. Finally, Chen et al. reported Pearson correlations between [^{11}C]UCB-J values in the hippocampus and cognitive measures. Notably, statistically significant correlations were found between SRTM2-derived hippocampus BP_{ND} and episodic memory on one hand (R = 0.56; p =0.01), and with clinical dementia rating–sum of boxes (CDR-SB) score on the other hand (R = −0.61; p =0.003). These results were later confirmed in a study conducted by the group of KU Leuven where prodromal AD patients (aMCI) were tested with both [^{11}C]UCB-J and [^{18}F]MK-6240 that allows the assessment of Tau protein deposition [58]. Within the median temporal lobe, there was a clear correlation between [^{11}C]UCB-J SUVR decrease and [^{18}F]MK-6240 SUVR increase (r = −0.076, p = 0.02). The authors corroborated previous results with SV2A PET imaging while highlighting the hippocampus as the most prominent region for synapse loss. Moreover, synaptic loss correlates with cognitive decline measured by MMSE score and episodic memory test (r = 0.77 and 0.7 respectively).

As a logical follow-up of the preclinical study that supported the development of the brivaracetam anti-epileptic drug, UCB Pharma and Yale University released a drug occupancy study in human for brivaracetam using [^{11}C]UCB-J PET [59]. Three cohorts with four, five, and four healthy subjects were tested for baseline and displacement, displacement and 4 h post-dose and steady state of oral dosing of brivaracetam respectively. The [^{11}C]UCB-J PET scanning procedure consisted of an intravenous bolus plus constant infusion. The regional V_T parameter was derived from 1T model and SV2A occupancy computed by regressing the change in V_T against the baseline V_T across regions. The main occupancy results were 66% and 70% for intravenous 100 mg of brivaracetam, 84% and 85% for intravenous 200 mg of brivaracetam, and 78–84% for intravenous 1500 mg of levetiracetam approximately 4 h post-dose. Moreover, the combined IC_{50} estimate across Cohorts 2 and 3 was 0.46 µg·mL^{-1} for brivaracetam and 4.02 µg·mL^{-1} for levetiracetam, leading the authors to state that brivaracetam has 8.7-fold higher affinity for SV2A than levetiracetam.

Recently, the group of Yale University published a PET imaging study with [^{11}C]UCB-J in a cohort of Parkinson's disease patients [60]. Based on current pathophysiological knowledge of PD, more precisely on the fact that synaptic changes are centrally involved in PD and characteristic of the disease pathogenesis, the authors seek to detect synaptic loss in vivo with [^{11}C]UCB J PET in individuals with mild bilateral PD [61]. Twelve PD subjects, assessed according to the Movement Disorders Society Unified PD Rating Scale, were compared to 12 matched normal controls. As previously described, the PET data were processed using SRTM2 kinetic modelling with BP_{ND} as imaging outcome. Statistical lower SV2A-specific binding values were found in PD subjects in several brain regions, the substantia nigra (−45%; p <0.001), red nucleus (−31%; p =0.03), locus coeruleus (−17%; p =0.03), and parahippocampal gyrus (−12%; p <0.01). The lower SV2A binding in the substantia nigra was further confirmed by in vitro autoradiography (−17%; p <0.005). The authors concluded that their results were in line with the hypothesis claiming that clinical symptomatology of PD may begin in the synapses of brainstem nuclei and remain present throughout the course of the disease [61].

Concerning psychiatric disorders, a recent article reported the in vivo investigation of synaptic density in schizophrenia [62]. This study is in line with evidences implicating synaptic dysfunction in the pathophysiology of schizophrenia. The authors included 18 individuals with a DSM-5 diagnosis of schizophrenia and compared them to 18 healthy volunteers using [^{11}C]UCB-J V_T, derived from the 1T model, as imaging outcome. Post-hoc analyses revealed that mean [^{11}C]UCB-J V_T (ml·cm^{-3}) was significantly reduced in schizophrenic patients relative to the control group in the frontal cortex (t = 2.51, df = 34.0, p =0.03), and in the anterior cingulate cortex (t = 2.83, df = 34.0, p =0.02) with large effect sizes (Cohen's d = 0.8 and 0.9 respectively). Interestingly, Onwordi et al. computed the distribution volume ratio (DVR, using the CS as reference region) in the selected ROIs and found a significant reduction in the hippocampus, in addition to the frontal cortex (FC) and the anterior cingulate cortex (ACC). This discrepancy was analyzed in the sense that DVR values show lower variability and so are likely to have greater sensitivity to detect group differences. Furthermore, the authors stated the absence of significant associations between [^{11}C]UCB-J V_T and corrected grey matter volume in any of the analyzed ROIs, as well as between [^{11}C]UCB-J V_T in the FC, ACC and the assessed clinical variables. The authors completed their clinical results by preclinical in vitro experiment that showed the absence of effect of chronic haloperidol administration on SV2A levels in the rat frontal cortex. This study raised the question whether these results rely on specific loss of SV2A protein or indicate lower synaptic terminal levels. So far, there is no evidence that antipsychotic drugs act on synaptic density, but it would be clinically relevant to compare longitudinally patients who do not respond to antipsychotic treatments and patients who do so.

In parallel, the group from Yale university investigated major depressive disorder (MDD) and post-traumatic stress disorder (PTSD) with the hypothesis that synaptic density, measured by [^{11}C]UCB-J PET, would be negatively associated with severity of depressive symptoms [63]. The study included twenty-six unmedicated clinical subjects and twenty-one age, sex, and smoking-matched healthy control subjects. The [^{11}C]UCB-J V_T derived from 1T model after partial volume correction used to correct for effects of tissue atrophy. Three regions of interest were considered: the hippocampus, the dorsolateral prefrontal cortex (dlPFC), and the anterior cingulate cortex (ACC). In each of these regions, a significant lower [^{11}C]UCB-J V_T in individuals with high severity depressive symptoms was detected in comparison to healthy controls. Strikingly, the authors reported significant negative correlations between the [^{11}C]UCB-J V_T and the severity of depressive symptoms (scored with the HAMD-17 scale) across all clinical subjects in dlPFC (r = −0.633, p = 0.001), ACC (r = −0.634, p = 0.001) and hippocampus (r = −0.487, p = 0.012). Finally, Holmes et al. looked at the functional connectivity in the study cohort and performed voxel-wise correlations to associate the seed connectivity data from the dlPFC to synaptic density. They found a significant negative correlation between dlPFC-posterior cingulate cortex (PCC) connectivity and SV2A density in the dlPFC in the clinical group (r = −0.60, p = 0.002). The authors concluded that their results prompt the discovery and evaluation of new treatments that should target synaptic density to increase synaptic connections.

Overall, the clinical outcomes of SV2A PET imaging suggest that this technic might bring new outcomes in clinical trials for disease monitoring and treatment response assessment, as well as in the refinement of the studies by providing new inclusion/exclusion criteria.

5. Conclusions

Currently, [^{11}C]UCB-J and [^{18}F]UCB-H are the two most widely used radiotracers for SV2A PET imaging in patients with neurological or psychiatric disorders, but they both present disadvantages. On one hand, the half-life of ^{11}C drastically limits the availability of [^{11}C]UCB-J, and [^{18}F]UCB-H provides a lower specific binding signal on the other hand. Notwithstanding, clinical outcomes obtained with both of them were very coherent with each other. The second generation of SV2A tracers is leaded by [^{18}F]SynVesT-1 and [^{18}F]SynVesT-2 which are both produced via labelling of an arylstannane precursor. The use of organotin compounds required ICP-MS analysis to ensure the injectability to the patients. Improvements of the radiochemical yield of iodonium precursors labelling could be an alternative to this stannic derivatives limitation. Alternatively, another radiotracer, a ^{18}F-difluoromethyl analogue of UCB-J that can be produced by C-H activation, does not require the synthesis of a specific precursor. However, for future PET clinical imaging applications, the radiochemical synthesis should also be improved. Nevertheless, and despite the quality of the newly designed fluorinated radiotracers, SV2A PET imaging still faces its biggest challenge, that is, in being included in a clinical routine for patient care and management.

Author Contributions: Conceptualization, G.B. and S.D.; writing—original draft preparation, G.B., S.D., M.A.B. and E.S.; writing—review and editing, G.B., S.D., M.A.B. and E.S. All authors have read and agreed to the published version of the manuscript.

Funding: This research received no external funding.

Acknowledgments: In this section you can acknowledge any support given which is not covered by the author contribution or funding sections. This may include administrative and technical support, or donations in kind (e.g., materials used for experiments).

Conflicts of Interest: The authors declare no conflict of interest.

References

1. Lynch, B.A.; Lambeng, N.; Nocka, K.; Kensel-Hammes, P.; Bajjalieh, S.M.; Matagne, A.; Fuks, B. The Synaptic Vesicle Protein Sv2a Is the Binding Site for the Antiepileptic Drug Levetiracetam. *Proc. Natl. Acad. Sci. USA* **2004**, *101*, 9861–9866. [CrossRef] [PubMed]
2. Bajjalieh, S.M.; Frantz, G.D.; Weimann, J.M.; McConnell, S.K.; Scheller, R.H. Differential Expression of Synaptic Vesicle Protein 2 (Sv2) Isoforms. *J. Neurosci.* **1994**, *9*, 5223–5235. [CrossRef]
3. Crowder, K.M.; Gunther, J.M.; Jones, T.A.; Hale, B.D.; Zhang, H.Z.; Peterson, M.R.; Scheller, R.H.; Chavkin, C.; Bajjalieh, S.M. Abnormal Neurotransmission in Mice Lacking Synaptic Vesicle Protein 2a (Sv2a). *Proc. Natl. Acad. Sci. USA* **1999**, *26*, 15268–15273. [CrossRef] [PubMed]
4. Bartholome, O.; Van den Ackerveken, P.; Sánchez Gil, J.; de la Brassinne Bonardeaux, O.; Leprince, P.; Franzen, R.; Rogister, B. Puzzling out Synaptic Vesicle 2 Family Members Functions. *Front. Mol. Neurosci.* **2017**, 148. [CrossRef]
5. Bakker, A.; Krauss, G.L.; Albert, M.S.; Speck, C.L.; Jones, L.R.; Stark, C.E.; Yassa, M.A.; Bassett, S.S.; Shelton, A.L.; Gallagher, M. Reduction of Hippocampal Hyperactivity Improves Cognition in Amnestic Mild Cognitive Impairment. *Neuron* **2012**, *3*, 467–474. [CrossRef]
6. Pascal, E.S.; Zhu, L.; Verret, L.; Vossel, K.A.; Orr, A.G.; Cirrito, J.R.; Devidze, N.; Ho, K.; Yu, G.-Q.; Palop, J.J.; et al. Levetiracetam Suppresses Neuronal Network Dysfunction and Reverses Synaptic and Cognitive Deficits in an Alzheimer's Disease Model. *Proc. Natl. Acad. Sci. USA* **2012**, *109*, E2895–E2903.
7. Terry, R.D.; Masliah, E.; Salmon, D.P.; Butters, N.; DeTeresa, R.; Hill, R.; Hansen, L.A.; Katzman, R. Physical Basis of Cognitive Alterations in Alzheimer's Disease: Synapse Loss Is the Major Correlate of Cognitive Impairment. *Ann. Neurol.* **1991**, *4*, 572–580. [CrossRef]
8. Scheff, S.W.; Price, D.A.; Schmitt, F.A.; Mufson, E.J. Hippocampal Synaptic Loss in Early Alzheimer's Disease and Mild Cognitive Impairment. *Neurobiol. Aging* **2006**, *10*, 1372–1384. [CrossRef]
9. Klitgaard, H.; Verdu, P. Levetiracetam: The First Sv2a Ligand for the Treatment of Epilepsy. *Expert Opin. Drug Discov.* **2007**, *2*, 11. [CrossRef]
10. Gillard, M.; Fuks, B.; Michel, P.; Vertongen, P.; Massingham, R.; Chatelain, P. Binding Characteristics of [3h]Ucb 30889 to Levetiracetam Binding Sites in Rat Brain. *Eur. J. Pharmacol.* **2003**, *478*, 1–9. [CrossRef]
11. Klitgaard, H.; Matagne, A.; Gobert, J.; Wulfert, E. Evidence for a Unique Profile of Levetiracetam in Rodent Models of Seizures and Epilepsy. *Eur. J. Pharmacol.* **1998**, *353*, 191–206. [CrossRef]
12. Gillard, M.; Fuks, B.; Leclercq, K.; Matagne, A. Binding Characteristics of Brivaracetam, a Selective, High Affinity Sv2a Ligand in Rat, Mouse and Human Brain: Relationship to Anti-Convulsant Properties. *Eur. J. Pharmacol.* **2011**, *664*, 36–44. [CrossRef]
13. Mercier, J.; Archen, L.; Bollu, V.; Carré, S.; Evrard, Y.; Jnoff, E.; Kenda, B.; Lallemand, B.; Michel, P.; Montel, F.; et al. Discovery of Heterocyclic Nonacetamide Synaptic Vesicle Protein 2a (Sv2a) Ligands with Single-Digit Nanomolar Potency: Opening Avenues Towards the First Sv2a Positron Emission Tomography (Pet) Ligands. *Chem. Med. Chem.* **2014**, *9*, 693–698. [CrossRef]
14. Zhang, L.; Villalobos, A. Strategies to Facilitate the Discovery of Novel Cns Pet Ligands. *EJNMMI Radiopharm. Chem.* **2017**, 13. [CrossRef] [PubMed]

15. Need, A.; Kant, N.; Jesudason, C.; Barth, V. Approaches for the Discovery of Novel Positron Emission Tomography Radiotracers for Brain Imaging. *Clin. Transl. Imaging* **2017**, *5*, 265–274. [CrossRef]
16. Pike, V.W. Considerations in the Development of Reversibly Binding Pet Radioligands for Brain Imaging. *Curr. Med. Chem.* **2016**, *18*, 1818–1869. [CrossRef] [PubMed]
17. Zhang, L.; Villalobos, A.; Beck, E.M.; Bocan, T.; Chappie, T.A.; Chen, L.; Grimwood, S.; Heck, S.D.; Helal, C.J.; Hou, X.; et al. Design and Selection Parameters to Accelerate the Discovery of Novel Central Nervous System Positron Emission Tomography (Pet) Ligands and Their Application in the Development of a Novel Phosphodiesterase 2a Pet Ligand. *J. Med. Chem.* **2013**, *56*, 4568–4579. [CrossRef]
18. Hancheng, C.; Mangner, T.J.; Muzik, O.; Wang, M.-W.; Chugani, D.C.; Chugani, H.T. Radiosynthesis of (11)C-Levetiracetam: A Potential Marker for Pet Imaging of Sv2a Expression. *ACS Med. Chem. Lett.* **2014**, *10*, 1152–1155.
19. Warnock, G.I.; Aerts, J.; Bahri, M.A.; Bretin, F.; Lemaire, C.; Giacomelli, F.; Mievis, F.; Mestdagh, N.; Buchanan, T.; Valade, A.; et al. Evaluation of 18f-Ucb-H as a Novel Pet Tracer for Synaptic Vesicle Protein 2a in the Brain. *J. Nucl. Med.* **2014**, *8*, 1336–1341. [CrossRef]
20. Aerts, J.; Otabashi, M.; Giacomelli, F.; Warnock, G.; Bahri, M.; Bretin, F.; Sauvage, X.; Thielen, C.; Lemaire, C.; Salmon, E.; et al. Radiosynthesis and First Small Animal Micropet Imaging of [18f]Ucb-H, a New Fluorine-18 Labelled Tracer Targeting Synaptic Vesicle Protein 2a (Sv2a). *EANM Abstr.* **2013**, *40*, S158.
21. Warnier, C.; Lemaire, C.; Becker, G.; Zaragoza, G.; Giacomelli, F.; Aerts, J.; Otabashi, M.; Bahri, M.A.; Mercier, J.; Plenevaux, A.; et al. Enabling Efficient Positron Emission Tomography (Pet) Imaging of Synaptic Vesicle Glycoprotein 2a (Sv2a) with a Robust and One-Step Radiosynthesis of a Highly Potent 18f-Labeled Ligand ([18f]Ucb-H). *J. Med. Chem.* **2016**, *59*, 8955–8966. [CrossRef]
22. Pike, V.W.; Aigbirhio, F.I. Reactions of Cyclotron-Produced [18f] Fluoride with Diaryliodonium Salts—A Novel Single-Step Route to No-Carrier-Added [18] Fluoroarenes. *J. Chem. Soc. Chem. Commun.* **1995**, *21*, 2215–2216. [CrossRef]
23. Estrada, S.; Lubberink, M.; Thibblin, A.; Sprycha, M.; Buchanan, T.; Mestdagh, N.; Kenda, B.; Mercier, J.; Provins, L.; Gillard, M.; et al. [11c]Ucb-a, a Novel Pet Tracer for Synaptic Vesicle Protein 2a. *Nucl. Med. Biol.* **2016**, *43*, 325–332. [CrossRef] [PubMed]
24. Nabulsi, N.B.; Mercier, J.; Holden, D.; Carre, S.; Najafzadeh, S.; Vandergeten, M.C.; Lin, S.F.; Deo, A.; Price, N.; Wood, M.; et al. Synthesis and Preclinical Evaluation of 11c-Ucb-J as a Pet Tracer for Imaging the Synaptic Vesicle Glycoprotein 2a in the Brain. *J. Nucl. Med.* **2016**, *57*, 777–784. [CrossRef] [PubMed]
25. Rokka, J.; Schlein, E.; Eriksson, J. Improved synthesis of SV2A targeting radiotracer [11C]UCB-J. *EJNMMI Radiopharm. Chem.* **2019**, *4*. [CrossRef] [PubMed]
26. Sephton, S.M.; Miklovicz, T.; Russell, J.J.; Doke, A.; Li, L.; Boros, I.; Aigbirhio, F.I. Automated radiosynthesis of [11C]UCB-J for imaging synaptic density by positron emission tomography. *J. Label. Compd. Radiopharm.* **2020**, *63*, 151–158. [CrossRef]
27. Li, S.; Cai, Z.; Zhang, W.; Holden, D.; Lin, S.-F.; Finnema, S.J.; Shirali, A.; Ropchan, J.; Carre, S.; Mercier, J.; et al. Synthesis and in vivo evaluation of [18F]UCB-J for PET imaging of synaptic vesicle glycoprotein 2A (SV2A). *Eur. J. Nucl. Med. Mol. Imaging* **2019**, *46*, 1952–1965. [CrossRef]
28. Li, S.; Cai, Z.; Wu, X.; Holden, D.; Pracitto, R.; Kapinos, M.; Gao, H.; Labaree, D.C.; Nabulsi, N.; Carson, R.E.; et al. Synthesis and in Vivo Evaluation of a Novel PET Radiotracer for Imaging of Synaptic Vesicle Glycoprotein 2A (SV2A) in Nonhuman Primates. *ACS Chem. Neurosci.* **2018**, *10*, 1544–1554. [CrossRef]
29. Constantinescu, C.C.; Tresse, C.; Zheng, M.; Gouasmat, A.; Carroll, V.M.; Mistico, L.; Alagille, D.; Sandiego, C.M.; Papin, C.; Marek, K.; et al. Development and In Vivo Preclinical Imaging of Fluorine-18-Labeled Synaptic Vesicle Protein 2A (SV2A) PET Tracers. *Mol. Imaging Boil.* **2018**, *21*, 509–518. [CrossRef]
30. Cai, Z.; Li, S.; Finnema, S.; Lin, S.; Zhang, W.; Holden, D.; Carson, R.; Huang, Y. Imaging Synaptic Density with Novel 18f-Labeled Radioligands for Synaptic Vesicle Protein-2a (Sv2a): Synthesis and Evaluation in Nonhuman Primates. *J. Nucl. Med.* **2017**, *58* (Suppl. S1), 547.
31. Cai, Z.; Li, S.; Zhang, W.; Pracitto, R.; Wu, X.; Baum, E.; Finnema, S.J.; Holden, D.; Toyonaga, T.; Lin, S.-F.; et al. Synthesis and Preclinical Evaluation of an 18F-Labeled Synaptic Vesicle Glycoprotein 2A PET Imaging Probe: [18F]SynVesT-2. *ACS Chem. Neurosci.* **2020**, *11*, 592–603. [CrossRef] [PubMed]

32. Patel, S.; Knight, A.; Krause, S.; Teceno, T.; Tresse, C.; Li, S.; Cai, Z.; Gouasmat, A.; Carroll, V.M.; Barret, O.; et al. Preclinical In Vitro and In Vivo Characterization of Synaptic Vesicle 2A-Targeting Compounds Amenable to F-18 Labeling as Potential PET Radioligands for Imaging of Synapse Integrity. *Mol. Imaging Biol.* **2019**, 1–10. [CrossRef] [PubMed]
33. Trump, L.; Lemos, A.; Lallemand, B.; Pasau, P.; Mercier, J.; Lemaire, C.; Luxen, A.; Genicot, C. Late-Stage 18f-Difluoromethyl Labeling of N-Heteroaromatics with High Molar Activity for Pet Imaging. *Angew. Chem. Int. Ed.* **2019**, *131*, 13283–13288. [CrossRef]
34. Trump, L.; Lemos, A.; Jacq, J.; Pasau, P.; Lallemand, B.; Mercier, J.; Genicot, C.; Luxen, A.; Lemaire, C. Development of a General Automated Flow Photoredox 18F-Difluoromethylation of N-Heteroaromatics in an AllinOne Synthesizer. *Org. Process. Res. Dev.* **2020**. [CrossRef]
35. Serrano, M.E.; Becker, G.; Bahri, M.A.; Seret, A.; Mestdagh, N.; Mercier, J.; Mievis, F.; Giacomelli, F.; Lemaire, C.; Salmon, E.; et al. Evaluating the In Vivo Specificity of [18F]UCB-H for the SV2A Protein, Compared with SV2B and SV2C in Rats Using microPET. *Molecules* **2019**, *24*, 1705. [CrossRef]
36. Bretin, F.; Warnock, G.; Bahri, M.A.; Aerts, J.; Mestdagh, N.; Buchanan, T.; Valade, A.; Mievis, F.; Giacomelli, F.; Lemaire, C.; et al. Preclinical radiation dosimetry for the novel SV2A radiotracer [18F]UCB-H. *EJNMMI Res.* **2013**, *3*, 35. [CrossRef]
37. Becker, G.; Warnier, C.; Serrano, M.E.; Bahri, M.A.; Mercier, J.; Lemaire, C.; Salmon, E.; Luxen, A.; Plenevaux, A. Pharmacokinetic Characterization of [18F]UCB-H PET Radiopharmaceutical in the Rat Brain. *Mol. Pharm.* **2017**, *14*, 2719–2725. [CrossRef]
38. Serrano, M.E.; Bahri, M.A.; Becker, G.; Seret, A.; Mievis, F.; Giacomelli, F.; Salmon, E.; Luxen, A.; Plenevaux, A. Quantification of [18F]UCB-H Binding in the Rat Brain: From Kinetic Modelling to Standardised Uptake Value. *Mol. Imaging Boil.* **2018**, *21*, 888–897. [CrossRef]
39. Serrano, M.E.; Bahri, M.A.; Becker, G.; Seret, A.; Germonpré, C.; Lemaire, C.; Giacomelli, F.; Mievis, F.; Luxen, A.; Salmon, E.; et al. Exploring with [18F]UCB-H the in vivo Variations in SV2A Expression through the Kainic Acid Rat Model of Temporal Lobe Epilepsy. *Mol. Imaging Boil.* **2020**, 1–11. [CrossRef]
40. Finnema, S.J.; Nabulsi, N.B.; Eid, T.; Detyniecki, K.; Lin, S.-F.; Chen, M.-K.; Dhaher, R.; Matuskey, D.; Baum, E.; Holden, D.; et al. Imaging synaptic density in the living human brain. *Sci. Transl. Med.* **2016**, *8*, 348ra96. [CrossRef]
41. Toyonaga, T.; Smith, L.M.; Finnema, S.J.; Gallezot, J.-D.; Naganawa, M.; Bini, J.; Mulnix, T.; Cai, Z.; Ropchan, J.; Huang, Y.; et al. In Vivo Synaptic Density Imaging with 11C-UCB-J Detects Treatment Effects of Saracatinib in a Mouse Model of Alzheimer Disease. *J. Nucl. Med.* **2019**, *60*, 1780–1786. [CrossRef] [PubMed]
42. Nicolas, J.-M.; Hannestad, J.; Holden, D.; Kervyn, S.; Nabulsi, N.; Tytgat, D.; Huang, Y.; Chanteux, H.; Staelens, L.; Matagne, A.; et al. Brivaracetam, a selective high-affinity synaptic vesicle protein 2A (SV2A) ligand with preclinical evidence of high brain permeability and fast onset of action. *Epilepsia* **2015**, *57*, 201–209. [CrossRef] [PubMed]
43. Bahri, M.A.; Plenevaux, A.; Aerts, J.; Bastin, C.; Becker, G.; Mercier, J.; Valade, A.; Buchanan, T.; Mestdagh, N.; LeDoux, D.; et al. Measuring brain synaptic vesicle protein 2A with positron emission tomography and [18 F]UCB-H. *Alzheimer's Dementia: Transl. Res. Clin. Interv.* **2017**, *3*, 481–486. [CrossRef] [PubMed]
44. Bastin, C.; Bahri, M.A.; Meyer, F.; Manard, M.; Delhaye, E.; Plenevaux, A.; Becker, G.; Seret, A.; Mella, C.; Giacomelli, F.; et al. In vivo imaging of synaptic loss in Alzheimer's disease with [18F]UCB-H positron emission tomography. *Eur. J. Nucl. Med. Mol. Imaging* **2019**, *47*, 390–402. [CrossRef]
45. Salinas, C.A.; Searle, G.E.; Gunn, R.N. The simplified reference tissue model: Model assumption violations and their impact on binding potential. *Br. J. Pharmacol.* **2014**, *35*, 304–311. [CrossRef]
46. Wahlund, L.O.; Barkhof, F.; Fazekas, F.; Bronge, L.; Augustin, M.; Sjögren, M.; Wallin, A.; Ader, H.; Leys, D.; Pantoni, L.; et al. A New Rating Scale for Age-Related White Matter Changes Applicable to MRI and CT. *Stroke* **2001**, *32*, 1318–1322. [CrossRef]
47. Schain, M.; Benjaminsson, S.; Varnäs, K.; Forsberg, A.; Halldin, C.; Lansner, A.; Farde, L.; Varrone, A. Arterial input function derived from pairwise correlations between PET-image voxels. *Br. J. Pharmacol.* **2013**, *33*, 1058–1065. [CrossRef]
48. Rousset, O.G.; Ma, Y.; Evans, A.C. Correction for partial volume effects in PET: Principle and validation. *J. Nucl. Med.* **1998**, *39*, 904–911.

49. Erlandsson, K.; Buvat, I.; Pretorius, P.H.; Thomas, B.A.; Hutton, B.F. A review of partial volume correction techniques for emission tomography and their applications in neurology, cardiology and oncology. *Phys. Med. Boil.* **2012**, *57*, R119–R159. [CrossRef]
50. Thomas, B.A.; Cuplov, V.; Bousse, A.; Mendes, A.; Thielemans, K.; Hutton, B.F.; Erlandsson, K. PETPVC: A toolbox for performing partial volume correction techniques in positron emission tomography. *Phys. Med. Boil.* **2016**, *61*, 7975–7993. [CrossRef]
51. Tzourio-Mazoyer, N.; Landeau, B.; Papathanassiou, D.; Crivello, F.; Etard, O.; Delcroix, N.; Mazoyer, B.; Joliot, M. Automated Anatomical Labeling of Activations in SPM Using a Macroscopic Anatomical Parcellation of the MNI MRI Single-Subject Brain. *NeuroImage* **2002**, *15*, 273–289. [CrossRef] [PubMed]
52. Wu, Y.; Carson, R.E. Noise Reduction in the Simplified Reference Tissue Model for Neuroreceptor Functional Imaging. *J. Cereb. Blood. Flow Metab.* **2002**, *22*, 1440–1452. [CrossRef] [PubMed]
53. Rossano, S.; Toyonaga, T.; Finnema, S.J.; Naganawa, M.; Lu, Y.; Nabulsi, N.; Ropchan, J.; De Bruyn, S.; Otoul, C.; Stockis, A.; et al. Assessment of a white matter reference region for 11C-UCB-J PET quantification. *Br. J. Pharmacol.* **2019**, 271678 19879230. [CrossRef]
54. Koole, M.; Van Aalst, J.; Devrome, M.; Mertens, N.; Serdons, K.; Lacroix, B.; Mercier, J.; Sciberras, D.; Maguire, R.P.; Van Laere, K. Quantifying SV2A density and drug occupancy in the human brain using [11C]UCB-J PET imaging and subcortical white matter as reference tissue. *Eur. J. Nucl. Med. Mol. Imaging* **2018**, *46*, 396–406. [CrossRef] [PubMed]
55. Varnäs, K.; Stepanov, V.; Halldin, C. Autoradiographic mapping of synaptic vesicle glycoprotein 2A in non-human primate and human brain. *Synapse* **2020**. [CrossRef]
56. Bretin, F.; Bahri, M.A.; Bernard, C.; Warnock, G.; Aerts, J.; Mestdagh, N.; Buchanan, T.; Otoul, C.; Koestler, F.; Mievis, F.; et al. Biodistribution and Radiation Dosimetry for the Novel SV2A Radiotracer [18F]UCB-H: First-in-Human Study. *Mol. Imaging Boil.* **2015**, *17*, 557–564. [CrossRef]
57. Chen, M.-K.; Mecca, A.P.; Naganawa, M.; Finnema, S.J.; Toyonaga, T.; Lin, S.-F.; Najafzadeh, S.; Ropchan, J.; Lu, Y.; McDonald, J.W.; et al. Assessing Synaptic Density in Alzheimer Disease With Synaptic Vesicle Glycoprotein 2A Positron Emission Tomographic Imaging. *JAMA Neurol.* **2018**, *75*, 1215–1224. [CrossRef]
58. Vanhaute, C.H.R.J.; Ceccarini, J.; Michiels, L.; Sunaert, S.; Lemmens, R.; Emsell, L.; Vandenbulcke, M.; Van Laere, K. Changes in Synaptic Density in Relation to Tau Deposition in Prodromal Alzheimer's Disease: A Dual Protocol Pet-Mr Study. In Proceedings of the European Association of Nuclear Medicine 2019, EANM, Barcelona, Spain, 12–16 October 2019; Volume 46, pp. S177–S178.
59. Finnema, S.J.; Rossano, S.; Naganawa, M.; Henry, S.; Gao, H.; Pracitto, R.; Maguire, R.P.; Mercier, J.; Kervyn, S.; Nicolas, J.; et al. A single-center, open-label positron emission tomography study to evaluate brivaracetam and levetiracetam synaptic vesicle glycoprotein 2A binding in healthy volunteers. *Epilepsia* **2019**, *60*, 958–967. [CrossRef]
60. Matuskey, D.; Tinaz, S.; Wilcox, K.C.; Naganawa, M.; Toyonaga, T.; Dias, M.; Henry, S.; Pittman, B.; Ropchan, J.; Nabulsi, N.; et al. Synaptic Changes in Parkinson Disease Assessed with in vivo Imaging. *Ann. Neurol.* **2020**, *87*, 329–338. [CrossRef]
61. Bellucci, A.; Mercuri, N.B.; Venneri, A.; Faustini, G.; Longhena, F.; Pizzi, M.; Missale, C.; Spano, P. Review: Parkinson's disease: From synaptic loss to connectome dysfunction. *Neuropathol. Appl. Neurobiol.* **2016**, *42*, 77–94. [CrossRef]
62. Onwordi, E.C.; Halff, E.F.; Whitehurst, T.; Mansur, A.; Cotel, M.C.; Wells, L.; Creeney, H.; Bonsall, D.; Rogdaki, M.; Shatalina, E.; et al. Synaptic Density Marker Sv2a Is Reduced in Schizophrenia Patients and Unaffected by Antipsychotics in Rats. *Nat Commun.* **2020**, *11*, 246. [CrossRef] [PubMed]
63. Holmes, S.E.; Scheinost, D.; Finnema, S.J.; Naganawa, M.; Davis, M.T.; DellaGioia, N.; Nabulsi, N.; Matuskey, D.; Angarita, G.A.; Pietrzak, R.H.; et al. Lower Synaptic Density Is Associated with Depression Severity and Network Alterations. *Nat. Commun.* **2019**, *10*, 1529. [CrossRef] [PubMed]

Sample Availability: Samples of the compounds......are available from the authors.

© 2020 by the authors. Licensee MDPI, Basel, Switzerland. This article is an open access article distributed under the terms and conditions of the Creative Commons Attribution (CC BY) license (http://creativecommons.org/licenses/by/4.0/).

Review

Positron Emission Tomography (PET) Radiopharmaceuticals in Multiple Myeloma

Christos Sachpekidis [1,*], Hartmut Goldschmidt [2] and Antonia Dimitrakopoulou-Strauss [1]

1. Clinical Cooperation Unit Nuclear Medicine, German Cancer Research Center, 69120 Heidelberg, Germany; a.dimitrakopoulou-strauss@dkfz.de
2. Department of Internal Medicine V, University Hospital Heidelberg and National Center for Tumor Diseases (NCT), 69120 Heidelberg, Germany; Hartmut.Goldschmidt@med.uni-heidelberg.de
* Correspondence: c.sachpekidis@dkfz-heidelberg.de or christos_saxpe@yahoo.gr; Tel.: +49-6221-42-2478; Fax: +49-6221-42-2476

Academic Editor: Anne Roivainen
Received: 16 December 2019; Accepted: 26 December 2019; Published: 29 December 2019

Abstract: Multiple myeloma (MM) is a plasma cell disorder, characterized by clonal proliferation of malignant plasma cells in the bone marrow. Bone disease is the most frequent feature and an end-organ defining indicator of MM. In this context, imaging plays a pivotal role in the management of the malignancy. For several decades whole-body X-ray survey (WBXR) has been applied for the diagnosis and staging of bone disease in MM. However, the serious drawbacks of WBXR have led to its gradual replacement from novel imaging modalities, such as computed tomography (CT), magnetic resonance imaging (MRI) and positron emission tomography/computed tomography (PET/CT). PET/CT, with the tracer ^{18}F-fluorodeoxyglucose (^{18}F-FDG), is now considered a powerful diagnostic tool for the detection of medullary and extramedullary disease at the time of diagnosis, a reliable predictor of survival as well as the most robust modality for treatment response evaluation in MM. On the other hand, ^{18}F-FDG carries its own limitations as a radiopharmaceutical, including a rather poor sensitivity for the detection of diffuse bone marrow infiltration, a relatively low specificity, and the lack of widely applied, established criteria for image interpretation. This has led to the development of several alternative PET tracers, some of which with promising results regarding MM detection. The aim of this review article is to outline the major applications of PET/CT with different radiopharmaceuticals in the clinical practice of MM.

Keywords: multiple myeloma; positron emission tomography/computed tomography; radiopharmaceuticals; ^{18}F-fluorodeoxyglucose

1. Introduction

Multiple myeloma (MM) is a neoplastic plasma cell disorder, characterized by the uncontrolled, clonal proliferation of plasma cells in the bone marrow. It is the second most common hematologic malignancy after non-Hodgkin's lymphoma accounting for approximately 1% of neoplastic diseases, and the most common primary tumor of the skeleton [1]. MM is almost always preceded from a premalignant precursor condition (monoclonal gammopathy of undetermined significance, MGUS), which then develops into asymptomatic or smoldering myeloma (SMM) and, finally, into symptomatic disease [2]. Bone involvement in the form of focal osteolytic lesions—the hallmark radiographic sign of MM—represents a marker of disease-related end-organ damage, necessitating immediate initiation of treatment [3]. Bone disease is a major cause of morbidity and mortality for patients suffering from MM. Since practically all patients develop bone involvement during the course of the disease [4], its reliable identification represents a pivotal diagnostic challenge. Historically, skeletal damage has been assessed by conventional, whole-body X-ray survey (WBXR), which was the standard imaging approach for

MM. Nevertheless, this modality carries several limitations, including a low sensitivity—requiring a more than 30% bone demineralization before an osteolytic lesion becomes evident—its failure to detect extramedullary disease (EMD), which is a significant adverse prognostic factor of MM, and its poor performance in treatment response assessment [5]. The drawbacks of planar radiography have been overcome in recent years with the development and introduction in clinical practice of myeloma of novel imaging modalities, namely whole-body computed tomography (CT), magnetic resonance imaging (MRI) and positron emission tomography/computed tomography (PET/CT). These techniques offer a higher sensitivity than WBXR, leading to its gradual substitution by them.

It is undisputable that the role of PET/CT with the radiotracer ^{18}F-fluorodeoxyglucose (^{18}F-FDG) in MM has been upgraded with an increasing amount of literature highlighting its value in diagnosis, prognosis and treatment response evaluation of the disease. According to the latest update of the International Myeloma Working group (IMWG), the detection of one or more osteolytic lesions on CT or PET/CT fulfills the criteria of bone disease and, therefore, of symptomatic MM requiring treatment [4].

This review article provides an overview of the position of PET/CT in MM management with focus on the most widely used tracer ^{18}F-FDG. In addition, the main data published on new PET tracers targeting different molecular pathways involved in MM pathogenesis are presented.

2. ^{18}F-FDG PET/CT in MM

PET/CT is a whole-body imaging technique combining the functional information of PET with the morphological assessment provided by CT. ^{18}F-FDG, the workhorse of PET imaging, is a biomarker of intracellular glucose metabolism. The tracer is actively transported into cells by the glucose transporter proteins (GLUT), which are expressed at a high degree in tumor cells due to their enhanced glucose demands. ^{18}F-FDG, as a glucose analogue, is taken up by the neoplastic cells, undergoes phosphorylation and then gets trapped intracellularly, since ^{18}F-FDG is not a substrate for further metabolic processing by either phosphohexose isomerase or glucose-6-phosphate dehydrogenase [6].

^{18}F-FDG PET/CT has become nowadays a standard imaging technique in several tumor entities. Due to its ability in providing whole-body evaluations in a single session, the modality can assess the extent of oncological disease in a satisfying manner. In MM in particular, PET/CT can detect with a high sensitivity and specificity both medullary and extramedullary lesions [7]. Another important advantage of PET is the potential of quantification of tracer uptake by means of the index standardized uptake value (SUV), which reflects the amount of tracer activity in a particular region of interest. This quantification of tracer uptake aids in objective interpretation of PET/CT scans in addition to obtaining cross-sectional imaging and assessing ^{18}F-FDG uptake visually, particularly in terms of patient follow-up. Furthermore—and most importantly—^{18}F-FDG PET/CT can assess the metabolic burden and activity of MM in different stages of the disease due to its ability in differentiating between metabolically active and inactive lesions, with significant implications in treatment response assessment [5,7].

2.1. ^{18}F-FDG PET/CT in the Diagnosis and Staging of MM

^{18}F-FDG PET/CT has been proven to be a very useful modality for the whole-body evaluation of the active burden of MM. Its reported sensitivity and specificity for assessment of medullary and extramedullary disease extent ranges from 80–100% [7–12]. The uptake pattern, SUV and different pharmacokinetic parameters of ^{18}F-FDG correlate with the percentage of bone marrow plasma cells [13] (Figure 1).

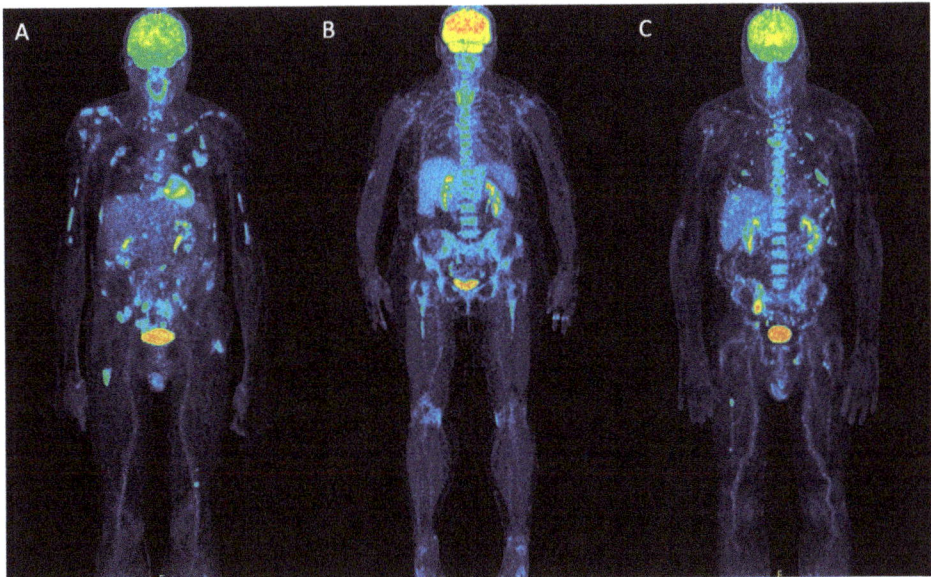

Figure 1. Maximum intensity projection (MIP) PET/CT images of newly diagnosed MM patients before treatment, representing examples of different pathologic patterns of ^{18}F-FDG uptake. (**A**) demonstrates a patient with multiple focal lesions in the skeleton. (**B**) depicts a patient with intense diffuse tracer uptake in the bone marrow of the axial skeleton and the proximal humeri and femora without clearly delineated focal lesions. (**C**) shows a patient with a mixed pattern of ^{18}F-FDG uptake with intense, diffuse uptake in the axial skeleton and multiple, focal bone marrow lesions.

PET/CT has been compared with other imaging modalities and has been shown to be superior to WBXR and comparable to MRI. In particular, a prospective study comparing ^{18}F-FDG PET/CT with WBXR and pelvic-spinal MRI highlighted the superiority of PET/CT to WBXR in 46% of cases (sensitivity 92% vs. 61%). The sensitivity of PET/CT in the spine was inferior to MRI, underestimating the disease in a third of the patients; however, ^{18}F-FDG PET/CT detected sites of active disease in areas outside the field of the MRI view [8]. Similarly, the results of a systematic review of 18 studies comparing the above-mentioned modalities showed a higher sensitivity of MRI at detecting diffuse disease of the spine, while ^{18}F-FDG PET/CT was more sensitive than WBXR with regard to detection of bone lesions [10]. In another systematic review of 17 studies no significant differences were found between ^{18}F-FDG PET/CT (sensitivity 91%, specificity 69%) and MRI (sensitivity 88%, specificity 68%) regarding detection rate of bone disease [11]. Recently, the prospective French IMAJEM study revealed no difference in the detection of bone lesions at diagnosis when comparing PET/CT and MRI with the former being positive in 95% and the latter in 91% of the patients [12].

Interestingly, there is a lack of studies regarding the comparison of ^{18}F-FDG PET/CT with whole-body CT. According to the recently published consensus statement by the IMWG, although whole-body low-dose CT is the preferred method for the detection of lytic bone lesions in MM, ^{18}F-FDG PET/CT should be considered as a valuable option, because of its ability to identify lytic lesions and extramedullary masses. Moreover, in cases of WBXR-negativity and whole-body MRI-unavailability, ^{18}F-FDG PET/CT is recommended for the differentiation between active and smoldering MM [7].

Further, the newly emerging, hybrid PET/MRI technique seems highly attractive in the diagnostic approach of MM since it combines two modalities with a high potential in myeloma evaluation in a single exam. The results of the only prospective study comparing PET/CT with PET/MRI demonstrated good image quality provided by PET/MRI and high correlation between the modalities regarding the number of detected active lesions and SUV values [14]. However, further studies are warranted to evaluate the potential role of this novel technique in the diagnostics and management of MM.

2.2. Prognostic Value of ^{18}F-FDG PET/CT in MM

^{18}F-FDG PET/CT is a reliable outcome predictor and is regarded as the elective technique for treatment response evaluation of MM due to its ability to distinguish active from inactive sites of disease [9,12,15]. In newly diagnosed, symptomatic MM patients, three independent PET factors have been recognized to affect both progression-free survival (PFS) and overall survival (OS) in different prospective studies. These parameters are the number of focal, ^{18}F-FDG-avid lesions, the SUV_{max} of the lesions, and the presence of EMD. Bartel et al. were the first to show in a group of 239 MM patients treated upfront with novel agents and double autologous stem-cell transplantation (ASCT) that the presence of more than three ^{18}F-FDG-avid focal lesions was related to fundamental features of myeloma biology and genomics and was the leading independent parameter associated with inferior PFS and OS [9]. A few years later, in a study by Zamagni et al., including 192 MM patients treated with thalidomide-dexamethasone induction therapy and double ASCT, it was shown that the presence at baseline of at least three focal lesions, a $SUV_{max} > 4.2$ of the hottest lesion, and the presence of EMD adversely affected 4-year estimates of PFS, while $SUV_{max} > 4.2$ and EMD were also correlated with shorter OS [15]. Further, the IMAJEM study highlighted the role of EMD as an independent, adverse prognostic factor for both PFS and OS in 134 patients receiving a combination of lenalidomide, bortezomib, and dexamethasone with or without ASCT, followed by lenalidomide maintenance [12]. The prognostic significance of the three established PET risk factors was recently confirmed in a prospective study of 48 MM patients treated with induction treatment and ASCT. In that study it was also shown that not only quantitative PET parameters from focal lesions, but also those from reference bone marrow samples, are associated with adverse PFS in the disease [16].

Apart from its predictive role in symptomatic MM, ^{18}F-FDG PET/CT has shown prognostic value in asymptomatic SMM patients. Although existing data are relatively limited, the first published results reflect the potential role of the modality in predicting the risk of progression from SMM to symptomatic disease. Siontis et al. studied a group of 122 SMM patients and found that the 2-year risk of progression to active MM was 75% in patients with a positive PET/CT (with or without lytic lesions), compared to 30% in patients with a negative PET/CT. The median time to progression (TTP) was 21 months for the PET/CT positive group, while the respective TTP for the PET/CT negative group was 60 months [17]. In another prospective, multicentric study of 120 SMM patients and a median follow-up of 2.2 years, patients with a positive PET study without underlying osteolysis had a higher risk of progression to active MM and a shorter TTP than patients who were PET-negative. In particular, 58% of the patients with a positive PET scan progressed to active myeloma in 2 years with a median TTP of 1.1 years, compared to those with a negative PET scan demonstrating a progression rate of 33% and a median TTP of 4.5 years [18].

2.3. The Value of ^{18}F-FDG PET/CT in Therapy Assessment

Due to its ability in distinguishing between active and inactive lesions, ^{18}F-FDG PET/CT is the best imaging tool for therapy response assessment and is considered the gold standard for treatment monitoring in MM [7] (Figure 2). Several studies have highlighted the role of the modality in the evaluation of the metabolic response to therapy in different stages of the treatment protocol, for example during induction treatment as well as after ASCT [9,12,15,19–23].

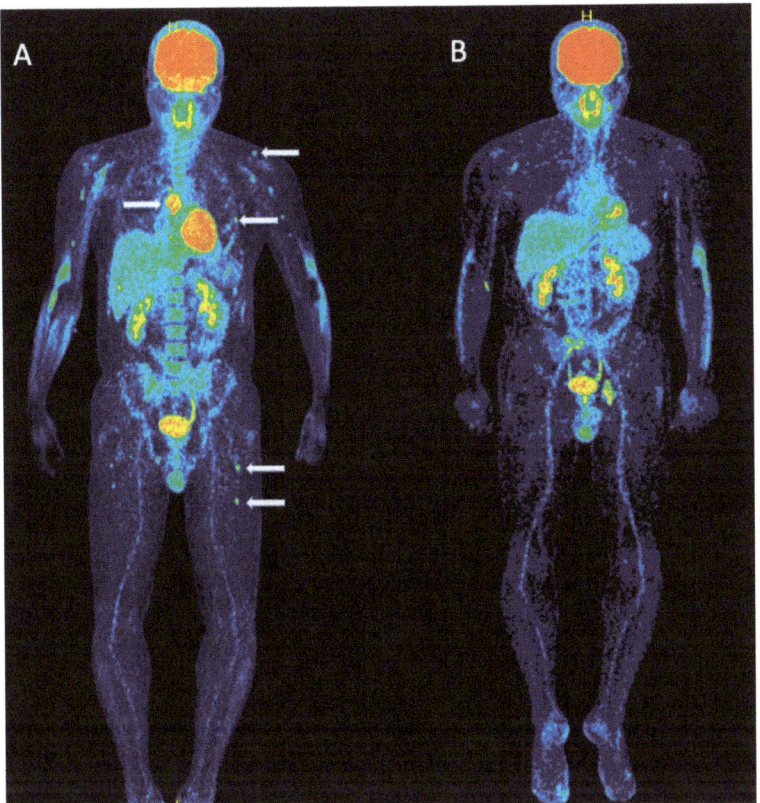

Figure 2. A 39-years old symptomatic MM patient scheduled for HDT and ASCT, undergoing ^{18}F-FDG PET/CT before and after therapy. Maximum intensity projection (MIP) ^{18}F-FDG PET/CT before therapy (**A**) revealed a mixed pattern of ^{18}F-FDG uptake with intense, diffuse uptake in the axial skeleton and multiple, focal bone marrow lesions for example in the sternum, ribs, humerus, scapula and femur (arrows). Follow-up ^{18}F-FDG PET/CT MIP after high-dose chemotherapy and ASCT (**B**) demonstrated a complete remission of both diffuse bone marrow uptake as well as focal MM lesions.

In a study published by the Little Rock group in 2009 involving 239 previously untreated MM patients, it was shown that complete ^{18}F-FDG suppression in focal lesions and EMD after induction treatment and before ASCT conferred superior OS and PFS, and was identified as an independent favorable prognostic variable [9]. A few years later, the same group published a study on a larger cohort involving 302 MM patients studied with PET/CT on day 7 of induction treatment. The authors showed that the persistence of more than three ^{18}F-FDG-avid lesions imparted inferior OS and PFS, suggesting a therapy change in patients with persistent findings on PET/CT early after induction therapy [19]. Most recently, this team published the findings of a trial in 596 patients examined with PET/CT at different time points (day 7 of induction, end of induction, post transplantation, and at maintenance treatment). They demonstrated that patients achieving complete suppression of ^{18}F-FDG activity in focal lesions following treatment at each studied time point had nonsignificant differences in their PFS and OS values than the patients with no lesions at baseline. Importantly, at each time point, patients with no detectable lesions had a significantly superior outcome compared to patients with at least one detectable lesion at that time point, irrespective of whether they had lesions at baseline [22].

The Bologna group has also highlighted the importance of ^{18}F-FDG PET/CT in assessment of response to therapy in MM in different time points. In particular, they have shown that the persistence of severe ^{18}F-FDG uptake—as reflected by the number of focal lesions, SUVmax and presence of EMD—after thalidomide/dexamethasone induction therapy is an early predictor of the worst long-term clinical outcomes. Moreover, a complete response (CR) on PET/CT after ASCT conferred superior PFS and OS in comparison with persistence of ^{18}F-FDG uptake, while the prognostic value of PET/CT was retained also at the time of relapse, with patients positive on PET/CT having a significantly shorter survival compared to those with a negative PET/CT scan [15]. A few years later, they showed in a group of 282 patients that attainment of PET/CT negativity by 3 months after the last cycle of first-line treatment (chemotherapy, novel agents with or without ASCT) significantly influenced both PFS and OS [21]. PET/CT has also been shown effective in response evaluation of patients undergoing allogeneic stem cell transplantation with persistence of EMD being an independent predictor of poor outcome and, on the other hand, achievement of CR on PET/CT after transplantation being associated with a significantly longer OS [23].

The French group (IMAJEM trial) recently evaluated the role of PET/CT after induction treatment (lenalidomide, bortezomib, and dexamethasone) as well as before lenalidomide maintenance in a group of 134 MM patients. The authors showed that normalization of PET/CT after three cycles of induction therapy was associated with improved PFS, and that normalization before maintenance resulted in longer PFS and OS, in comparison to patients without normalization of their PET findings [12]. They could, moreover, show that change in SUV after three cycles of induction therapy was an independent prognostic factor for PFS, rendering SUV a potentially powerful tool for the prediction of long-term outcome in MM [24].

Other groups have also studied ^{18}F-FDG PET/CT in the treatment response evaluation of MM, using different therapeutic agents and protocols. Most of them have confirmed the benefit of applying the modality in the workup of MM patients [25–27].

2.4. The Value of ^{18}F-FDG PET/CT in Minimal Residual Disease (MRD) Diagnostics

A field that is constantly drawing more attention in MM therapy assessment is that of standardization and optimization of minimal residual disease (MRD) detection, which is becoming standard diagnostic care. This is driven by the need to improve the definition of disease remission due to the unprecedented rates of CR brought in recent years by the incorporation of novel agents in the treatment of MM patients. It is clear that in MM there is a direct correlation between the depth of response and prolonged survival rates [28]. At present, MRD is detected within the bone marrow, either by multicolor flow cytometry (MFC) or by next generation sequencing technologies [29].

Data on the potential role of ^{18}F-FDG PET/CT in evaluation of the depth of response—beyond the level of conventionally defined CR- are limited but growing. Zamagni et al. retrospectively analyzed 282 MM patients who were evaluated at baseline and during posttreatment follow-up with serial PET/CT scans. They found that the modality could provide a more accurate definition of CR, allowing to stratify patients in conventional CR after up-front therapy into different prognostic subgroups, according to the persistence or absence of ^{18}F-FDG metabolic activity. In particular, the achievement of PET-negativity after treatment was an independent predictor of prolonged PFS and OS for patients with conventionally defined CR [21]. Furthermore, the complementary role of PET/CT and MRD diagnostics with MFC in predicting patient outcome has been supported by some studies. A subanalysis of the IMAJEM trial in 86 patients before maintenance evaluated for both PET/CT and MRD, assessed by MFC, revealed a higher PFS for the group of patients with both a normalized PET/CT and a negative MRD versus patients with either PET positivity and/or MRD positivity before maintenance [12]. In line with these results, the Little Rock group showed in 83 MM patients in CR with available MRD and functional imaging data (in this case PET/CT and/or diffusion weighted MRI) that double-positive and double-negative features defined groups with dismal and excellent PFS, respectively [30]. Most recently, a retrospective study analyzed the prediction of outcome with the combination of ^{18}F-FDG

PET/CT and MRD, assessed by MFC, in 103 patients with newly diagnosed MM. Apart from confirming the benefit—in terms of PFS—linked to the achievement of negativity by MFC and ^{18}F-FDG PET/CT individually, the authors showed that the combination of negativity by both techniques conferred significantly higher PFS than each technique alone, also supporting the potential complementarity between PET/CT and MFC in MRD detection [31].

2.5. Limitations of ^{18}F-FDG PET/CT

Limitations of ^{18}F-FDG PET/CT include its limited availability in comparison to conventional radiological modalities as well as its higher cost. Moreover, the poor sensitivity for the detection of diffuse bone marrow infiltration or skull lesions, due to masking of their activity by the underlying physiological tracer uptake in the brain, is an important drawback. In a report of 227 MM patients the incidence of PET false-negativity was 11% in these patients, a finding attributed to the significantly lower expression of the gene coding for hexokinase-2, which catalyzes the first step of glycolysis [32]. However, this explanation warrants further validation [33]. Further, ^{18}F-FDG, as a glucose analog, is generally restricted in oncological imaging by both false positive (inflammation, post-surgical areas, recent use of chemotherapy, fractures, etc.) and false negative results (hyperglycemia, recent administration of high-dose steroids, etc.). Finally, issues are raised due to the lack of established criteria for image interpretation of ^{18}F-FDG PET/CT scans in MM, resulting in poor interobserver reproducibility in interpreting results. In an attempt to standardize the interpretation of ^{18}F-FDG PET/CT, the Bologna group has recently proposed the Italian Myeloma criteria for PET Use (IMPeTUs) based on the standard Deauville five-point system [34]. These descriptive criteria take into account the number and site of focal lesions, the presence of EMD, as well as the diffuse bone marrow involvement. The first results from the application of IMPeTUs seem to improve the interobserver reproducibility in scan interpretation; however, this needs to be confirmed in further studies.

3. Non ^{18}F-FDG PET Tracers in MM

Due to the limitations of ^{18}F-FDG as an imaging biomarker of MM, several other PET tracers have been proposed and tested in patients with the malignancy. Although some of them have given promising results regarding detection of MM lesions, most studies were performed in rather small patient cohorts and, thus, require validation in further prospective clinical trials. The most important of them will be addressed in the following paragraphs.

3.1. ^{18}F-Choline and ^{11}C-Choline

Choline is a component of phosphatidylcholine and, as such, functions as a substrate for cell membrane biosynthesis. The uptake of radiolabeled choline is increased in proliferating cells because it is involved in membrane metabolism and growth. Choline PET imaging has been traditionally used in the diagnostics of prostate cancer.

The first report of ^{11}C-choline uptake in myeloma lesions was an incidental finding of a solitary plasmacytoma in a patient being re-staged for prostate cancer [35]. Based on this finding, a comparison study of ^{11}C-choline vs. ^{18}F-FDG PET/CT in assessing bone involvement was performed by the Bologna group in a heterogeneous group of 10 MM patients (4 patients at completion of initial therapy, 2 during follow-up and 4 at disease relapse). In 2/10 patients with suspicion of disease relapse, both the ^{11}C-Choline and ^{18}F-FDG PET/CT scans were positive and identified the same number and sites of bone lesions. In 4/10 patients, both techniques were positive, but ^{11}C-choline identified a nonsignificant higher number of lesions than ^{18}F-FDG. Finally, 4 patients were negative with both tracers, a finding consistent with clinical, laboratory and radiological data indicating a CR at the time of imaging [36]. Almost ten years later, another pilot study on choline PET was published on a larger MM patient cohort. Twenty-one patients with suspected progressive or relapsing MM were studied with ^{18}F-choline and ^{18}F-FDG PET/CT. No myeloma lesions were detected in two cases, while uncountable foci were observed in four patients. In the rest, 15 patients with countable bone foci, ^{18}F-choline PET/CT depicted a significantly higher number of lesions than ^{18}F-FDG PET/CT [37]. Further, the performance of ^{18}F-choline and ^{18}F-FDG PET/CT in the detection of skeletal involvement was compared in a case series of five MM patients in a pairwise fashion. Skeletal lesions were detected in all five ^{18}F-choline PET/CT scans compared to four out of five ^{18}F-FDG PET/CT scans. Altogether ^{18}F-choline PET/CT detected a total of 134 bone lesions compared to 64 lesions detected by ^{18}F-FDG PET/CT. Interestingly, the vast majority of the missed lesions in ^{18}F-FDG PET/CT were in the axial skeleton including the skull vault [38].

To summarize, choline PET seems to have a better detection rate of focal lesions than ^{18}F-FDG PET. However, no comparison studies between the two PET tracers in previously untreated MM patients have been performed. A limitation of choline PET is its unfavorable physiological distribution involving increased uptake in the bone marrow and the liver parenchyma potentially masking lesions in these organs; although hepatic lesions are rare in MM and can be reliably detected with MRI, the increased activity in the bone marrow compartment may pose significant diagnostic challenges, in particular in patients showing a diffuse bone marrow infiltration pattern. Moreover, the use of ^{11}C-choline is limited in centres with an on-site cyclotron and radiopharmacy facilities, because of the very short half-life of the radioisotope (20 min).

3.2. ^{11}C-Acetate

^{11}C-acetate is rapidly picked-up by cells and metabolized into acetyl-CoA by the key enzyme acetyl-CoA synthase, which is overexpressed in certain cancer cells [39]. The use of ^{11}C-acetate in MM can be justified by the elevated lipid synthesis in proliferating abnormal plasma cells as reported by studies with myeloma cell lines [40].

Similarly to radiolabeled choline, the first report of ^{11}C-acetate uptake in myeloma lesions was an incidental finding [41]. In total, two comparative studies of ^{11}C-acetate with ^{18}F-FDG have been published thus far. Ho et al. evaluated a heterogeneous group of 35 untreated patients (26 with symptomatic MM, 5 with SMM, and 4 with MGUS), 9 of which undergoing also dual tracer follow-up PET/CT. The authors reported a significantly higher overall sensitivity for symptomatic MM with ^{11}C-acetate than with ^{18}F-FDG (84.6% vs. 57.7%), while the specificity for ^{11}C-acetate and ^{18}F-FDG PET/CT was 100% and, 93.1% respectively. Furthermore, all indolent plasma cell neoplasms (SMM and MUGS) were negative by ^{11}C-acetate PET, whereas 2 cases of MGUS were false-positive by ^{18}F-FDG [42]. A similar study was published a few months later by Lin et al. in 15 untreated MM patients examined with both tracers at diagnosis, 13 of which being evaluated with a repeated dual-tracer examination after completion of induction treatment. They found a higher detection rate for both diffuse and focal myeloma lesions at initial staging using ^{11}C-acetate than ^{18}F-FDG. Moreover, after treatment the diffuse bone marrow ^{11}C-acetate uptake showed a statistically significant difference in SUV$_{max}$ reductions between patients with at least a very good partial response and those with at most a partial response. Such a difference between patients in these two response groups was not observed with ^{18}F-FDG PET/CT [43].

In summary, these preliminary findings imply a potential role for ^{11}C-acetate PET/CT for the evaluation of patients with MM. Nevertheless, practical and logistical considerations are raised due to the fact that the synthesis of the tracer requires technical expertise and an on-site cyclotron.

3.3. ^{11}C-Methionine

^{11}C-Methionine is an aminoacidic PET tracer mainly employed in the diagnosis of central nervous system tumors. The uptake of the tracer primarily reflects its transmembrane transport by the sodium-independent L-transporter into cells. This transport is driven by concentration gradient and is thus influenced by the intracellular metabolism of the amino acid, which in turn reflects proliferation activity [44]. The concept of applying ^{11}C-methionine in MM is mainly based on the knowledge that radiolabeled amino acids show a rapid uptake and metabolic incorporation into newly synthesized immunoglobulins [45]. Moreover, the uptake of 35S-methionine into myeloma cells is higher as compared with other hematopoietic cells [46]. Despite the limited literature on the topic, ^{11}C-methionine PET/CT concordantly appears to perform better than ^{18}F-FDG in detection of myeloma lesions.

Dankerl et al. were the first to apply this PET tracer for imaging of MM in a group of 19 patients with active disease. The authors detected disseminated multifocal ^{11}C-methionine–positive bone marrow lesions in all patients, except two, a finding suggesting widespread dissemination of MM in the hematopoietic bone marrow. The two patients without extensive disease on ^{11}C-methionine PET showed exclusive EMD and monofocal medullary MM, respectively [46]. The first comparative study was published in 2013 by Nakamoto et al. in 20 patients with MM (n = 15) and plasmacytoma (n = 5) who underwent ^{18}F-FDG PET/CT and ^{11}C-methionine PET/CT scans. On a patient basis, two patients were accurately diagnosed only by ^{11}C-methionine PET/CT, while in the remaining 18 patients consistent results were obtained. However, the potential upgrade of staging or restaging was necessary in 6 of 11 positive patients because more abnormal lesions were demonstrated by ^{11}C-methionine PET/CT. The patient-based sensitivity, specificity and accuracy of ^{11}C-methionine PET/CT for restaging were 89%, 100% and 93%, respectively, while those of ^{18}F-FDG PET/CT were 78%, 100% and 86%, respectively [47]. Two years later, Okasaki et al. studied 64 patients with MM or MGUS (21 previously untreated, 43 restaged after treatment) undergoing PET/CT with the tracers ^{11}C-4'-thiothymidine (^{11}C-4DST), ^{11}C-methionine, and ^{18}F-FDG. The main findings of the study were the following: firstly, the number of equivocal lesions observed using ^{18}F-FDG was larger compared to using ^{11}C-methionine or ^{11}C-4DST both before and after therapy. Secondly, ^{11}C-methionine and ^{11}C-4DST were superior to ^{18}F-FDG in clearly detecting skull lesions because of their low physiological accumulation in the brain [48].

The Würzburg group has also highlighted the superiority of ^{11}C-methionine over ^{18}F-FDG for staging and re-staging of both intra- and extramedullary MM lesions [49,50]. These results were further confirmed in both patient- and lesion-based analyses in the largest so far, dual-center study of 78 patients (4 solitary plasmacytoma, 5 SMM, 69 symptomatic MM) published in 2017 [51]. Moreover, the same group has recently performed the first head-to-head comparison of ^{11}C-methionine and ^{11}C-choline for metabolic imaging of MM in 19 patients with a history of MM (n = 18) or solitary bone plasmacytoma (n = 1). ^{11}C-methionine provided advantages over ^{11}C-choline in terms of higher sensitivity by detecting a higher number of intramedullary lesions in approximately 40% of patients, as well as by achieving higher lesion-to-background contrast [52].

Drawbacks of ^{11}C-Methionine PET are considered to be its increased physiological biodistribution in the liver parenchyma and the bone marrow, potentially reducing the detection rate of MM lesions. Moreover, the ^{11}C labeling of the tracer prevents a relatively massive production and distribution of ^{11}C-Methionine [53].

3.4. ^{18}F-Fluorothymidine (^{18}F FLT)

^{18}F-Fluorothymidine (^{18}F-FLT) is the most studied cellular proliferation PET agent [54]. ^{18}F-FLT is taken up by cells and phosphorylated by thymidine kinase 1, which is upregulated by about tenfold during the S-phase of the cell cycle, producing ^{18}F-FLT monophosphate (^{18}F-FLT-MP), which can then be sequentially phosphorylated to form ^{18}F-FLT diphosphate (^{18}F-FLT-DP) and ^{18}F-FLT triphosphate (^{18}F-FLT-TP). These phosphorylated products are metabolically trapped intracellularly without being incorporated into DNA. The tracer retention within cells reflects, in part, thymidine kinase activity and is often positively correlated with cellular proliferation [55].

The knowledge regarding application and performance of ^{18}F-FLT PET in MM is limited. Agool et al. studied a group of 18 patients with different hematologic disorders, among which were two patients with MM. The authors found that the affected osteolytic areas in these two MM patients demonstrated a low ^{18}F-FLT uptake [56]. In a pilot study on combined ^{18}F-FDG and ^{18}F-FLT PET/CT imaging in 8 myeloma patients (4 patients with symptomatic MM, 4 patients with SMM) the number of myeloma-indicative lesions was significantly higher for ^{18}F-FDG PET/CT than for ^{18}F-FLT

PET/CT. A common finding of the study was a mismatch of focally increased ^{18}F-FDG uptake and reduced ^{18}F-FLT uptake (lower than the surrounding bone marrow) in myeloma lesions. Moreover, ^{18}F-FLT PET/CT was characterized by high background activity in the bone marrow compartment, complicating the evaluation of bone marrow lesions [57].

In conclusion, despite the limited number of patients studied so far, the preliminary results indicate that ^{18}F-FLT does not seem suitable as a single PET tracer in MM diagnostics.

3.5. ^{68}Ga-Pentixafor

Chemokine receptor 4 (CXCR4) is a pleiotropic, G-protein coupled chemokine receptor expressed on hematopoeitic stem and progenitor cells in the bone marrow niche. CXCR4 can mediate the migration as well as the homing process of these cells in the bone marrow in response to its ligand, stromal cell-derived factor 1 (SDF-1) [58]. In MM, CXCR4 is involved in myeloma cell homing, bone marrow retention, angiogenesis and metastasis, while collective evidence from several studies support the pivotal role of CXCR4 in different stages of MM, disease progression, development of therapeutic resistance and MRD, as well as poor prognosis [59–66].

^{68}Ga-pentixafor is a radiolabeled peptide that shows high affinity for CXCR4. The major advantage of the tracer is its potential use in a thera(g)nostic approach in combination with the ^{177}Lu- or ^{90}Y-labeled agent pentixather in progressive MM patients with CXCR4-positive tumor cells, as confirmed by a ^{68}Ga-pentixafor PET scan. Preliminary results of the CXCR4-directed endoradiotherapy with pentixather in three heavily pretreated MM patients were relatively encouraging with low levels of toxicity, good tolerance of the treatment and high initial response rates [67].

Two studies have investigated the diagnostic performance of ^{68}Ga-pentixafor in comparison to ^{18}F-FDG in patients with advanced MM. The initial results in 14 MM patients showed a slight superiority of the novel tracer over ^{18}F-FDG in the relapsed disease setting, with 10/14 patients showing MM manifestations on ^{68}Ga-pentixafor PET, while 9/14 were positive on ^{18}F-FDG PET [68]. The larger second study included 35 patients undergoing ^{68}Ga-pentixafor PET/CT for evaluation of eligibility for endoradiotherapy. In 19 patients, ^{18}F-FDG PET/CT was also available for correlation. ^{68}Ga-pentixafor PET detected CXCR4-positive disease in 23/35 subjects (66%). Importantly, in the 19 patients in whom a comparison to ^{18}F-FDG PET was available, 8/19 (42%) patients had an equal number of lesions with both tracers, in 4/19 (21%) subjects ^{68}Ga-pentixafor PET detected more lesions, while ^{18}F-FDG PET proved superior in 7/19 (37%) of them [69].

Most recently, the first comparative study of ^{18}F-FDG and ^{68}Ga-pentixafor PET/CT in 30 patients with newly diagnosed MM was published. ^{68}Ga-Pentixafor PET/CT had a significantly higher positive rate than ^{18}F-FDG PET/CT in detection of myeloma lesions (93.3% vs. 53.3%). In quantitative analysis, bone marrow uptake values in ^{68}Ga-Pentixafor were positively correlated with end organ damage, staging, and laboratory biomarkers related to tumor burden including serum β2-microglobulin, serum free light chain, and 24-h urine light chain. In contrary, in ^{18}F-FDG PET/CT, only the SUV mean of total bone marrow was positively correlated with serum free light chain and 24-h urine light chain [70]. These results indicate that ^{68}Ga-pentixafor PET might be a promising biomarker in assessing the tumor burden of newly diagnosed MM patients.

3.6. ^{18}F-Sodium Fluoride (^{18}F-NaF)

^{18}F-NaF is a highly sensitive biomarker of bone reconstruction, with potential indications in a wide range of bone disease [71–74]. The uptake of the tracer in bone occurs by chemisorption onto hydroxyapatite, followed by exchange with hydroxyl groups in the hydroxyapatite, resulting in formation of fluoroapatite. The tracer accumulates in nearly all sites of increased new bone formation, reflecting regional blood flow, osteoblastic activity and bone turnover [71,75,76].

An increasing interest has been raised in the last years on the potential application of ^{18}F-NaF PET/CT in MM diagnostics and management. This interest was based, however, on a very small number of studied MM patients without comparison with a robust reference imaging method [77–80].

Despite this initial enthusiasm, subsequent publications demonstrated rather discouraging results. In particular, ^{18}F-NaF PET/CT did not confer any superiority or complementarity to ^{18}F-FDG PET/CT in detection of MM lesions, showing both lower sensitivity and specificity [81–83]. Moreover, ^{18}F-NaF PET/CT does not seem to add significantly to ^{18}F-FDG PET/CT in the treatment response evaluation of MM patients, as shown in a study of 34 patients undergoing high-dose chemotherapy and ASCT [84].

The low sensitivity of ^{18}F-NaF PET/CT in detecting myeloma lesions is mainly attributed to the fact that the tracer indicates osteoblastic activity. However, since the hallmark of MM is the osteolytic lesion, the accumulation of ^{18}F-NaF takes place only in the accompanying, sometimes minimal, reactive osteoblastic changes [85]. Further, being a very sensitive radiopharmaceutical for osteoblastic activity, ^{18}F-NaF accumulates in practically in every site of newly mineralizing bone, irrelevant of its aetiology. This means that any cause of bone reconstruction, such as traumatic or degenerative bone lesions, will lead to tracer accumulation, significantly decreasing its specificity as a myeloma tracer [86].

3.7. ^{18}F-FAZA

One of the reasons leading to an increased metabolic activity detected with ^{18}F-FDG PET/CT is tumor hypoxia. Tumor hypoxia leads to enhanced production of several hypoxia inducible factors, resulting in increased microvessel density (MVD) around the malignant plasma cells [6]. MVD has been proven to be correlated with disease progression in MM [87]. Based on this approach, de Waal et al. applied the PET tracer 1-α-D: -(5-deoxy-5-[^{18}F]-fluoroarabinofuranosyl)-2-nitroimidazole (^{18}F-FAZA), which accumulates in tumor hypoxia. The authors studied 5 patients with relapsed MM with ^{18}F-FDG PET and ^{18}F-FAZA PET. Although all patients had a positive ^{18}F-FDG PET scan, no lesions were demonstrated on ^{18}F-FAZA PET, reflecting a limited performance of this tracer in the workup of MM patients [88].

3.8. ^{89}Zr-Daratumumab

The membrane glycoprotein cluster of differentiation 38 (CD38) is expressed at a high density by almost all myeloma cells, and at relatively low levels on normal hematopoietic cells. CD38 is an established therapeutic target in MM. Daratumumab is an FDA-approved therapeutic monoclonal antibody that binds directly to CD38, offering a clinical benefit in MM patients [89–91]. Recently, daratumumab was radiolabeled with ^{89}Zr through deferoxamine (DFO), producing the PET agent ^{89}Zr-DFO-daratumumab. The results of a Phase I first-in-human ^{89}Zr-DFO-daratumumab PET/CT imaging study in six MM patients demonstrated successful whole-body PET visualization of MM with focal tracer uptake in previously known as well as unknown sites of osseous myeloma, consistent with successful CD38-targeted immunoPET imaging of myeloma in human patients [92]. Although these results warrant validation in further prospective studies, they are highly promising for the usage of this PET antibody in diagnosis and staging of MM. Moreover, it could be applied in terms of a personalized, daratumumab-directed imaging in order to identify those MM patients who would benefit from daratumumab and thus predict the effectiveness of therapy in the context of a thera(g)nostic approach in MM.

4. Conclusions

PET/CT with ^{18}F-FDG is increasingly gaining acceptance in the management of MM patients, and is considered a powerful diagnostic tool for the detection of medullary and extramedullary disease at initial diagnosis, a reliable predictor of survival, as well as the most robust modality for treatment response evaluation in the disease. On the other hand, ^{18}F-FDG carries the limitations of a rather poor sensitivity for the detection of diffuse bone marrow infiltration, a relatively low specificity, and the lack of widely applied, established criteria for image interpretation. These drawbacks have led to the development of several alternative PET tracers for MM detection. Some of these radiotracers have provided promising results—such as ^{18}F-choline and ^{11}C-choline, ^{11}C-acetate, ^{11}C-methionine,

^{68}Ga-pentixafor and ^{89}Zr-Daratumumab—but most studies were performed in small patient cohorts and require validation in further prospective clinical trials.

Funding: This research received no external funding.

Conflicts of Interest: The authors declare no conflict of interest.

References

1. Altekruse, S.F.; Kosary, C.L.; Krapcho, M.; Neyman, N.; Aminou, R.; Waldron, W.; Ruhl, J.; Howlader, N.; Tatalovich, Z.; Cho, H.; et al. (Eds.) *SEER Cancer Statistics Review, 1975–2007*; National Cancer Institute: Bethesda, MD, USA, 2009. Available online: http://seer.cancer.gov/csr/1975_2007/index.html (accessed on 5 December 2019).
2. Landgren, O.; Kyle, R.A.; Rajkumar, S.V. From myeloma precursor disease to multiple myeloma: New diagnostic concepts and opportunities for early intervention. *Clin. Cancer Res.* **2011**, *17*, 1243–1252. [CrossRef]
3. Zamagni, E.; Tacchetti, P.; Cavo, M. Imaging in multiple myeloma: How? When? *Blood* **2019**, *133*, 644–651. [CrossRef]
4. Rajkumar, S.V.; Dimopoulos, M.A.; Palumbo, A.; Blade, J.; Merlini, G.; Mateos, M.V.; Kumar, S.; Hillengass, J.; Kastritis, E.; Richardson, P.; et al. International Myeloma Working Group updated criteria for the diagnosis of multiple myeloma. *Lancet Oncol.* **2014**, *15*, e538–e548. [CrossRef]
5. Zamagni, E.; Cavo, M. The role of imaging techniques in the management of multiple myeloma. *Br. J. Haematol.* **2012**, *159*, 499–513. [CrossRef]
6. De Waal, E.G.M.; Glaudemans, A.W.J.M.; Schröder, C.P.; Vellenga, E.; Slart, R.H.J.A. Nuclear medicine imaging of multiple myeloma, particularly in the relapsed setting. *Eur. J. Nucl. Med. Mol. Imaging* **2017**, *44*, 332–341. [CrossRef]
7. Cavo, M.; Terpos, E.; Nanni, C.; Moreau, P.; Lentzsch, S.; Zweegman, S.; Hillengass, J.; Engelhardt, M.; Usmani, S.Z.; Vesole, D.H.; et al. Role of 18F-FDG PET/CT in the diagnosis and management of multiple myeloma and other plasma cell disorders: A consensus statement by the International Myeloma Working Group. *Lancet Oncol.* **2017**, *18*, e206–e217. [CrossRef]
8. Zamagni, E.; Nanni, C.; Patriarca, F.; Englaro, E.; Castellucci, P.; Geatti, O.; Tosi, P.; Tacchetti, P.; Cangini, D.; Perrone, G.; et al. A prospective comparison of 18F-fluorodeoxyglucose positron emission tomography-computed tomography, magnetic resonance imaging and whole-body planar radiographs in the assessment of bone disease in newly diagnosed multiple myeloma. *Haematologica* **2007**, *92*, 50–55. [CrossRef] [PubMed]
9. Bartel, T.B.; Haessler, J.; Brown, T.L.; Shaughnessy, J.D., Jr.; van Rhee, F.; Anaissie, E.; Alpe, T.; Angtuaco, E.; Walker, R.; Epstein, J.; et al. F18-fluorodeoxyglucose positron emission tomography in the context of other imaging techniques and prognostic factors in multiple myeloma. *Blood* **2009**, *114*, 2068–2076. [CrossRef]
10. Van Lammeren-Venema, D.; Regelink, J.C.; Riphagen, I.I.; Zweegman, S.; Hoekstra, O.S.; Zijlstra, J.M. 18F-fluoro-deoxyglucose positron emission tomography in assessment of myeloma-related bone disease: A systematic review. *Cancer* **2012**, *118*, 1971–1981. [CrossRef] [PubMed]
11. Lu, Y.Y.; Chen, J.H.; Lin, W.Y.; Liang, J.A.; Wang, H.Y.; Tsai, S.C.; Kao, C.H. FDG PET or PET/CT for detecting intramedullary and extramedullary lesions in multiple Myeloma: A systematic review and meta-analysis. *Clin. Nucl. Med.* **2012**, *37*, 833–837. [CrossRef] [PubMed]
12. Moreau, P.; Attal, M.; Caillot, D.; Macro, M.; Karlin, L.; Garderet, L.; Facon, T.; Benboubker, L.; Escoffre-Barbe, M.; Stoppa, A.M.; et al. Prospective evaluation of magnetic resonance imaging and [18F]fluorodeoxyglucose positron emission tomography-computed tomography at diagnosis and before maintenance therapy in symptomatic patients with multiple myeloma included in the IFM/DFCI 2009 Trial: Results of the IMAJEM study. *J. Clin. Oncol.* **2017**, *35*, 2911–2918. [CrossRef] [PubMed]
13. Sachpekidis, C.; Mai, E.K.; Goldschmidt, H.; Hillengass, J.; Hose, D.; Pan, L.; Haberkorn, U.; Dimitrakopoulou-Strauss, A. (18)F-FDG dynamic PET/CT in patients with multiple myeloma: Patterns of tracer uptake and correlation with bone marrow plasma cell infiltration rate. *Clin. Nucl. Med.* **2015**, *40*, e300–e307. [CrossRef] [PubMed]

14. Sachpekidis, C.; Hillengass, J.; Goldschmidt, H.; Mosebach, J.; Pan, L.; Schlemmer, H.P.; Haberkorn, U.; Dimitrakopoulou-Strauss, A. Comparison of (18)F-FDG PET/CT and PET/MRI in patients with multiple myeloma. *Am. J. Nucl. Med. Mol. Imaging* **2015**, *5*, 469–478. [PubMed]
15. Zamagni, E.; Patriarca, F.; Nanni, C.; Zannetti, B.; Englaro, E.; Pezzi, A.; Tacchetti, P.; Buttignol, S.; Perrone, G.; Brioli, A.; et al. Prognostic relevance of 18-F FDG PET/CT in newly diagnosed multiple myeloma patients treated with up-front autologous transplantation. *Blood* **2011**, *118*, 5989–5995. [CrossRef]
16. Sachpekidis, C.; Merz, M.; Kopp-Schneider, A.; Jauch, A.; Raab, M.S.; Sauer, S.; Hillengass, J.; Goldschmidt, H.; Dimitrakopoulou-Strauss, A. Quantitative dynamic 18F-fluorodeoxyglucose positron emission tomography/computed tomography before autologous stem cell transplantation predicts survival in multiple myeloma. *Haematologica* **2019**, *104*, e420–e423. [CrossRef]
17. Siontis, B.; Kumar, S.; Dispenzieri, A.; Drake, M.T.; Lacy, M.Q.; Buadi, F.; Dingli, D.; Kapoor, P.; Gonsalves, W.; Gertz, M.A.; et al. Positron emission tomography-computed tomography in the diagnostic evaluation of smoldering multiple myeloma: Identification of patients needing therapy. *Blood Cancer J.* **2015**, *5*, e364. [CrossRef]
18. Zamagni, E.; Nanni, C.; Gay, F.; Pezzi, A.; Patriarca, F.; Bellò, M.; Rambaldi, I.; Tacchetti, P.; Hillengass, J.; Gamberi, B.; et al. 18F-FDG PET/CT focal, but not osteolytic, lesions predict the progression of smoldering myeloma to active disease. *Leukemia* **2016**, *30*, 417–422. [CrossRef]
19. Usmani, S.Z.; Mitchell, A.; Waheed, S.; Crowley, J.; Hoering, A.; Petty, N.; Brown, T.; Bartel, T.; Anaissie, E.; van Rhee, F.; et al. Prognostic implications of serial 18-fluoro-deoxyglucose emission tomography in multiple myeloma treated with total therapy 3. *Blood* **2013**, *121*, 1819–1823. [CrossRef]
20. Nanni, C.; Zamagni, E.; Celli, M.; Caroli, P.; Ambrosini, V.; Tacchetti, P.; Brioli, A.; Zannetti, B.; Pezzi, A.; Pantani, L.; et al. The value of 18F-FDG PET/CT after autologous stem cell transplantation (ASCT) in patients affected by multiple myeloma (MM): Experience with 77 patients. *Clin. Nucl. Med.* **2013**, *38*, e74–e79. [CrossRef]
21. Zamagni, E.; Nanni, C.; Mancuso, K.; Tacchetti, P.; Pezzi, A.; Pantani, L.; Zannetti, B.; Rambaldi, I.; Brioli, A.; Rocchi, S.; et al. PET/CT improves the definition of complete response and allows to detect otherwise unidentifiable skeletal progression in multiple Myeloma. *Clin. Cancer Res.* **2015**, *21*, 4384–4390. [CrossRef]
22. Davies, F.E.; Rosenthal, A.; Rasche, L.; Petty, N.M.; McDonald, J.E.; Ntambi, J.A.; Steward, D.M.; Panozzo, S.B.; van Rhee, F.; Zangari, M.; et al. Treatment to suppression of focal lesions on positron emission tomography-computed tomography is a therapeutic goal in newly diagnosed multiple myeloma. *Haematologica* **2018**, *103*, 1047–1053. [CrossRef] [PubMed]
23. Patriarca, F.; Carobolante, F.; Zamagni, E.; Montefusco, V.; Bruno, B.; Englaro, E.; Nanni, C.; Geatti, O.; Isola, M.; Sperotto, A.; et al. The role of positron emission tomography with 18F-fluorodeoxyglucose integrated with computed tomography in the evaluation of patients with multiple myeloma undergoing allogeneic stem cell transplantation. *Biol. Blood Marrow Transplant.* **2015**, *21*, 1068–1073. [CrossRef] [PubMed]
24. Bailly, C.; Carlier, T.; Jamet, B.; Eugene, T.; Touzeau, C.; Attal, M.; Hulin, C.; Facon, T.; Leleu, X.; Perrot, A.; et al. Interim PET Analysis in First-Line Therapy of Multiple Myeloma: Prognostic Value of ΔSUVmax in the FDG-Avid Patients of the IMAJEM Study. *Clin. Cancer Res.* **2018**, *24*, 5219–5224. [CrossRef] [PubMed]
25. Dimitrakopoulou-Strauss, A.; Hoffmann, M.; Bergner, R.; Uppenkamp, M.; Haberkorn, U.; Strauss, L.G. Prediction of progression-free survival in patients with multiple myeloma following anthracycline-based chemotherapy based on dynamic FDG-PET. *Clin. Nucl. Med.* **2009**, *34*, 576–584. [CrossRef]
26. Elliott, B.M.; Peti, S.; Osman, K.; Scigliano, E.; Lee, D.; Isola, L.; Kostakoglu, L. Combining FDG-PET/CT with laboratory data yields superior results for prediction of relapse in multiple myeloma. *Eur. J. Haematol.* **2011**, *86*, 289–298. [CrossRef]
27. Korde, N.; Roschewski, M.; Zingone, A.; Kwok, M.; Manasanch, E.E.; Bhutani, M.; Tageja, N.; Kazandjian, D.; Mailankody, S.; Wu, P.; et al. Treatment with carfilzomib-lenalidomide-dexamethasone with lenalidomide extension in patients with smoldering or newly diagnosed multiple myeloma. *JAMA Oncol.* **2015**, *1*, 746–754. [CrossRef]
28. Paiva, B.; García-Sanz, R.; San Miguel, J.F. Multiple Myeloma Minimal Residual Disease. In *Plasma Cell Dyscrasias. Cancer Treatment and Research*; Roccaro, A., Ghobrial, I., Eds.; Springer: Cham, Switzerland, 2016; Volume 169, pp. 103–122. [CrossRef]

29. Kumar, S.; Paiva, B.; Anderson, K.C.; Durie, B.; Landgren, O.; Moreau, P.; Munshi, N.; Lonial, S.; Bladé, J.; Mateos, M.V.; et al. International Myeloma Working Group consensus criteria for response and minimal residual disease assessment in multiple myeloma. *Lancet Oncol.* **2016**, *17*, e328–e346. [CrossRef]
30. Rasche, L.; Alapat, D.; Kumar, M.; Gershner, G.; McDonald, J.; Wardell, C.P.; Samant, R.; Van Hemert, R.; Epstein, J.; Williams, A.F.; et al. Combination of flow cytometry and functional imaging for monitoring of residual disease in myeloma. *Leukemia* **2019**, *33*, 1713–1722. [CrossRef]
31. Alonso, R.; Cedena, M.T.; Gómez-Grande, A.; Ríos, R.; Moraleda, J.M.; Cabañas, V.; Moreno, M.J.; López-Jiménez, J.; Martín, F.; Sanz, A.; et al. Imaging and bone marrow assessments improve minimal residual disease prediction in multiple myeloma. *Am. J. Hematol.* **2019**, *94*, 853–861. [CrossRef]
32. Rasche, L.; Angtuaco, E.; McDonald, J.E.; Buros, A.; Stein, C.; Pawlyn, C.; Thanendrarajan, S.; Schinke, C.; Samant, R.; Yaccoby, S.; et al. Low expression of hexokinase-2 is associated with false-negative FDG-positron emission tomography in multiple myeloma. *Blood* **2017**, *130*, 30–34. [CrossRef]
33. Kircher, S.; Stolzenburg, A.; Kortuem, K.M.; Kircher, M.; Da Via, M.; Samnick, S.; Buck, A.; Einsele, H.; Rosenwald, A.; Lapa, C. Hexokinase-2 expression in MET-positive FDG-negative multiple myeloma. *J. Nucl. Med.* **2019**, *60*, 348–352. [CrossRef] [PubMed]
34. Nanni, C.; Versari, A.; Chauvie, S.; Bertone, E.; Bianchi, A.; Rensi, M.; Bellò, M.; Gallamini, A.; Patriarca, F.; Gay, F.; et al. Interpretation criteria for FDG PET/CT in multiple myeloma (IMPeTUs): Final results. IMPeTUs (Italian myeloma criteria for PET USe). *Eur. J. Nucl. Med. Mol. Imaging* **2018**, *45*, 712–719. [CrossRef]
35. Ambrosini, V.; Farsad, M.; Nanni, C.; Schiavina, R.; Rubello, D.; Castellucci, P.; Pasquini, E.; Franchi, R.; Cavo, M.; Fanti, S. Incidental finding of an 11C-choline PET positive solitary plasmacytoma lesion. *Eur. J. Nucl. Med. Mol. Imaging* **2006**, *33*, 1522. [CrossRef] [PubMed]
36. Nanni, C.; Zamagni, E.; Cavo, M.; Rubello, D.; Tacchetti, P.; Pettinato, C.; Farsad, M.; Castellucci, P.; Ambrosini, V.; Montini, G.C.; et al. 11C-choline vs. 18F-FDG PET/ CT in assessing bone involvement in patients with multiple myeloma. *World J. Surg. Oncol.* **2007**, *5*, 68. [CrossRef] [PubMed]
37. Cassou-Mounat, T.; Balogova, S.; Nataf, V.; Calzada, M.; Huchet, V.; Kerrou, K.; Devaux, J.Y.; Mohty, M.; Talbot, J.N.; Garderet, L. 18F-fluorocholine versus 18F-fluorodeoxyglucose for PET/CT imaging in patients with suspected relapsing or progressive multiple myeloma: A pilot study. *Eur. J. Nucl. Med. Mol. Imaging* **2016**, *43*, 1995–2004. [CrossRef] [PubMed]
38. Meckova, Z.; Lambert, L.; Spicka, I.; Kubinyi, J.; Burgetova, A. Is fluorine-18-fluorocholine PET/CT suitable for the detection of skeletal involvement of multiple myeloma? *Hell. J. Nucl. Med.* **2018**, *21*, 167–168. [CrossRef] [PubMed]
39. Grassi, I.; Nanni, C.; Allegri, V.; Morigi, J.J.; Montini, G.C.; Castellucci, P.; Fanti, S. The clinical use of PET with (11)C-acetate. *Am. J. Nucl. Med. Mol. Imaging* **2012**, *2*, 33–47.
40. Khoo, S.H.; Al-Rubeai, M. Metabolic characterization of a hyper-productive state in an antibody producing NS0 myeloma cell line. *Metab. Eng.* **2009**, *11*, 199–211. [CrossRef]
41. Lee, S.M.; Kim, T.S.; Lee, J.W.; Kwon, H.W.; Kim, Y.I.; Kang, S.H.; Kim, S.K. Incidental finding of an ^{11}C-acetate PET-positive multiple myeloma. *Ann. Nucl. Med.* **2010**, *24*, 41–44. [CrossRef]
42. Ho, C.L.; Chen, S.; Leung, Y.L.; Cheng, T.; Wong, K.N.; Cheung, S.K.; Liang, R.; Chim, C.S. ^{11}C-Acetate PET/CT for metabolic characterization of multiple myeloma: A comparative study with 18F-FDG PET/CT. *J. Nucl. Med.* **2014**, *55*, 749–752. [CrossRef]
43. Lin, C.; Ho, C.L.; Ng, S.H.; Wang, P.N.; Huang, Y.; Lin, Y.C.; Tang, T.C.; Tsai, S.F.; Rahmouni, A.; Yen, T.C. (11)C-Acetate as a new biomarker for PET/CT in patients with multiple myeloma: Initial staging and postinduction response assessment. *Eur. J. Nucl. Med. Mol. Imaging* **2014**, *41*, 41–49. [CrossRef] [PubMed]
44. Glaudemans, A.W.; Enting, R.H.; Heesters, M.A.; Dierckx, R.A.; van Rheenen, R.W.; Walenkamp, A.M.; Slart, R.H. Value of ^{11}C-methionine PET in imaging brain tumours and metastases. *Eur. J. Nucl. Med. Mol. Imaging* **2013**, *40*, 615–635. [CrossRef] [PubMed]
45. Hammerton, K.; Cooper, D.A.; Duckett, M.; Penny, R. Biosynthesis of immunoglobulin in human immunoproliferative diseases. I. Kinetics of synthesis and secretion of immunoglobulin and protein by bone marrow cells in myeloma. *J. Immunol.* **1978**, *121*, 409–417. [PubMed]
46. Dankerl, A.; Liebisch, P.; Glatting, G.; Friesen, C.; Blumstein, N.M.; Kocot, D.; Wendl, C.; Bunjes, D.; Reske, S.N. Multiple Myeloma: Molecular Imaging with C-Methionine PET/CT—Initial Experience. *Radiology* **2007**, *242*, 498–508. [CrossRef] [PubMed]

47. Nakamoto, Y.; Kurihara, K.; Nishizawa, M.; Yamashita, K.; Nakatani, K.; Kondo, T.; Takaori-Kondo, A.; Togashi, K. Clinical value of ^{11}C-methionine PET/CT in patients with plasma cell malignancy: Comparison with ^{18}F-FDG PET/CT. *Eur. J. Nucl. Med. Mol. Imaging* **2013**, *40*, 708–715. [CrossRef] [PubMed]
48. Okasaki, M.; Kubota, K.; Minamimoto, R.; Miyata, Y.; Morooka, M.; Ito, K.; Ishiwata, K.; Toyohara, J.; Inoue, T.; Hirai, R.; et al. Comparison of (11)C-4′-thiothymidine, (11)C-methionine, and (18)F-FDG PET/CT for the detection of active lesions of multiple myeloma. *Ann. Nucl. Med.* **2015**, *29*, 224–232. [CrossRef] [PubMed]
49. Lapa, C.; Knop, S.; Schreder, M.; Rudelius, M.; Knott, M.; Jörg, G.; Samnick, S.; Herrmann, K.; Buck, A.K.; Einsele, H.; et al. 11C-Methionine-PET in Multiple Myeloma: Correlation with Clinical Parameters and Bone Marrow Involvement. *Theranostics* **2016**, *6*, 254–261. [CrossRef]
50. Lapa, C.; Schreder, M.; Lückerath, K.; Samnick, S.; Rudelius, M.; Buck, A.K.; Kortüm, K.M.; Einsele, H.; Rosenwald, A.; Knop, S. [11 C]Methionine emerges as a new biomarker for tracking active myeloma lesions. *Br. J. Haematol.* **2018**, *181*, 701–703. [CrossRef]
51. Lapa, C.; Garcia-Velloso, M.J.; Lückerath, K.; Samnick, S.; Schreder, M.; Otero, P.R.; Schmid, J.S.; Herrmann, K.; Knop, S.; Buck, A.K.; et al. 11C-Methionine-PET in Multiple Myeloma: A Combined Study from Two Different Institutions. *Theranostics* **2017**, *7*, 2956–2964. [CrossRef]
52. Lapa, C.; Kircher, M.; Da Via, M.; Schreder, M.; Rasche, L.; Kortüm, K.M.; Einsele, H.; Buck, A.K.; Hänscheid, H.; Samnick, S. Comparison of 11C-Choline and 11C-Methionine PET/CT in Multiple Myeloma. *Clin. Nucl. Med.* **2019**, *44*, 620–624. [CrossRef]
53. Nanni, C. PET/CT with Standard Non-FDG Tracers in Multiple Myeloma. In *Molecular Imaging in Multiple Myeloma*; Nanni, C., Fanti, S., Zanoni, L., Eds.; Springer: Cham, Switzerland, 2019; pp. 93–98.
54. Peck, M.; Pollack, H.A.; Friesen, A.; Muzi, M.; Shoner, S.C.; Shankland, E.G.; Fink, J.R.; Armstrong, J.O.; Link, J.M.; Krohn, K.A. Applications of PET imaging with the proliferation marker [18F]-FLT. *Q. J. Nucl. Med. Mol. Imaging* **2015**, *59*, 95–104. [PubMed]
55. Lodge, M.A.; Holdhoff, M.; Leal, J.P.; Bag, A.K.; Nabors, L.B.; Mintz, A.; Lesser, G.J.; Mankoff, D.A.; Desai, A.S.; Mountz, J.M.; et al. Repeatability of 18F-FLT PET in a multicenter study of patients with high-grade glioma. *J. Nucl. Med.* **2017**, *58*, 393–398. [CrossRef] [PubMed]
56. Agool, A.; Slart, R.H.; Kluin, P.M.; de Wolf, J.T.; Dierckx, R.A.; Vellenga, E. F-18 FLT PET: A noninvasive diagnostic tool for visualization of the bone marrow compartment in patients with aplastic anemia: A pilot study. *Clin. Nucl. Med.* **2011**, *36*, 286–289. [CrossRef] [PubMed]
57. Sachpekidis, C.; Goldschmidt, H.; Kopka, K.; Kopp-Schneider, A.; Dimitrakopoulou-Strauss, A. Assessment of glucose metabolism and cellular proliferation in multiple myeloma: A first report on combined 18F-FDG and 18F-FLT PET/CT imaging. *EJNMMI Res.* **2018**, *8*, 28. [CrossRef] [PubMed]
58. Zou, Y.R.; Kottmann, A.H.; Kuroda, M.; Taniuchi, I.; Littman, D.R. Function of the chemokine receptor CXCR4 in haematopoiesis and in cerebellar development. *Nature* **1998**, *393*, 595–599. [CrossRef] [PubMed]
59. Alsayed, Y.; Ngo, H.; Runnels, J.; Leleu, X.; Singha, U.K.; Pitsillides, C.M.; Spencer, J.A.; Kimlinger, T.; Ghobrial, J.M.; Jia, X.; et al. Mechanisms of regulation of CXCR4/SDF-1 (CXCL12)–dependent migration and homing in multiple myeloma. *Blood* **2007**, *109*, 2708–2717. [CrossRef] [PubMed]
60. Paiva, B.; Corchete, L.A.; Vidriales, M.B.; Puig, N.; Maiso, P.; Rodriguez, I.; Alignani, D.; Burgos, L.; Sanchez, M.L.; Barcena, P.; et al. Phenotypic and genomic analysis of multiple myeloma minimal residual disease tumor cells: A new model to understand chemoresistance. *Blood* **2016**, *127*, 1896–1906. [CrossRef]
61. Peled, A.; Klein, S.; Beider, K.; Burger, J.A.; Abraham, M. Role of CXCL12 and CXCR4 in the pathogenesis of hematological malignancies. *Cytokine* **2018**, *109*, 11–16. [CrossRef]
62. Chatterjee, S.; Behnam Azad, B.; Nimmagadda, S. The intricate role of CXCR4 in cancer. *Adv. Cancer Res.* **2014**, *124*, 31–82. [CrossRef]
63. Guo, F.; Wang, Y.; Liu, J.; Mok, S.C.; Xue, F.; Zhang, W. CXCL12/CXCR4: A symbiotic bridge linking cancer cells and their stromal neighbors in oncogenic communication networks. *Oncogene* **2016**, *35*, 816–826. [CrossRef]
64. Liu, S.H.; Gu, Y.; Pascual, B.; Yan, Z.; Hallin, M.; Zhang, C.; Fan, C.; Wang, W.; Lam, J.; Spilker, M.E.; et al. A novel CXCR4 antagonist IgG1 antibody (PF-06747143) for the treatment of hematologic malignancies. *Blood Adv.* **2017**, *1*, 1088–1100. [CrossRef] [PubMed]
65. Coniglio, S.J. Role of Tumor-Derived Chemokines in Osteolytic Bone Metastasis. *Front Endocrinol.* **2018**, *9*, 313. [CrossRef] [PubMed]

66. Ullah, T.R. The role of CXCR4 in multiple myeloma: Cells' journey from bone marrow to beyond. *J. Bone Oncol.* **2019**, *17*, 100253. [CrossRef] [PubMed]
67. Herrmann, K.; Schottelius, M.; Lapa, C.; Osl, T.; Poschenrieder, A.; Hänscheid, H.; Lückerath, K.; Schreder, M.; Bluemel, C.; Knott, M.; et al. First-in-Human Experience of CXCR4-Directed Endoradiotherapy with 177Lu- and 90Y-Labeled Pentixather in Advanced-Stage Multiple Myeloma with Extensive Intra- and Extramedullary Disease. *J. Nucl. Med.* **2016**, *57*, 248–251. [CrossRef]
68. Philipp-Abbrederis, K.; Herrmann, K.; Knop, S.; Schottelius, M.; Eiber, M.; Lückerath, K.; Pietschmann, E.; Habringer, S.; Gerngroß, C.; Franke, K.; et al. In Vivo molecular imaging of chemokine receptor CXCR4 expression in patients with advanced multiple myeloma. *EMBO Mol. Med.* **2015**, *7*, 477–487. [CrossRef]
69. Lapa, C.; Schreder, M.; Schirbel, A.; Samnick, S.; Kortüm, K.M.; Herrmann, K.; Kropf, S.; Einsele, H.; Buck, A.K.; Wester, H.J.; et al. [68Ga]Pentixafor-PET/CT for imaging of chemokine receptor CXCR4 expression in multiple myeloma-Comparison to [18F]FDG and laboratory values. *Theranostics* **2017**, *7*, 205–212. [CrossRef]
70. Pan, Q.; Cao, X.; Luo, Y.; Li, J.; Feng, J.; Li, F. Chemokine receptor-4 targeted PET/CT with 68Ga-Pentixafor in assessment of newly diagnosed multiple myeloma: Comparison to 18F-FDG PET/CT. *Eur. J. Nucl. Med. Mol. Imaging* **2019**. [CrossRef]
71. Czernin, J.; Satyamurthy, N.; Schiepers, C. Molecular mechanisms of bone 18F-NaF deposition. *J. Nucl. Med.* **2010**, *51*, 1826–1829. [CrossRef]
72. Segall, G.; Delbeke, D.; Stabin, M.G.; Even-Sapir, E.; Fair, J.; Sajdak, R.; Smith, G.T. SNM. SNM practice guideline for sodium 18F-fluoride PET/CT bone scans 1.0. *J. Nucl. Med.* **2010**, *51*, 1813–1820. [CrossRef]
73. Beheshti, M.; Mottaghy, F.M.; Payche, F.; Behrendt, F.F.; Van den Wyngaert, T.; Fogelman, I.; Strobel, K.; Celli, M.; Fanti, S.; Giammarile, F.; et al. (18)F-NaF PET/CT: EANM procedure guidelines for bone imaging. *Eur. J. Nucl. Med. Mol. Imaging* **2015**, *42*, 1767–1777. [CrossRef]
74. Hillner, B.E.; Siegel, B.A.; Hanna, L.; Duan, F.; Quinn, B.; Shields, A.F. 18F-fluoride PET used for treatment monitoring of systemic cancer therapy: Results from the National Oncologic PET Registry. *J. Nucl. Med.* **2015**, *56*, 222–228. [CrossRef] [PubMed]
75. Hawkins, R.A.; Choi, Y.; Huang, S.C.; Hoh, C.K.; Dahlbom, M.; Schiepers, C.; Satyamurthy, N.; Barrio, J.R.; Phelps, M.E. Evaluation of the skeletal kinetics of fluorine-18-fluoride ion with PET. *J. Nucl. Med.* **1992**, *33*, 633–642. [PubMed]
76. Grant, F.D.; Fahey, F.H.; Packard, A.B.; Davis, R.T.; Alavi, A.; Treves, S.T. Skeletal PET with 18 F-fluoride: Applying new technology to an old tracer. *J. Nucl. Med.* **2008**, *49*, 68–78. [CrossRef] [PubMed]
77. Kurdziel, K.A.; Shih, J.H.; Apolo, A.B.; Lindenberg, L.; Mena, E.; McKinney, Y.Y.; Adler, S.S.; Turkbey, B.; Dahut, W.; Gulley, J.L.; et al. The kinetics and reproducibility of 18F-sodium fluoride for oncology using current PET camera technology. *J. Nucl. Med.* **2012**, *53*, 1175–1184. [CrossRef] [PubMed]
78. Nishiyama, Y.; Tateishi, U.; Shizukuishi, K.; Shishikura, A.; Yamazaki, E.; Shibata, H.; Yoneyama, T.; Ishigatsubo, Y.; Inoue, T. Role of 18F-fluoride PET/CT in the assessment of multiple myeloma: Initial experience. *Ann. Nucl. Med.* **2013**, *27*, 78–83. [CrossRef] [PubMed]
79. Xu, F.; Liu, F.; Pastakia, B. Different lesions revealed by 18F-FDG PET/CT and 18F-NaF PET/CT in patients with multiple myeloma. *Clin. Nucl. Med.* **2014**, *39*, e407–e409. [CrossRef] [PubMed]
80. Oral, A.; Yazici, B.; Ömür, Ö.; Comert, M.; Saydam, G. 18F-FDG and 18F-NaF PET/CT Findings of a Multiple Myeloma Patient With Thyroid Cartilage Involvement. *Clin. Nucl. Med.* **2015**, *40*, 873–876. [CrossRef]
81. Sachpekidis, C.; Goldschmidt, H.; Hose, D.; Pan, L.; Cheng, C.; Kopka, K.; Haberkorn, U.; Dimitrakopoulou-Strauss, A. PET/CT studies of multiple myeloma using (18) F-FDG and (18) F-NaF: Comparison of distribution patterns and tracers' pharmacokinetics. *Eur. J. Nucl. Med. Mol. Imaging* **2014**, *41*, 1343–1353. [CrossRef]
82. Ak, İ.; Onner, H.; Akay, O.M. Is there any complimentary role of F-18 NaF PET/CT in detecting of osseous involvement of multiple myeloma? A comparative study for F-18 FDG PET/CT and F-18 FDG NaF PET/CT. *Ann. Hematol.* **2015**, *94*, 1567–1575. [CrossRef]
83. Sachpekidis, C.; Hillengass, J.; Goldschmidt, H.; Anwar, H.; Haberkorn, U.; Dimitrakopoulou-Strauss, A. Quantitative analysis of 18F-NaF dynamic PET/CT cannot differentiate malignant from benign lesions in multiple myeloma. *Am. J. Nucl. Med. Mol. Imaging* **2017**, *7*, 148–156.

84. Sachpekidis, C.; Hillengass, J.; Goldschmidt, H.; Wagner, B.; Haberkorn, U.; Kopka, K.; Dimitrakopoulou-Strauss, A. Treatment response evaluation with 18F-FDG PET/CT and 18F-NaF PET/CT in multiple myeloma patients undergoing high-dose chemotherapy and autologous stem cell transplantation. *Eur. J. Nucl. Med. Mol. Imaging* **2017**, *44*, 50–62. [CrossRef] [PubMed]
85. Even-Sapir, E.; Mishani, E.; Flusser, G.; Metser, U. 18F-Fluoride positron emission tomography and positron emission tomography/computed tomography. *Semin. Nucl. Med.* **2007**, *37*, 462–469. [CrossRef] [PubMed]
86. Dimitrakopoulou-Strauss, A. PET-CT in der nuklearmedizinischen Diagnostik des multiplen Myeloms [PET-CT for nuclear medicine diagnostics of multiple myeloma]. *Radiologe* **2014**, *54*, 564–571. [CrossRef] [PubMed]
87. Rajkumar, S.V.; Mesa, R.A.; Fonseca, R.; Schroeder, G.; Plevak, M.F.; Dispenzieri, A.; Lacy, M.Q.; Lust, J.A.; Witzig, T.E.; Gertz, M.A.; et al. Bone marrow angiogenesis in 400 patients with monoclonal gammopathy of undetermined significance, multiple myeloma, and primary amyloidosis. *Clin. Cancer Res.* **2002**, *8*, 2210–2216. [PubMed]
88. De Waal, E.G.; Slart, R.H.; Leene, M.J.; Kluin, P.M.; Vellenga, E. 18F-FDG PET increases visibility of bone lesions in relapsed multiple myeloma: Is this hypoxia-driven? *Clin. Nucl. Med.* **2015**, *40*, 291–296. [CrossRef] [PubMed]
89. Dimopoulos, M.A.; Oriol, A.; Nahi, H.; San-Miguel, J.; Bahlis, N.J.; Usmani, S.Z.; Rabin, N.; Orlowski, R.Z.; Komarnicki, M.; Suzuki, K.; et al. Daratumumab, Lenalidomide, and Dexamethasone for Multiple Myeloma. *N. Engl. J. Med.* **2016**, *375*, 1319–1331. [CrossRef] [PubMed]
90. Facon, T.; Kumar, S.; Plesner, T.; Orlowski, R.Z.; Moreau, P.; Bahlis, N.; Basu, S.; Nahi, H.; Hulin, C.; Quach, H.; et al. Daratumumab plus Lenalidomide and Dexamethasone for Untreated Myeloma. *N. Engl. J. Med.* **2019**, *380*, 2104–2115. [CrossRef]
91. Moreau, P.; Attal, M.; Hulin, C.; Arnulf, B.; Belhadj, K.; Benboubker, L.; Béné, M.C.; Broijl, A.; Caillon, H.; Caillot, D.; et al. Bortezomib, thalidomide, and dexamethasone with or without daratumumab before and after autologous stem-cell transplantation for newly diagnosed multiple myeloma (CASSIOPEIA): A randomised, open-label, phase 3 study. *Lancet* **2019**, *394*, 29–38. [CrossRef]
92. Ulaner, G.; Sobol, N.; O'Donoghue, J.; Burnazi, E.; Staton, K.; Weber, W.; Lyashchenko, S.; Lewis, J.; Landgren, C.O. Preclinical development and First-in-human imaging of 89Zr-Daratumumab for CD38 targeted imaging of myeloma. *J. Nucl. Med.* **2019**, *60*, 203.

© 2019 by the authors. Licensee MDPI, Basel, Switzerland. This article is an open access article distributed under the terms and conditions of the Creative Commons Attribution (CC BY) license (http://creativecommons.org/licenses/by/4.0/).

MDPI
St. Alban-Anlage 66
4052 Basel
Switzerland
Tel. +41 61 683 77 34
Fax +41 61 302 89 18
www.mdpi.com

Molecules Editorial Office
E-mail: molecules@mdpi.com
www.mdpi.com/journal/molecules

www.ingramcontent.com/pod-product-compliance
Lightning Source LLC
LaVergne TN
LVHW070743100526
838202LV00013B/1292